T0181220

Progress in IS

More information about this series at http://www.springer.com/series/10440

Jorge Marx Gómez • Michael Sonnenschein •
Ute Vogel • Andreas Winter • Barbara Rapp •
Nils Giesen
Editors

Advances and New Trends in Environmental and Energy Informatics

Selected and Extended Contributions from the 28th International Conference on Informatics for Environmental Protection

Editors
Jorge Marx Gómez
University of Oldenburg
Oldenburg, Germany

Michael Sonnenschein
University of Oldenburg
Oldenburg, Germany

Ute Vogel
University of Oldenburg
Oldenburg, Germany

Andreas Winter
University of Oldenburg
Oldenburg, Germany

Barbara Rapp
University of Oldenburg
Oldenburg, Germany

Nils Giesen
University of Oldenburg
Oldenburg, Germany

Progress in IS
ISBN 978-3-319-79486-0 ISBN 978-3-319-23455-7 (eBook)
DOI 10.1007/978-3-319-23455-7

Springer Cham Heidelberg New York Dordrecht London

Softcover reprint of the hardcover 1st edition 2016

Printed on acid-free paper

Springer International Publishing AG Switzerland is part of Springer Science+Business Media (www.springer.com)

We dedicate this book to all people whose lives were taken and affected by terror, repression and violence inside and outside of universities around the world.

Foreword

This book is a collection of selected outstanding scientific papers based on contributions to the 28th International Conference on Informatics for Environmental Protection – EnviroInfo 2014 – which took place in September 2014 at the Oldenburg University (Germany). The EnviroInfo conference series aims at presenting and discussing the latest state-of-the-art development on information and communications technology (ICT) in environmental-related fields. Selected papers out of the EnviroInfo 2014 conference have been extended to full book chapters by the authors and were reviewed each by three members of a 37-person program committee to improve the contributions. This book outlines some of the major topics in the field of ICT for energy efficiency and further relevant environmental topics addressed by the EnviroInfo 2014. It includes concepts, methods, approaches, and applications of emerging fields of research and practice in ICT dedicated to environmental and sustainability topics and problems.

The book provides and covers a broad range of topics. It is structured in six application-oriented main clusters comprising all chapters:

- **Green IT**, aiming to design, implement, and use ICT in an environment-friendly and resource-preserving (efficient) way considering their whole life cycle
- **From Smart Grids to Smart Homes**, providing technical approaches and systems with the aim of an efficient management of power generation from renewable energy sources as well as an efficient power consumption
- **Smart Transportation**, aiming to provide innovative services to various modes of transport and traffic management
- **Sustainable Enterprises and Management**, aiming to provide various elements involved in managing and organization's operations, with a strong emphasis on maintaining socio-environmental integrity
- **Environmental Decision Support**, aiming to inform environmental and natural resource management based on techniques for objective assessments of options and their outcomes

- **Social Media for Sustainability**, becoming an indispensable tool for value creation in education and sustainable business with community engagement for sustainability

Crucial elements in the book clusters are the criteria environmental protection, sustainability, and energy as well as resource efficiency which are spread all over the book chapters.

Chapters allocated in the "Green IT" cluster deal mainly with the topics energy awareness, resource optimization, energy efficiency, and carbon footprint improvement as well as simulation approaches. Three chapters are dedicated to an optimized operation of data centers with respect to energy consumption. Modeling and simulation of data center's operations are essential tools for this purpose. Two chapters are oriented towards the energy efficiency of mobile and context-aware applications.

The cluster "From Smart Grids to Smart Homes" comprises chapters concerning the management of renewables in power grids, IT architecture analysis for smart grids, control algorithms in smart grids, as well as energy efficiency of buildings and eco-support features at home. First, an approach for increasing distribution grid capacity for connecting renewable generators is presented. The next two chapters address architectures and control algorithms for smart grids. After dealing with an approach to urban energy planning demonstrated on the example of a district in Vienna, a new method for reducing energy consumption at home is presented.

In the "Smart Transportation" cluster, chapters present and discuss themes on sustainable mobility supported by mobile technologies to inform on air quality, safer bicycling through spatial information on road traffic, and system dynamics modeling to understand rebound effects in air traffic.

Authors of chapters allocated to "Sustainable Enterprises and Management" argue on web-based software tools for sustainability management for small enterprises, while the others focus on green business process management and IT for green improvements for bridging the gap from strategic planning to everyday work.

Contributions of chapters in the cluster "Environmental Decision Support" are assigned to the thematic fields of supporting environmental policy making by specific software tools, enrichment of environmental data streams by use of semantic web technologies, and quality improvement of river flood prediction by enhancing flood simulation.

The remaining "Social Media for Sustainability" cluster includes a contribution to sustainable practices in educational environments by means of social media and a contribution on sustainable development in rural areas by neighborhood effects using ICT. Web technologies are the key element to improve environmental awareness and communication between different stakeholders.

We strongly believe that this book will contribute to the increasing awareness of researchers and practitioners in the field of ICT and sustainability, and we hope that it can reach a wide international audience and readership.

Reviewing Committee

- Hans-Knud Arndt, University of Magdeburg, Germany
- Jörg Bremer, University of Oldenburg, Germany
- Thomas Brinkhoff, Jade-Hochschule, Wilhelmshaven/Oldenburg/Elsfleth, Germany
- Christian Bunse, Fachhochschule Stralsund, Germany
- Luis Rafael Canali, National University of Technology Córdoba, Argentina
- Lester Cowley, Nelson Mandela Metropolitan University, Port Elizabeth & George, South Africa
- Clemens Düpmeier, Karlsruhe Institute of Technology, Germany
- Amr Eltaher, University of Duisburg, Germany
- Luis Ferreira, Polytechnic Institute of Cávado and Ave, Portugal
- Werner Geiger, Karlsruhe Institute of Technology, Germany
- Nils Giesen, University of Oldenburg, Germany
- Albrecht Gnauck, HTW Berlin, Germany
- Johannes Göbel, University of Hamburg, Germany
- Paulina Golinska, University of Poznan, Poland
- Marion Gottschalk, OFFIS Institute for Information Technology, Oldenburg, Germany
- Klaus Greve, University of Bonn, Germany
- Axel Hahn, University of Oldenburg, Germany
- Oliver Kramer, University of Oldenburg, Germany
- Nuno Lopes, Polytechnic Institute of Cávado and Ave, Portugal
- Somayeh Malakuti, University of Technology, Dresden, Germany
- Jorge Marx Gómez, University of Oldenburg, Germany
- Ammar Memari, University of Oldenburg, Germany
- Andreas Möller, Leuphana University, Lüneburg, Germany
- Stefan Naumann, Hochschule Trier, Germany
- Alexandra Pehlken, University of Oldenburg, Germany
- Joachim Peinke, University of Oldenburg, Germany

- Werner Pillmann, International Society for Environmental Protection, Vienna, Austria
- Barbara Rapp, University of Oldenburg, Germany
- Wolf-Fritz Riekert, HDM Stuttgart, Germany
- Brenda Scholtz, Nelson Mandela Metropolitan University, Port Elizabeth & George, South Africa
- Karl-Heinz Simon, University of Kassel, Germany
- Michael Sonnenschein, University of Oldenburg, Germany
- Frank Teuteberg, University of Osnabrück, Germany
- Ute Vogel, University of Oldenburg, Germany
- Benjamin Wagner vom Berg, University of Oldenburg, Germany
- Andreas Winter, University of Oldenburg, Germany
- Volker Wohlgemuth, HTW Berlin, Germany

Acknowledgments

First of all, we would like to thank the authors for elaborating and providing their book chapters in time and in high quality. The editors would like to express their sincere gratitude to all reviewers who devoted their valuable time and expertise in order to evaluate each individual chapter. Furthermore, we would like to extend our gratitude to all sponsors of the EnviroInfo 2014 conference. Without their financial support, this book would not have been possible. Finally, our thanks go to Springer Publishing House for their confidence and trust in our work.

Oldenburg, May 2015.

Jorge Marx Gómez
Michael Sonnenschein
Ute Vogel
Andreas Winter
Barbara Rapp
Nils Giesen

Contents

Part I
Green IT

Chapter 1
Extending Energetic Potentials of Data Centers by Resource Optimization to Improve Carbon Footprint

Alexander Borgerding and Gunnar Schomaker

Abstract The electric power is one of the major operating expenses in data centers. Rising and varying energy costs induce the need of further solutions to use energy efficiently. The first steps to improve efficiency have already been accomplished by applying virtualization technologies. However, a practical approach for data center power control mechanisms is still missing.

In this paper, we address the problem of energy efficiency in data centers. Efficient and scalable power usage for data centers is needed. We present different approaches to improve efficiency and carbon footprint as background information. We propose an in-progress idea to extend the possibilities of power control in data centers and to improve efficiency. Our approach is based on virtualization technologies and live-migration to improve resource utilization by comparing different effects on virtual machine permutation on physical servers. It delivers an efficiency-aware VM placement by assessing different virtual machine permutation. In our approach, the applications are untouched and the technology is non-invasive regarding the applications. This is a crucial requirement in the context of Infrastructure-as-a-Service (IaaS) environments.

Keywords Data center • VM placement • Energy efficiency • Power-aware • Resource management • Server virtualization

1 Introduction

The IP traffic increases year by year worldwide. New Information and Communication Technology (ICT) services are coming up and existing services are migrating to IP technology, for example, VoIP, TV, radio and video streaming. Following

A. Borgerding (✉)
University of Oldenburg, 26111 Oldenburg, Germany
e-mail: alexander.borgerding@uni-oldenburg.de

G. Schomaker
Software Innovation Campus Paderborn, Zukunftsmeile 1, 33102 Paderborn, Germany
e-mail: schomaker@sicp.de

J. Marx Gómez et al. (eds.), *Advances and New Trends in Environmental and Energy Informatics*, Progress in IS, DOI 10.1007/978-3-319-23455-7_1

these trends, the power consumption of ICT obtains a more and more significant value. In the same way, data centers are growing in number and size in order to comply with the increasing demand. As a result, their share of electric power consumption increases too, e.g. it has doubled in the period 2000–2006 [16]. In addition, energy costs rise continuously and the data center operators are faced with customer questions about sustainability and carbon footprint while economical operation is an all-over goal. The electric power consumption has become one of the major expenses in data centers.

A high performance server in idle-state consumes up to 60 % of its peak power [11]. To reduce the quantity of servers in idle-state, virtualization technologies are used. Virtualization technologies allow several virtual machines (VMs) to be operated on one physical server or machine (PM). In this way the number of servers in idle-state can be reduced to save energy [6]. However, the rising energy costs lead to a rising cost pressure and further solutions are needed as they will be proposed in the following.

This paper extends our contribution to EnviroInfo 2014 – 28th International Conference on Informatics for Environmental Protection [3] and is organized as follows: Sect. 2 motivates and defines the problem of energy efficiency and integrating renewable energy in data centers. Section 3 gives background on approaches relevant to energy efficiency, virtualization technology and improving the carbon footprint. In Sect. 4, we present the resource-efficient and energy-adaptive approach. The paper is concluded by comments on our progressing work in Sect. 5.

2 Problem Definition

The share of volatile renewable power sources is increasing. This leads to volatile energy availability and lastly to varying energy price models. To deal with the variable availability, we need an approach that ensures controllable power consumption beyond general energy efficiency. Thus, we need to improve the efficiency of the data center using an intelligent and efficient VM placement in order to adapt to volatile energy availability and improve carbon footprint while keeping the overall goal to use the invested energy as efficient as possible.

The increasing amount of IT services combined with steadily raising energy costs place great demands on data centers. These conditions induce the need to operate a maximum number of IT services with minimal employment of resources, since the aim is an economical service operation. Therefore, the effectiveness of the invested power should be at a maximum level. In this paper, we focus on the server's power consumption and define the efficiency of a server as the work done per energy unit [5].

In the related work part of this paper, we analyze different kinds of approaches in the context of energy consumption, energy efficiency and integrating renewable power. In this research approach, we want to explore which further options exist to use energy efficient and how we can take effect on the data center's power consumption and, finally, to adapt it to available volatile renewable energy.

To take advantage of current developments, power consumption should be increasable in times of low energy prices and reducible otherwise while we stick to a high efficiency level in both cases. In Service Level Agreements (SLAs) for instance, a specific application throughput within a time frame is defined. Due to these agreements, we can use periods of low energy prices to produce the throughput far before the time frame exceeds. In periods of high energy prices, a scheduled decrease of the previously built buffer can be used to save energy costs.

Some approaches [4, 10, 15] use geographically-distributed data centers to schedule the workload across data centers with high renewable energy availability. The methodology is only suitable in big, geographically-spread scenarios and the overall power consumption is not affected. Hence, we do not pursue these approaches. In general, many approaches are based on strategies with focus on CPU utilization because CPU utilization correlates with the server's power consumption directly [5]. The utilization of other server components does not have such an effect on the server's power consumption. However, the application's performance depends not only on CPU usage, but all required resources are needed for optimal application performance. Hence, the performance relies on other components too and we also want to focus on these other components such as Network Interface Card (NIC), Random Access Memory (RAM) and Hard Disk Drive (HDD) to improve the efficiency, especially if their utilization does not have an adverse effect on the server's power consumption. Our assumption is that the optimized utilization of these resources is not increasing the power consumption, but it can be used to improve the efficiency and application performance.

There are different types of applications; some applications work stand-alone while others rely on several components running on different VMs. Components of the latter communicate via network and the network utilization takes effect on such distributed applications. In our approach, we want to include these communication topology topics. However, the applications' requirements are changing during operation, sometimes in large scale and in short intervals. Therefore, we need an online algorithm that acts at runtime to respond to changing values. We need to keep obstacles at a low level by acting agnostic to the applications. The capable approach should be applicable without the need to change the operating applications. This is a crucial requirement in the context of Infrastructure-as-a-Service (IaaS) environments.

Being agnostic to applications means to influence their performance without they become aware of our methodology. For example, if an application intends to write a file on the hard disk, it has to wait until it gets access to the hard disk. This is a usual situation an application can handle. In the wait state, the application cannot distinguish whether the wait was caused by another application writing on the hard disk or by our methodology.

The problem of determining an efficient VM placement can be formulated as an extended bin-packing problem, where VMs (objects) must be allocated to the PMs (bins). In the bin-packing problem, objects of different volumes must be fitted into a finite number of bins, each of the same volume, in a way that minimizes the number of bins used. The bin-packing problem has an NP-hard complexity. Compared to the VM allocating problem, we have a multidimensional bin packing problem. Instead of the object size, we have to deal with several resource requirements of VMs.

In a data center with k PMs and n VMs operated on the PMs, the number of configuration possibilities is described by partioning a set of n elements into k partitions while the k sets are disjoint and nonempty. This is described by the Stirling numbers of the second kind:

$$S_{n,k} = \frac{1}{k!} \sum_{j=0}^{k} (-1)^{k-j} \binom{k}{j} j^{n}$$

In case of a data center with 10 VMs and 3 PMs, we have $S_{10,3} = 9330$ different and possible VM allocations to the PMs that are named as configurations in this paper.

Hence, a global bin-packing solver will not be able to deliver a VM placement for a fast acting online approach.

The formal description of the VM placing problem relating to the bin-packing problem is as follows: A set of virtual machines $\mathrm{V} = \{\mathrm{VM}_1, \ldots, \mathrm{VM}_n\}$ and a set of physical machines $\mathrm{P} = \{\mathrm{PM}_1, \ldots, \mathrm{PM}_k\}$ is given. The VMs are represented by their resource demand vectors d_i. The PMs are represented by their resource capacity vectors c_s. The resource capacity vector of a PM describes the available resources that can be requested by VMs. The goal is to find a configuration so that for all PMs in P:

$$\sum_{i=1}^{j} d_i \leq c_s$$

while j is the total number of VMs on the PM.

To measure the quality of an allocated configuration C, the efficiency E defined by:

$$E(C) = \frac{work\,done}{unit\,energy}$$

is a suitable metric [5]. The aggregated idle times of the PMs may also indicate the quality of the configuration.

To the best of our knowledge, this is the first approach that researches on agnostic methodologies, without scheduling components, to control the data centers power consumption with the aim of efficiency and the possibility to increase and decrease the power consumption as well.

3 Related Work

Power consumption and energy efficiency in data centers is a topic, on which a lot of work has already been done. In this section, we give an overview of different approaches.

The usage of low-power components seems to offer solutions for lower energy consumption. Meisner et al. [12] handled the question whether low power consumption correlates with energy efficiency in the data center context. They discovered that the usage of low power components is not the solution. They compared low power servers with high power servers and defined the energy efficiency of a system as the work done per energy unit. They achieved better efficiency with the high power servers and found that modern servers are only maximally efficient at 100 % utilization.

Another potential for improvement is to let IT requirements follow energy availability. There are some approaches [4, 10, 14] that use local energy conditions. They migrate the server workload to data center destinations with available renewable power. These ideas are finally only suitable for distributed and widespread data centers. Data center locations at close quarters typically have the same or not significantly different energy conditions. In the latter scenario, the consumption of renewable energy can be increased, but the efficient power usage is not taken into consideration.

A different idea is mentioned by Krioukov et al. [9]. In this work, a scheduler has access to a task list, where the task with the earliest deadline is at the top. This is an earliest deadline first (EDF) schedule. If renewable energy is available, the EDF scheduler starts tasks from the top of the task list to use the renewable energy. If less energy is available, tasks will be terminated. In such approaches, we have to deal with application-specific topics. To build a graded list of tasks to schedule, we determine the duration a task needs to be processed and we need a deadline for each task to be processed. Terminated tasks lead to application-specific issues that need to be resolved afterwards.

The approach of Hoyer [8] bases on prediction models to calculate the needed server capacity in advance to reduce unused server capacity. Optimistic, pessimistic and dynamic resource strategies were presented. This approach offers methodologies to improve efficiency, but controlling the data centers power consumption is not focused.

Tang et al. [17] propose a thermal-aware task scheduling. The ambition is to minimize cooling requirements and to improve the data center efficiency in this way. They set up a central database with server information, especially server heat information. An EDF scheduler is placing tasks with the earliest deadline on the coldest server. Thus, they avoid hot spots and cooling requirement can be decreased to improve efficiency. The usage of a graded task list comes with the same disadvantages as described before. To avoid dealing with application-specific topics, the virtual machine is a useful container to place IT loads instead of explicit application tasks. In many approaches, for example Corradi et al. [6], power consumption is reduced by concentrating VMs on a fewer number of servers and powering down unused ones to save energy. Chen et al. [5] describe the power consumption of a server as the sum of its static power consumption and its dynamic power consumption. The static power consumption is the consumption of the server in power-on state without workload. This amount of power can be saved with this approach. The dynamic part of server's power consumption correlates with its CPU utilization, as described by Pelley et al. [13]. Thus, most methodologies are only focused on CPU utilization.

Dalvanadi et al. [7] and Vu et al. [18] pointed out that network communication can also influence the overall performance of an IT service and network-aware VM placement is also an important and challenging issue. Hence, they embrace network traffic to minimize power consumption.

As described, many approaches [1, 15, 19] use virtualization technologies to concentrate VMs on a small number of PMs. While migrating VMs onto a PM, the size of the RAM is a limiting factor. If the RAM-size of the PM is exhausted, further VMs cannot be migrated onto this PM. This can be an adverse effect, especially if resources such as CPUs are still underutilized or completely idling. The memory sharing technology offers the possibility to condense redundant memory pages on a PM to one page. Unneeded physical memory can be freed to improve the VMs memory footprint. The VMs run on top of a hypervisor, which is responsible for allocating the physical resources to individual VMs. The hypervisor identifies identical memory pages on the different VMs on a PM and shares them among the VMs with pointers. This frees up memory for new pages. If a VM's information on that shared page changes, the hypervisor writes the memory to a new page and re-addresses a pointer. The capacity of the PM can be increased to concentrate further VMs on the PM and to achieve higher server utilization. Wood et al. [19] present a memory sharing-aware placement approach for virtual machines that includes a memory fingerprinting system to determine the sharing potential among a set of VMs. In addition, it makes use of live migration to optimize the VM placement.

In summary, the state of the art approaches deliver several solutions in the context of energy efficiency, but an efficiency-aware approach with combined data center power control mechanisms is still missing.

4 Resource-Efficient and Energy-Adaptive VM Placement Approach

In this section the in-progress idea for resource-efficient and energy-adaptive VM placement in data centers is proposed. To optimize the server utilization, many data center operators already use server virtualization technologies and operate several virtual machines on one physical server. This technology is the base for our further optimizations. In our approach, we are at the point that the first steps of optimizations have already been done. Hence, we are running a set of VMs concentrated on a small number of potential servers. Unused servers are already switched off. As further input, we get a target power consumption value.

It is generally accepted that applications operate ideally if they have access to all required server resources. With the aim of improving the data center's efficiency, resource-competing VMs should not be operated on the same physical server together. Our approach is to create a VM allocation that concentrates VMs with suitable resource requirements on the same physical server for ideal application performance and efficiency. In this constellation, each application has access to the required server resources and operates ideally. Finally, the overall server resources are more utilized than before and the efficiency rises. Beside the increased efficiency, this situation also leads to a higher power consumption and application performance. This scenario is suitable for times of high energy availability. Following the idea of green energy usage, this technology is also capable of reducing the data center's power consumption in situations of less green power availability. Therefore, the methodology can be used to explicitly reduce resource utilization by combining resource-competing applications, leading to lower power consumption but also to a potentially reduced application performance.

In data centers, applications induce specific power consumptions by their evoked server load. This required amount of power is so far understood as a fixed and restricted value. Our concept is to let this amount of power become a controllable value by applying a corresponding VM allocation.

The power consumption PC_{dc} of a data center breaks down as follows:

$$PC_{dc} = PC_{Support} + PC_{Servers}$$

The total power consumption is the sum of the power consumption of all data center components. Beside the power consumption of all PMs $PC_{Servers}$, we have the power consumption of the support infrastructure $PC_{Support}$ i.e. network components, cooling components, UPS, lights, etc.

Chen et al. [5] describe the power consumption of a server as the sum of its static (idle, without workload) power consumption $PC_{Servers\,idle}$ and its dynamic power consumption $PC_{servers\,dyn.}$:

$$PC_{Servers} = PC_{Servers\,idle} + PC_{servers\,dyn.}$$

$PC_{servers\,dyn.}$ is the amount of power we directly take influence on. It reflects the amount of power consumption deviance between 100 % server utilization and idle mode. Idle servers still consume 60 % of their peak power draw [11].

Hence, a sustainable part (up to 40 %) of the server's power consumption is controllable; it can be increased in times of high energy availability and decreased otherwise. Our approach is based on virtualization technology and the possibility to live-migrate VMs. The methodology is agnostic to the operating applications. This is an advantage compared to other task scheduling-based algorithms, since these have to deal with task execution times and other application-specific topics. In our approach, the applications are untouched and the technology is non-invasive regarding the applications; it only takes effect on the availability of server resources. The variable availability of server resources is a usual setting that applications are confronted with.

As described in the related work part of this paper, the PM's RAM can be a limiting factor while migrating further VMs to the PM. In addition, we make use of the technology to share RAM across the VMs to increase the number of VMs operated on a PM.

The following diagrams illustrate the practice, how the methodology's strategy migrates VMs between physical servers.

In Fig. 1.1, the initial, non-optimized situation is displayed showing a set of VMs operated on three physical servers. The resource utilization is highlighted (lighter colors meaning low, darker colors high utilizations). On PM2, for example the performance is affected by high network utilization.

Our methodology achieves an equilibrium allocation regarding the resource utilization, as shown in Fig. 1.2. VMs to migrate are chosen depending on their RAM size and their fraction of scarce resource utilization. The subsequent VM permutation leads to an average utilization of all involved resources. Hence, the approach increases efficiency and power consumption by resource usage optimization.

The configuration is suitable for times of high energy availability and low energy prices. In periods of less available renewable energy or high energy prices, we need to reduce the power consumption while keeping a high efficiency level.

The situation, as shown in Fig. 1.3, is the result with reduced power consumption objectives. The CPU utilization is reduced to likewise reduce the power consumption as well while the utilization of other resources is balanced. The result is the most effective constellation at reduced power conditions. The Dynamic Voltage

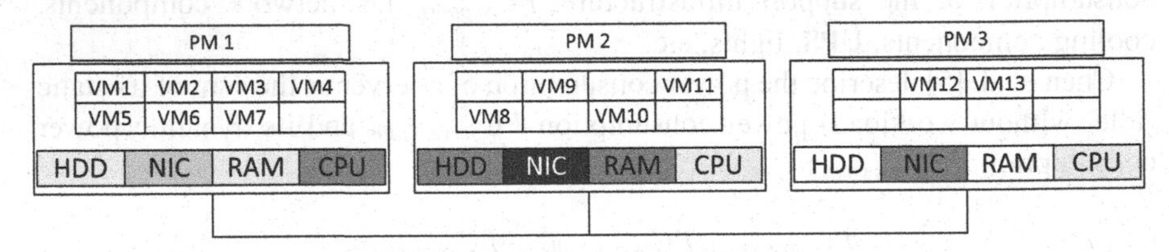

Fig. 1.1 Schematic VM on physical server diagram: initial situation

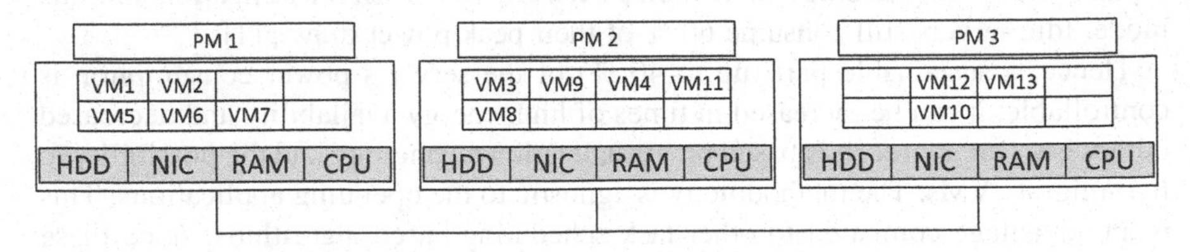

Fig. 1.2 Schematic VM on physical server diagram: optimized situation

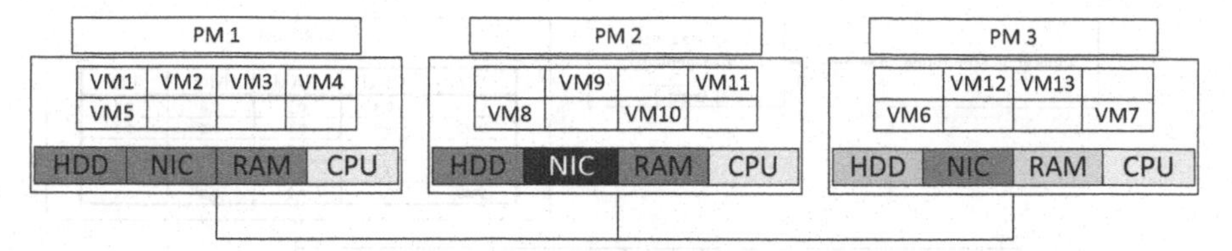

Fig. 1.3 Schematic VM on physical server diagram: aim of reduced power consumption

and Frequency Scaling (DVFS) technique is used to adapt the power consumption to the actual CPU utilization. DVFS allows a dynamical adaption of CPU voltage and CPU frequency according to the current resource demand.

4.1 System Model

As described, we have an NP-hard complexity if we stick to methodologies that involve all possible configurations to find the best suitable configuration for our actual requirements. To reduce the complexity and the long computation time, we change from an all-embracing global solution to local solving strategies. On the one hand, this is required for online acting approaches and on the other hand, we assume that we will not get significantly better overall solutions if we include all VMs to find a suitable configuration.

In Fig. 1.4, a component model of the entire system is shown. We have an application-monitoring component that delivers information about the applications and servers to the service level management (SLM). The SLM component contains all service level agreements (SLAs) and calculates new power target values for the data center to observe the SLAs. These values are propagated to all optimizers, working on every physical server. The optimizer compares the new incoming target values with its own actual value. If the difference is in range of a predefined hysteresis, the optimizer does not take any action. Otherwise it starts optimization. If the target is not in the predefined range and the actual value is lower than the target, the optimizer resolves resource competing constellations and hosts additional VMs from the offer pool. In the offer pool, all distributed optimizers can announce VMs, for example, if they do not fit to their actual placement strategy. The VMs in the offer pool are represented with their resource requirements that are the base for later VM placement swaps. If the actual value is higher than the target, the optimizer arranges a resource competing allocation to reduce the power consumption.

The energy availability, energy prices and service level values are independent and global values to aggregate to a target power consumption value. This is a task for the central service level management (SLM) component of our system. Here we do not have any local issues to attend, so we can calculate these values globally. As an additional effect of the globally defined target power consumption value, we

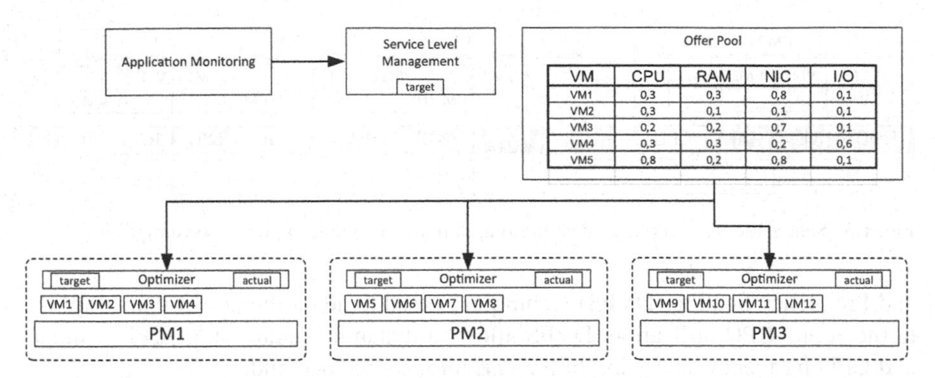

VM	CPU	RAM	NIC	I/O
VM1	0,3	0,3	0,8	0,1
VM2	0,3	0,1	0,1	0,1
VM3	0,2	0,2	0,7	0,1
VM4	0,3	0,3	0,2	0,6
VM5	0,8	0,2	0,8	0,1

Fig. 1.4 Schematic system model

have evenly distributed server utilization. This reduces the occurrence of hot spots, similar to the approach mentioned by Tang et al. [17].

We use a local optimizer component working on a single PM that focuses on a solution for its own PM. This component has to find a solution for just one PM and the set of possible VMs is reduced to the actual operated ones and to a subset of those in the offer pool. As input, the optimizer receives a defined target power consumption, which has to be reached with best possible efficiency.

4.2 Algorithm

CPU utilization is the most effective value regarding power consumption as mentioned before. In other words, the overall CPU utilization is the value to increase or decrease to take effect on the data center's power consumption. Our approach uses competing resource allocations to slow down applications and in series the CPU utilization. Consolidating VMs on a PM that utilize the same resources except the CPU can accomplish this. Consequently, the CPU utilization and power consumption decreases. This practice affects the application's performance and we need a feedback that is sent from the application-monitoring component to the SLM component to ensure the SLAs. With the information about the SLAs and actual application performance, the SLM component is able to calculate power consumption target values that achieve the economic data center objectives.

The target power consumption is broadcasted to all PMs. The PM has got an optimizer component that receives the target and compares it with its actual value. If the target is similar to the actual value, the optimizer does not interfere. Otherwise it starts optimizing. While doing this, the focus is kept on balanced resource utilization and efficiency. Hence, the overall CPU utilization is reduced or increased but all other resources are used as efficiently as possible. Balanced resource utilization is always the goal except for CPU utilization and resources that are used to build the competing resource situation. Merely the attainable CPU utilization is a variable and implicit value that corresponds to the power consumption target value.

Every PM's optimizer strives to reach the target value by optimizing its own situation. We have an offer pool of VMs, which can be accessed by every PM's optimizer. The optimizer is able to read the offered VMs from other PMs or even to offer VMs. If the target value is greater than the actual value, the optimizer removes suitable VMs from the pool to host until the target value is reached. If the target is lower than the actual value, the optimizer offers VMs to the pool to reduce the own value. Furthermore, additional VMs can be hosted from the pool to create competing resource situations to reduce the CPU utilization and to reach the target value. Developing a reduced power consumption VM allocation can be done in three ways:

(i) Migrate VMs to other PMs. This reduces the CPU utilization and the power consumption by DVFS technology.
(ii) The optimizer arranges a resource competing allocation, which reduces the CPU utilization and-as a result-decreases the power consumption by DVFS technology.
(iii) The optimizer arranges CPU overprovisioning. CPU utilization is already at 100 % and further VMs will be hosted. The additional VMs do not increase the PM's power consumption but reduce the power consumption of the PM they came from. Hence, the overall power consumption is decreasing.

The strategy to reduce the power consumption starts with (i) and is cascading down to the methodology of (iii). At first, the target is strived with (i), if this is not leading to the required results, we go on with (ii) and lastly with (iii). Using the methodology of (i) means, we have no further risks of SLA-violation because the application's performance is not influenced. In (ii) und (iii) we potentially slow down the applications, probably increasing the risk of SLA violations. Hence, the methodology always starts in step (i).

The formal description of the efficiency and power consumption problem is a follows: A set of virtual machines $V = \{VM_1, \ldots, VM_n\}$ and a set of physical machines $P = \{PM_1, \ldots, PM_k\}$ is given. The VMs are represented by their resource demand vectors d_i. The PMs are represented by their resource capacity vectors c_i. The goal is to find a configuration C so that for all PMs in P:

$$\sum_{i=1}^{j} d_i \leq c_s + x_s$$

where the vector x_s is an offset to control under- and overprovisioning of the server resources on PM_s while j is the total number of VMs on the PM_s. We use x_s to control the resource utilization on the PMs to induce the intended server utilization and thereby their power consumption.

To measure the quality of an allocated configuration C, we have now two different metrics. On the one hand, we have the efficiency E:

$$E(C) = \frac{work\,done}{unit\,energy}$$

On the other hand, we have the difference Δ between the PMs power consumption PC_{server} and the target power consumption PC_{target}:

$$\Delta_{(target,\,C)} = \left|PC_{target} - PC_{server}(C)\right|$$

The Δ represents the deviance (positive) from the target power consumption. In case of lower target power consumptions, a lasting deviance is the indicator to go on with the next step (ii) or (iii).

The process of reaching a suitable VM placement and the behaviour of the locally executed optimizer is demonstrated by the following pseudo code:

Inputs: ***t*** target power consumption for local PM, ***p*** actual PM's power consumption, resource utilization
Output: VM placement for local PM that evokes target power consumption

1. receive new target ***t*** given by SLM component
2. ***if t > p*** and the PM's CPU utilization is 100 %, offer VMs to other PMs via offer pool
3. ***if t > p*** and the PM's CPU utilization is lower than 100 % and all other resources are underutilized, the PM invites VMs to shelter from other PMs with high CPU utilization
4. ***if t > p*** and the PM's CPU utilization is lower than 100 % and other resources are strong utilized, offer VMs to other PMs to solve the competing resource situation
5. ***if t < p*** and the PM's CPU utilization is lower than 100 % and other resources are strong utilized, invite VMs to shelter from other PMs with high CPU utilization
6. ***if t < p*** and the PM's CPU utilization is 100 %, invite VMs to shelter from other PMs to create resource competing situation
7. ***if t = p*** do nothing

In addition to the event of changing power consumption targets, we have further events to deal with. During the operation a host can become over- or underloaded. A PM's overload might lead to SLA violations and an underload means that the efficiency is not at optimum level. Depending on the actual power consumption targets, we have to act in different ways. If the power consumption target is higher than the PM's actual value and an underload is detected, the condition of:

$$\sum_{i=1}^{j} d_i \leq a_s + x_s$$

implies to resolve resource competing allocations or to host further VMs.

If an overload is detected, the condition might be fulfilled. If x_s is used to reduce the power consumption and to strive the power consumption target, we have a pseudo-overloaded PM. The SLA component decides weather to risk SLA violations and stick to the values of x_s, or to consume expensive energy, for example and to adapt x_s for using additional CPU resources.

Another point is to find candidates for VM migrations. There are some key properties, which suitable candidates should fulfill. In general, the target PM must provide the required RAM space to avoid page swapping. The migration costs for a new VM allocation are a substantial topic we have to look at. Instead of choosing randomized candidates, we sort the set of candidates by their RAM size at first to build an ordered list. The RAM size indicates the duration of the migration because copying the RAM pages to the target PM is a major part in the migration process. In a second step, the resource requirements will be taken into account, reducing the set of potential candidates and leading to a graded list of suitable VMs.

Finally, the local optimizer initializes the migration of the best fitting VM.

4.3 Future Work and Experiments

In experiments, we evaluated scenarios with different VMs to validate our approach, based on affecting server's power consumption and application performance by applying various VM allocations. The test-VMs are running benchmarks, simulating applications that rely on different server resources (VM1 is running RAM benchmarks, VM2 is running file benchmarks and VM3 is running CPU benchmarks). The results show that VM placement strategies can improve the performance up to 34 % and increases the power consumption up to 16 %. Vice versa, VM permutations can decrease the power consumption up to 16 % (up to 50 % if idle PMs are switched off) and even decrease the application performance up to 34 %.

The test results are shown in Table 1.1. The first columns illustrate the VMs allocated to the physical servers PM1 and PM2, followed by the corresponding power consumption of the PMs. The performance columns are containing information about the achieved performance per VM. We normalized the performances with the achieved performance operating the specific VM on the PM separately. VM2 reaching lightly more than 100 % performance is caused by measurement

Table 1.1 Test results

Case	PM1	PM2	PM1 [W]	PM2 [W]	Perf. VM1 [%]	Perf. VM2 [%]	Perf. VM3 [%]	Ø Perf. [%]	Σ PC [W]	Ø Perf./Σ PC
1	VM1, VM2, VM3	–	54.1	38.0	43.62	100.00	44.80	62.81	92.1	0.6840
2	VM1, VM2	VM3	55.9	53.4	89.22	97.56	100.00	95.59	109.3	0.8692
3	VM1, VM3	VM2	55.0	39.9	49.51	100.00	50.32	66.61	94.9	0.7060
4	VM2, VM3	VM1	54.6	55.0	100.00	101.88	86.39	96.09	109.6	0.8759

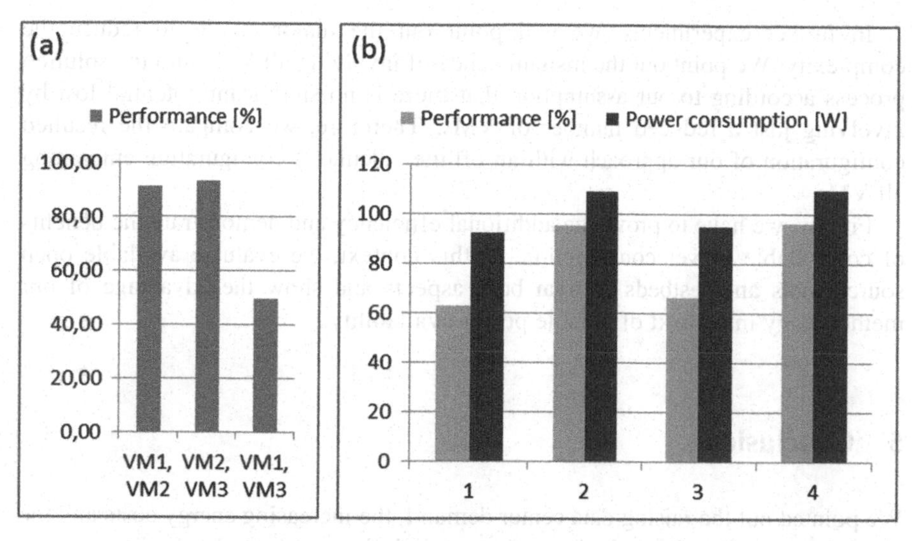

Fig. 1.5 Performance dependencies (**a**); Power consumption and performance per test-case (**b**)

uncertainty. The last columns summarize the average performance and power consumption (PC) results.

The last column shows the performance per power consumption as degree of efficiency. The fourth test-case attains the best efficiency. Figure 1.5b gives an overview of the achieved performance and power consumption. In Fig. 1.5a, the performance depending on the combined VMs is displayed. We got lightly different performance combining VM1, VM2 and VM2, VM3, but operating VM1 and VM3 together causes massive performance losses. The reached performance indicates variations of more than 40 % and shows high potential of our approach.

Our primary goal is to increase the data center's power efficiency. The essential research work is to analyze the different reachable effects by combining further methodologies, for example RAM-sharing, network aware approaches and integrating further resources (such as RAM, NIC and HDD) into the approach as described before.

In our ongoing work, we evaluate performance metrics to estimate the quality of VM allocations. The aggregated idle times of a PM just indicate the efficiency of the PM usage, as described. In order to be able to make sound statements on the quality of our VM allocations, we need further metrics, which allow us to involve the power consumption targets and to get a weighted estimation of the quality on efficiency and deviance to target power consumption.

To ensure comparability with other dynamic VM placing approaches, we implement our methodology into a standard framework such as OpenStack. It is planed to evaluate the methodology in an OpenStack testbed by applying it to different kinds of applications and different workloads. To generate reliable and comparable benchmark results, we execute benchmarks with standard applications and real world workload traces as mentioned in [2].

In further experiments, we will point out the major effects to reduce the complexity. We point out the insignificance of involving all VMs into the solution process according to our assumption that there is no significant potential lost by involving just a reduced number of VMs. Therefore, we compare the resulted configuration of our approach with an offline calculated configuration embracing all VMs.

Finally, we have to prove the additional efficiency and demonstrate the benefits of controllable power consumption. In this context, we evaluate available open source tools and testbeds to treat both aspects and show the advantage of our methodology in context of volatile power availability.

5 Conclusion

We pointed out the raising data center demand, the increasing energy costs and the requirement to handle volatile energy availability respectively. In Sect. 3, we presented different approaches related to energy efficiency, power consumption and usage of renewable power in data centers. We defined the problem of energy efficiency and proposed a resource-optimization approach that improves overall energy efficiency and also allows controlling actual data center power consumption without application-invasive measures. Our approach is an instrument to increase efficiency and to adapt to renewable power availability, both having a positive effect on the carbon footprint.

References

1. Beloglazov, A. and Buyya, R.: Optimal online deterministic algorithms and adaptive heuristics for energy and performance efficient dynamic consolidation of virtual machines in Cloud data centers. Concurrency and Computation: Practice and Experience, 2012.
2. Beloglazov, A. and Buyya, R.: OpenStack Neat: a framework for dynamic and energy-efficient consolidation of virtual machines in OpenStack clouds, Concurrency and Computation: Practice and Experience, 2014.
3. Borgerding, A. and Schomaker, G.: Extending Energetic Potentials of Data Centers by Resource Optimization to Improve Carbon Footprint. EnviroInfo 2014 - 28th International Conference on Informatics for Environmental Protection, pages 661–668, 2014.
4. Chen, C., et al.: Green-aware workload scheduling in geographically distributed data centers. 4th IEEE International Conference on Cloud Computing Technology and Science Proceedings, pages 82–89, 2012.
5. Chen, H., Caramanis, M.C. and Coskun, A.K.: The data center as a grid load stabilizer, Design Automation Conference (ASP-DAC), 2014 19th Asia and South Pacific, pages. 105-112, 20–23, Jan. 2014.
6. Corradi, A., Fanelli, M. and Foschini, L.: VM consolidation: A real case based on OpenStack Cloud, Future Generation Comp. Syst., 32, pages 118–127, 2014.
7. Dalvandi, A., et al.: Time-Aware VM-Placement and Routing with Bandwidth Guarantees in Green Cloud Data Centers. In Proceedings of the 2013 I.E. International Conference on Cloud

Computing Technology and Science - Volume 01 (CLOUDCOM '13), Vol. 1. IEEE Computer Society, Washington, DC, USA, pages 212–217, 2013.
8. Hoyer, M.: Resource management in virtualized data centers regarding performance and energy aspects, Doctoral dissertation, Oldenburg, Univ., Diss., 2011.
9. Krioukov, A., et al.: Integrating renewable energy using data analytics systems: Challenges and opportunities. IEEE Data Eng. Bull., 34(1), pages 3–11, 2011.
10. Liu, Z., et al.: Greening geographical load balancing. In Proceedings of the ACM SIGMETRICS Joint International Conference on Measurement and Modeling of Computer Systems, SIGMETRICS '11, pages 233–244, New York, NY, USA, 2011.
11. Meisner, D., et al.: PowerNap: eliminating server idle power. In Proceedings of the 14th international conference on Architectural support for programming languages and operating systems (ASPLOS XIV). ACM, New York, NY, USA, pages 205–216, 2009.
12. Meisner, D., et al.: Does low-power design imply energy efficiency for data centers? In Naehyuck Chang, Hiroshi Nakamura, Koji Inoue, Kenichi Osada and Massimo Poncino, editors, ISLPED, pages 109–114. IEEE/ACM, 2011.
13. Pelley, S., et al.: Understanding and abstracting total data center power. Proc. of the 2009 Workshop on Energy Efficient Design (WEED), Jun. 2009.
14. Qureshi, A., et al.: Cutting the electric bill for internet-scale systems. SIGCOMM Comput. Commun. Rev., 39(4), pages 123–134, August 2009.
15. Schröder, K., Schlitt, D., Hoyer, M. and Nebel, W.: Power and cost aware distributed load management. In Proceedings of the 1st International Conference on Energy-Efficient Computing and Networking (e-Energy '10). ACM, New York, NY, USA, pages 123–126, 2010.
16. Smith, J. W. and Sommerville, I.: Workload Classification & Software Energy Measurement for Efficient Scheduling on Private Cloud Platforms, CoRR abs/1105.2584, 2011.
17. Tang, Q., et al.: Thermal-Aware Task Scheduling for Data centers through Minimizing Heat Recirculation, The IMPACT Laboratory School of Computing and Informatics Arizona State University Tempe, AZ 85287, 2008.
18. Vu, H. and Hwang, S.: A Traffic and Power-aware Algorithm for Virtual Machine Placement in Cloud Data Center. International Journal of Grid & Distributed Computing 7.1, 2014.
19. Wood, T., Tarasuk-Levin, G., Shenoy, P., Desnoyers, P., Cecchet, E. and Corner, M. D.: Memory buddies: exploiting page sharing for smart colocation in virtualized data centers. In Proceedings of the 2009 ACM SIGPLAN/SIGOPS international conference on Virtual execution environments (VEE '09). ACM, New York, NY, USA, pages 31–40, 2009.

Chapter 2
Expansion of Data Centers' Energetic Degrees of Freedom to Employ Green Energy Sources

Stefan Janacek and Wolfgang Nebel

Abstract Rising power consumption of data centers is a topic of great concern and therefore several power saving technologies exist. This paper describes the idea of a data center overall power saving and controlling strategy, allowing the data center to enter optimized minimum power states while still keeping up flexibility but also to control its own power consumption to apply demand response management. Therefore, the degrees of freedom a virtualized data center has are modeled and the methodology used to control its energy state is described, taking into account the IT hardware like servers and network gear as well as the influence of cooling devices and power distribution devices. This knowledge can also be used to react to data center emergencies like cooling failures. We describe our simulation models, the methodology and the power saving potential of our system. We formulate the problem to control the data center's power consumption by applying different consolidation strategies as an extended bin packing optimization problem, where virtual machines must be packed on a specific number of servers. External constraints like the time-flexibility of the solution and the influence on supporting devices are applied by using cost functions. We present a solving approach for this problem in the form of the data center multi-criteria aware allocation heuristic *DaMucA* and compare it to the traditional first fit decreasing heuristic.

Keywords Data center • Power efficiency • Power regulation • ICT simulation • Bin packing • Multi-criteria optimization

S. Janacek (✉)
OFFIS e.V., 26121 Oldenburg, Germany
e-mail: stefan.janacek@offis.de

W. Nebel
University of Oldenburg, 26111 Oldenburg, Germany
e-mail: wolfgang.nebel@uni-oldenburg.de

J. Marx Gómez et al. (eds.), *Advances and New Trends in Environmental and Energy Informatics*, Progress in IS, DOI 10.1007/978-3-319-23455-7_2

1 Introduction

Information and Communication Technologies (ICT) and especially data centers play a significant role in our today's world. Growing markets as cloud computing and on-demand services fortify this trend. As a result, the power demand of ICT kept on rising during the last few years. Since energy costs have also become a major economical factor, power saving and efficiency technologies for data centers have emerged. Among them are technologies like virtualization [3], server consolidation [12], and application load scheduling to times of lower energy prices [9, 20]. A fairly recent trend is to enable the data center to benefit from renewable energy sources [11, 25, 32], allowing it to operate at full load in times of high availability and cutting its load otherwise. Unfortunately, this methodology needs to alter the running applications, stopping their execution in the worst case or it needs a network of connected data centers in different geographical locations. This approach may be viable for some scenarios; however, often this is not possible. Instead, this paper proposes the idea of expanding the degrees of freedom a data center already has without altering any of its running applications. The goal of the presented methodology is to let the data center mostly operate in a minimum energy state; however, to allow demand response management (DRM), it should be able to enter a specific energy state, hence be able to control its power consumption. This could, for example, be used to follow an external power profile induced by a Smart Grid. If the data center wants to participate in DRM, hence react to demand response requests (DRR), it has to fulfill certain qualifications. One of them is the need to be able to control its power up to a certain error threshold [6]. If this fails, the data center operator may even have to pay penalties or lose its participation qualification. The technique described in this paper can be used to enable the data center to predict its possibilities to apply its power consumption to a certain request. This allows the operator to decide for each DRR if the data center should apply to it or not. Reacting to DRR can also help improving the overall grid stability [29].

There are also motivations internal to the data center to control the power consumption of a data center subspace (for example a room or a cage). Especially when load balancing techniques are applied, a data center may have significant diverse power states in different rooms, leading to inefficient global device states or severe cooling inequalities. In these cases, the possibility to control the power consumption of a subset of the data center's devices can become necessary. In the case of a co-location data center,[1] the operator cannot influence the server's load and hence its power consumption. Instead he has to assure that a sufficient amount of electric power is available for his customer's services. If the running services need fewer resources as expected, it may occur that already purchased power contingents at the electricity market are not used, leading to penalties for the operator [29]. To prevent this, many data center operators use load banks to reach

[1] Co-location data center business model means that customers provide their own hardware (servers) and the data center only provides the infrastructure (power, cooling, network).

the desired power consumption. Taking advantage of a power consumption control mechanism as represented by the methodology described in this paper can replace these activities.

To achieve these advantages, the data center's existing degrees of freedom are identified and expanded to be able to reach a high power consumption variability, while still keeping the applications unchanged. A key aspect is the modeling and description of interdependencies between different device categories in data centers. A base technology for the methodology proposed is server virtualization that enables the data center to live-migrate virtual machines (VM) across different physical machines (PM, the terms physical machine and server in this paper mean the same thing). By using this technology, the migration of running applications encapsulated in VMs to different PMs can be used to intensely influence the server's power consumption, also affecting the amount of cooling and uninterruptible power supply (UPS) load needed, thus changing the entire power consumption of the data center. Here, very dense states with minimal power consumption are possible, as well as loosely packed states with a higher consumption but also with increased flexibility and thermal safety, for example, regarding possible reactions to Computer Room Air Handler (CRAH) failures or similar incidents.

The methodology assumes a virtualized data center, where the following operations are allowed as adjustments: (1) VM migrations, (2) server switch-ons, (3) server switch-offs. Each action takes a specific amount of time. In this paper we address the problem of finding a suitable VM allocation on the existing servers to either enter a minimal power state or to enter a state approaching a specific power demand while taking into account possible side effects that may occur in combination with the data center's hardware devices and the time it needs to enter this state. We formulate the problem as an NP-hard extended bin packing problem [10] with a global cost function, where VMs (items) must be allocated to the PMs (bins). This approach is not new [5]; however in the approach presented here it is not always optimal to just minimize the number of active PMs to reach the desired state. Traditional heuristics like first fit decreasing (FFD) lead to non-optimal solutions. In addition to the conference paper [15], we therefore introduce the *DaMucA* heuristic. In short, the problem that needs to be solved is to find the best tradeoff between a minimal (specific in case of DRR) energy state and an operational secure state that also assures a safe operation while still accounting for possible future load boosts or possible technical infrastructure failures. To the best of our knowledge, this is the first approach that researches a methodology that is able to control the data center's power demand using these adjustments while also taking into account interdependencies of the data center's hardware devices. To be able to evaluate the specific load states of a data center in terms of power demand, a data center simulation is presented that models the server's power demand, the efficiency of UPS devices, cooling power demands via approximating meta-models and network flows in a sample simulated data center.

The rest of this paper is organized as follows: Sect. 2 lists the related work, in Sect. 3 the models and architecture of the simulation is described, while Sect. 4

shows the problem formulation and the methodology (*DaMucA* heuristic) used. In Sect. 5, we present evaluation results and analyze the potential of the approach. We conclude in Sect. 6 and describe our next steps and further research.

2 Related Work

The area of research this paper addresses is also focus of other publications. General server power models can be found in [14, 24], while [21] already proposes additional models for racks and cooling units. Energy models for data centers are found in [1, 20]. Our research partly bases on these results.

In [11], the authors propose the idea to combine a data center with a local power network that includes renewable energy sources. Such a power network is, however, less complex than a smart grid, since it only consists of power producers. The authors also cover the aspect of the intermittency of these power producers. They propose to shift the work load to other data center locations, each profiting from individual energy advantages. A similar approach is covered in [32], including weather conditions at different locations. Mohsenian-Rad et al. [17] proposes a service request routing for data centers to distribute the load according to the electric grid in a smart grid. In [25], the authors present the idea of a carbon-aware data center operation. They propose three key ideas to implement this concept: on-site and off-site renewable energies and Renewable Energy Certificates (REC). In our research, the usage of RECs is, however, not a legitimate concept.

In [4], the authors discuss the possibilities of integrating a data center into demand response schemes. They also present financial evaluations to determine, at which point the data center can benefit from these methods.

Modeling of thermal behavior of data center components, especially of servers, has been researched before. In [30], the thermal load of processors and micro controllers is considered. Cochran et al. [8] handles thermal predictions of processors and combines it with a Dynamic Voltage and Frequency Scaling (DVFS) technique.

Much of the fundamental research regarding thermal modeling of data centers was done in [19, 23]. The authors provided cooling device models and a data center room computational fluid dynamics (CFD) model. In [18], the authors describe how to achieve a thermal modeling using machine learning strategies and the integrated sensor data of IT equipment. A similar approach is presented in [28] where the authors present an abstract heat flow model that uses the onboard thermal sensors. The model can predict the temperature distribution as well as air recirculation effects. Parolini et al. [22] describes another approach of thermal device modeling using a constrained Markov decision process. Sharma et al. [26] proposes to create a balanced thermal situation in the data center with a focus on cooling energy efficiency by migrating services to different PMs in the data center. The authors use detailed CFD simulations to model the thermal data center. They also investigate the consequences of possible cooling device failures and how to react to them, however, the time it needs to migrate VMs is not considered, only the thermal basis is researched. Thereby, the work of the authors forms a base research regarding thermal data center simulation. They also propose heuristics to approximate thermal behavior. The thermal modelling proposed in this paper bases on the thermal models and heuristics found in this work. In [27] a thermal aware task scheduling is presented that aims at minimizing the energy consumption while still considering the heat situation of the servers. The research addresses data centers with blade servers. The thermal models are also based on CFD simulations.

Thermal modeling of a server rack is arranged in [7, 33] presents a dynamic model for the temperature and cooling demand of server racks that are enclosed in hot aisle containment. The correlation of power consumption and temperature of server internal coolers is investigated in [31]. As a result, the authors state that it is possible to save power under certain conditions, when the cooling chain adapts itself to a higher temperature level and the server coolers compensate this by applying a higher rotation frequency. They also model the time that cool air needs to travel from CRAH units to a specific server rack. However, a detailed correlation to server load is not handled. Al-Haj et al. [2] handles the planning of VM migrations under consideration of VM interdependencies like communication, security aspects and other SLAs. These are not considered in this paper, since it aims at showing the concept to maximize the degrees of freedom. However, the methodology proposed here can easily be adopted to also support VM interdependencies, if needed.

3 Models and Simulation Architecture

The methodology described in this paper uses a data center simulation that is able to model the power consumption of IT hardware, in this case the PMs, and the supporting devices such as cooling and UPS devices. Figure 2.1 shows the architecture of the information flow of the models used for the simulation of the hardware devices in the data center.

3.1 Server Models and Application Load

The simulated data center consists of heterogeneous server models, the model data is based on the publicly available results of the SPEC power benchmark[2] and on own measurements [14]. These models are generally transferable and characterizable for other or future devices. The total power consumption of a PM is split into two parts: the minimum static power consumption P_{st} that describes the

[2] For benchmark details and results, see https://www.spec.org/power_ssj2008/.

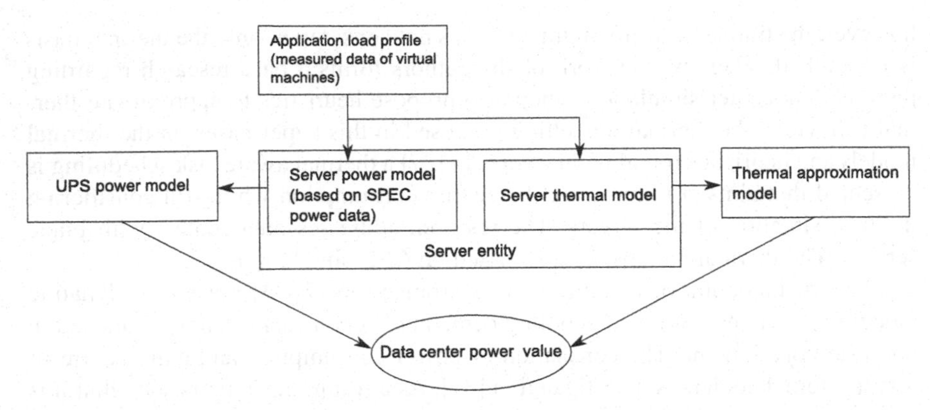

Fig. 2.1 Simulation model architecture

consumption in idle mode and the dynamic power consumption P_{dy} that is influenced by the utilization of the PM. As an indicator of the utilization, the CPU load is used as the only value; it has already been shown that it has strong correlations with the power consumption [13, 14]. The total power consumption of a PM is given by:

$$P_t = P_{st} + P_{dy}$$

The CPU utilization of a PM is calculated by adding all of the VM's utilizations at each instant of time. Let n be the total number of VMs on a server at an instant of time and $C_{VMi}(t)$ the CPU utilization of the VM i, the total CPU utilization $C_{PM}(t)$ of the PM at the time t is calculated as:

$$C_{PM}(t) = \sum_{i=1}^{n} C_{VMi}(t)$$

Additionally, a flexibility grade is introduced that defines the additional percentage of CPU load each VM can possess without overloading the hosting server. For example, a global flexibility grade of $fl = 0.2$ allows each VM to have a load boost of $C_{VMi}(t) * (1 + fl)$. Increasing the flexibility grade inevitably leads to a greater number of active servers in an allocation.

Our measurements showed that the variability in RAM allocations is very small; hence it is assumed that each VM has a static memory allocation. This value is retrieved by finding the maximum RAM allocation the VM had during the measurement duration. Each PM can operate a maximum number of VMs at each instant of time; this number is limited by the resource usage of each VM. Relevant values are the CPU load $C_{VMi}(t)$ at each time t and the RAM allocation M_{VMi} (as this value is static, it has no reference to time), where these in sum must not exceed the PM's physical resources C_{PM} and M_{PM}:

$$\forall t: C_{PM} \geq \sum_{i=1}^{n} C_{VMi}(t) \text{ and } M_{PM} \geq \sum_{i=1}^{n} M_{VMi}$$

The RAM allocation of VMs forms a hard and static boundary regarding the maximum number of VMs of each PM. Over-provisioning of RAM is not assumed. Finally, the total power consumption of a PM $P_{PM}(t)$ is calculated using the power models published in [13, 14] using the CPU utilization of the PM at the time t.

3.2 VM Allocation State

A VM allocation state A defines the power state of each PM (on or off) and for each PM that is powered on the list of VMs hosted on this PM. A state is legal, if all VMs can access the resources they need for their operation at the current time. To cross from one state to another, VMs will be migrated and PMs can be switched on or off respectively.

3.3 UPS Models

The data center simulation uses a basic UPS model scheme that evaluates the efficiency for a specific UPS device. For most UPS, the efficiency increases with rising load. Hence, the UPS should always be operated with the best efficiency factor, for example, at least with 80 % load. The methodology proposed in this paper uses the UPS model to find an allocation that leads to an improved UPS efficiency factor, compared to other methodologies that do not consider UPS power consumption. It is assumed that each UPS device has at least a minimum power consumption P_{Umin}, even if the devices (servers) attached to it are powered off. It is also assumed that UPS devices are not turned off if unused. Regarding this information, we formulate the following UPS power model that is used for the data center simulation: Let P_U be the total power consumption of all devices the UPS powers (servers) including the UPS device's own consumption and P_D the power consumption of all devices attached to the UPS. The efficiency factor function $i(P_D)$ defines the UPS efficiency at the power load P_D. Then $P_U(P_D)$ can be calculated as:

$$P_U(P_D) = \begin{cases} P_{Umin}, \text{ if } P_D < P_{Umin} \\ P_D + (1 - i(P_D))^* P_D, \text{ else} \end{cases}$$

3.4 Thermal Models

The thermal models in this simulation are used to evaluate the power consumption of the cooling devices depending on the workload of servers in different data center locations (room, racks, cages). Furthermore, they need to provide a thermal performance index for each rack and server that identifies its cooling efficiency. Using this index, the algorithm can prioritize efficient areas in the data center. Accurate and complex thermal simulations use CFD to model the airflow, pressure and temperature in a data center room [26]. However, these simulations need to be prepared for each data center room, meaning that the room needs to be modeled as a 3D space and the simulation itself takes a substantially amount of time, often several hours. The methodology proposed in this paper needs fast decisions, because it has to immediately react to DRR or to emergencies. The models needed here do not need to offer best accuracy, but instead, they need to give an approximation about a possible resulting situation. Another drawback of the CFD modelling is the need to create a specific simulation for each single data center room, small changes in the real room need a remodeling of the CFD 3D model. It is more practical to develop a simulation scenario, where static models can be automatically characterized to a specific real world situation. In the case of thermal data center simulation, this can be achieved by feeding a prepared model with thermal measurement data of a server room [18]. Many servers or racks in today's data centers already provide temperature data. Most important are the server's inlet and outlet air temperatures, or, if the inlet temperature is not known, the temperature of the air entering the room out of the CRAH units. Using this data, a thermal model of the data center room and its devices (servers) can be derived that provides sufficient accuracy for the needs of the methodology presented. This measurement data can also be used to account for air recirculation effects.

As presented in [26], the server's thermal sensors are used to measure the exhaust temperature T_{ex} and the cold air temperature entering the room T_c (or the server's inlet temperature respectively, if such a sensor exists). The difference between these temperatures marks the excess temperature rise and defines the heat dissipation effectiveness. Using this value, the thermal performance index can be derived and it can be determined, which regions are more effective than others and thus should be preferably used.

As heat energy models, a similar methodology as described in [16] is used. It is assumed that the heat produced by the servers Q is equal to the power consumption of these devices so that $Q = P_{servers}$. Based on these models, we define the function $P_{th}(P_{servers})$ that calculates the needed cooling power for a given server power consumption at the time t.

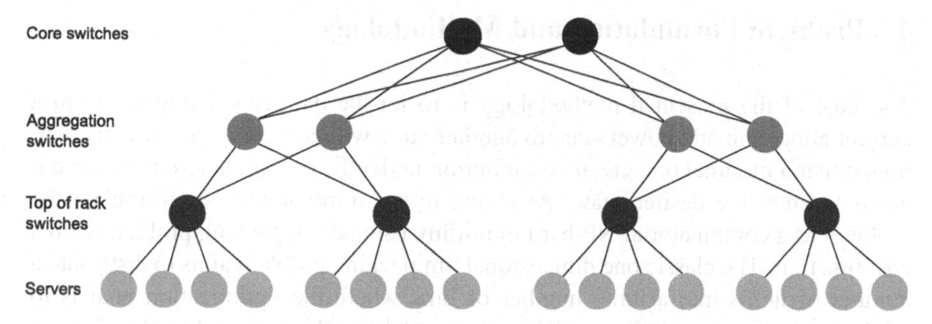

Fig. 2.2 Network topology model (VL2) with redundant switches on each level

3.5 Network Topology Model

The network of the simulated data center is modeled as a graph while the used topology is VL2 (see Fig. 2.2). It is assumed that the network connections between the different switch layers have different bandwidth sizes, allowing different amounts of parallel network traffic. In this paper, the network graph is used to determine the amount of live-migrations of VMs that can be performed in parallel. To be able to reach a different VM allocation state, often several migrations will occur; if most of them can be run in parallel, the target allocation state can be reached in less time.

The following rules apply to parallel migrations: (1) Each PM may only be source or target of one migration at the same time. (2) Each switch node in the network graph can only handle as much migrations so that the maximum bandwidth is not exceeded for more than 50 % (this value can be defined by the user/data center operator). In our model, this ensures that the running applications can still access the network safely; in other network scenarios this value might be changed according to the real conditions. When a new allocation state should be entered or evaluated, our algorithm calculates the needed migrations to cross from the current state to the new state and finds its involved PMs and their network paths respectively. It also evaluates which PMs can be switched off or have to be switched on. It then creates a migration plan where as much parallel migrations as possible are scheduled. Based on this information, the algorithm calculates the amount of steps $s(A_{current}, A_{target})$ that is needed to migrate from the current allocation state $A_{current}$ to the new state A_{target} where each step takes a constant amount of time (defined by the duration of migrations and server switches).

4 Problem Formulation and Methodology

The goal of the presented methodology is to let the data center migrate from a current allocation and power state to another state with a specific power consumption, either a minimal or a given consumption under the consideration of the time it needs to enter the desired state. As stated in the introduction, we formulate the problem as a combinatorial NP-hard multidimensional bin packing problem with a cost function. The classic one dimensional bin packing problem aims to distribute a number of items into a finite number of bins where the optimization goal is to minimize the number of bins used. However, applying this approach to the problem described here may lead to inefficient solutions. If the methodology just minimizes the number of servers, power savings will occur for the IT hardware but not for the supporting devices like UPS and cooling. These may run into significantly inefficient states, destroying the savings achieved by switching off servers. Similarly, if a specific power consumption should be approached, the modifications caused by the reactions of the supporting devices may lead to severe deviations. To eliminate these problems, a new approach is presented that still uses the bin packing representation of the problem; however, instead of trying to minimize the amount of bins used, a cost function is used to rate the effectiveness of the entire solution regarding power consumption and the time needed to reach the new state (see Fig. 2.3). The formal definition of the problem is as follows:

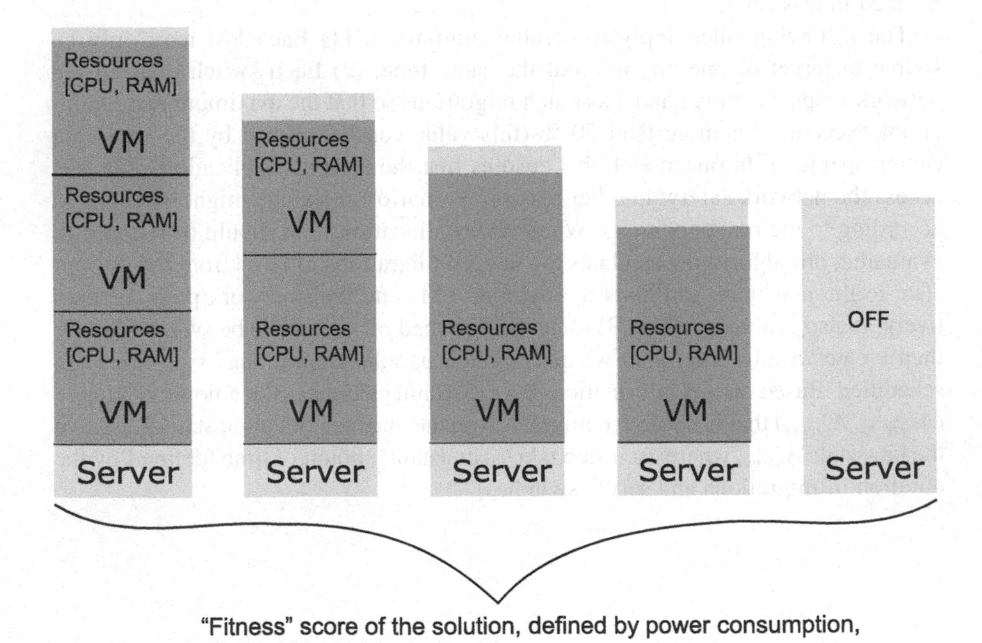

Fig. 2.3 Extended bin packing problem using a fitness function to evaluate each solution

Given is a set $V = \{v_1, \ldots, v_m\}$ of VMs in the data center with resource demand vectors $r_{v1}, r_{v2}, r_{v3}, \ldots, r_{vm}$ and a set $S = \{s_1, \ldots, s_k\}$ of PMs available with resource capacity vectors of $x_{s1}, x_{s2}, x_{s3}, \ldots, x_{sk}$.

Find an allocation A of all elements in V to an arbitrary number δ of elements in S so that for each $s \in S$:

$$\sum_{i=1}^{j} r_{vi} \leq x_s + b$$

where b is a buffer value used to prevent overloading a PM and j is the number of VMs on the PM s. The optimization goal is, in contrast to the classic bin packing problem not to minimize the number of used PMs, but instead to maximize the fitness of the allocation $f(A)$. This function evaluates the allocation A in terms of the proximity towards the desired power consumption (minimal or target value); the time it needs to enter this allocation is then considered when a new solution is chosen.

4.1 The DaMucA *Heuristic*

The algorithm consists of two main parts: the monitoring component that gathers information from the sensors in the data center and the *DaMucA* heuristic that uses the information to calculate a new allocation; either because consolidation is possible or because the current state induces problems like resource shortages, limited flexibility or emergencies indicated by alarms (see Fig. 2.4; *DaMucA* specific parts are located inside the dashed rectangle).

The heuristic distinguishes between three different cases: (1) server consolidation to save energy, inevitably reducing the grade of flexibility; (2) increasing flexibility, which normally leads to more energy consumption; (3) react to emergencies, which generally means a fast increase of flexibility for the global situation and keeping a blacklist of servers that for example became too hot. An entire allocation is built stepwise; before each step the heuristic updates the ratings for each server. These ratings evaluate the quality of each server regarding each specific criterion on the basis of the described models in Sect. 3. The ratings are consulted each time an action is planned. For example when a server should be switched off, the heuristic uses the ratings to determine which server fits best for the current action. Consequently after each step, the ratings need to be updated in order to reflect the changed situation. The following criteria for each server are taken into account: (1) the basic efficiency of the server's model, calculated by the relation of the power consumption and the performance; (2) the current utilization and the difference to the server's optimal state; (3) the efficiency of the connected UPS devices; (4) the thermal performance index; (5) the predicted network utilization of connected switches. All calculated rating values are normalized to the range [0, 1]. The heuristic suggests a fitting server for each requested action. However, to create

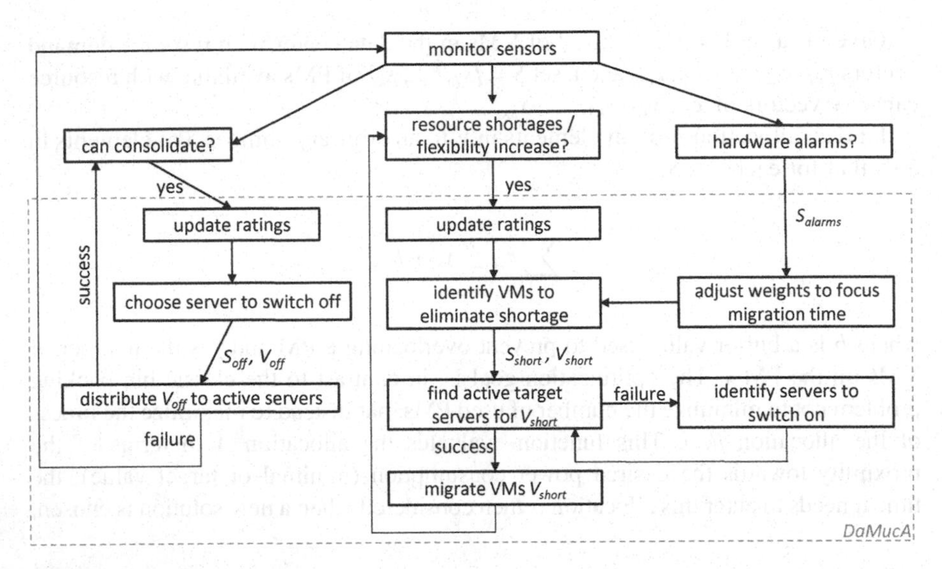

Fig. 2.4 Algorithm overview with a dedicated monitoring module and the DaMucA heuristic that suggests possible actions. All parts related to DaMucA are located inside the dashed rectangle. The variables S_{off}, S_{short} and S_{alarms} represent sets of servers that are subject of the corresponding action, V_{off} and V_{short} represent the set of VMs on the according servers

neighbor allocation states, it is possible to choose between different suggestions. To find an optimal allocation, a search algorithm based on the simulated annealing meta-heuristic is used that builds its solution candidates using the *DaMucA* heuristic and evaluates each candidate based on the following fitness function.

To measure the fitness of each allocation, first the total data center power consumption $P_{DC}(t, A)$ under the allocation A is calculated.

$$P_{DC}(A) = P_U\left(\sum_{i=1}^{\delta} P_{PMi}\right) + P_{th}\left(\sum_{i=1}^{\delta} P_{PMi}\right)$$

Next, the duration in steps to migrate from the data center's current allocation state $A_{current}$ to the solutions state $A_{solution}$ is retrieved using the network graph.

$$d = s(A_{current}, A_{solution})$$

In normal operation state, the methodology tries to let the data center operate in an energy efficient state, hence the optimization goal is to minimize $P_{DC}(A)$. The fitness function is then defined as follows:

$$f(A) = \frac{1}{P_{DC}(A)}$$

If the methodology is used to apply demand response management, target power consumption for the data center is given as $P_{DCtarget}$. In that case, the optimization goal is to minimize the deviance to the given consumption a:

$$a = \left| P_{DCtarget} - P_{DC} \right|$$

In this case, the fitness function uses a instead of $P_{DC}(A)$.

The second optimization goal is always to minimize the amount of steps d needed to reach the new allocation state, since the new state should always be reached with as few operations as possible. When two allocations are compared, first the fitness value is used and as a second condition the number of steps d is compared, for example if a solution needs a significantly lower amount of steps and the fitness is only marginally worse, this solution is preferred.

5 Analysis and Results

We evaluate a sample scenario with the data center models and architecture described in Sect. 3. The simulated data center consists of two rooms, 1228 PMs in 96 racks (48 racks per room) and 8 UPS devices. For each simulation, the number of VMs is static, meaning that there are no VMs coming into the simulation or leaving it.

The application load profiles of the VMs consist of load measurements of real applications hosted in a mid-sized data center. The main simulation scenario has 12621 VMs. At first, the operation state of the data center is in the initial non-optimized state where each PM is powered on and the VMs are distributed in an equilibrium allocation. In this case, the data center had a power consumption of about 270 kW (see Fig. 2.5).

The *DaMucA* solution is compared to a traditional FFD solution, which simply tries to minimize the number of active servers. It already considers the energy efficiency of each server model. The data center had a power consumption of about 148 kW using this solution, however, indicated by the rack power standard deviation, the distribution of the power consumption inside the data center is significantly differing. This harms the flexibility of the allocation and leads to potential hot spots. Figure 2.6 (bottom) shows this problem in detail. Examining the average power consumption of each rack, it can be observed that there are some racks with all servers powered off and some racks with power spikes. The *DaMucA* solution results in almost the same data center power consumption (achieved by savings in supporting devices), but it has more active servers, leading to enhanced flexibility and it creates a solution with an almost equilibrium power deviation regarding the racks (see Fig. 2.6, top).

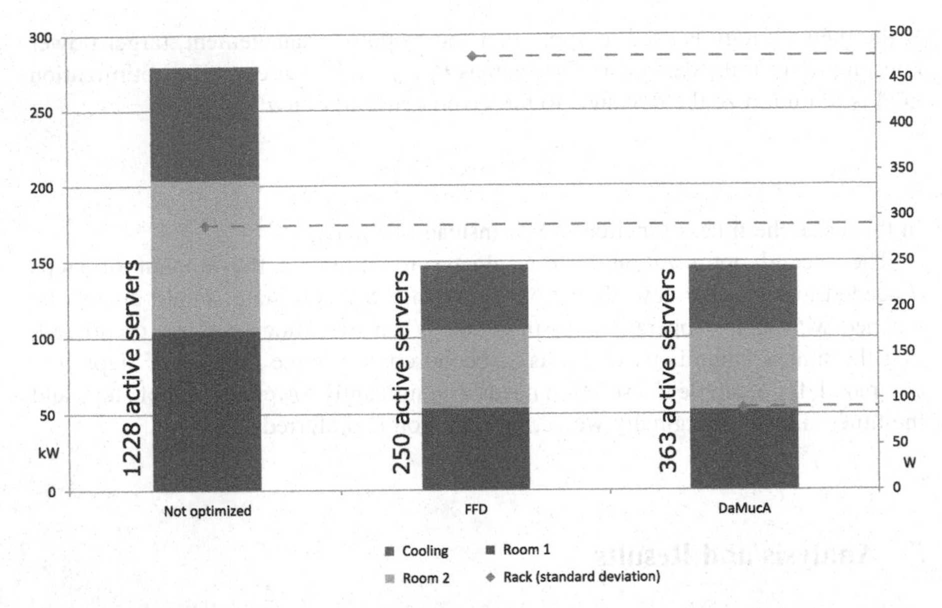

Fig. 2.5 Comparison of the power consumption of the entire data center by optimization method; each split up into the power used by each room and the cooling overhead power. CRAH devices located inside the room count for the room power consumption. The *diamonds* show the rack power standard deviation of all racks; a smaller value indicates a better solution

Figure 2.7 shows the scalability of data center power controlled by *DaMucA*. By increasing the flexibility grade, a continuous power regulation is possible. The methodology is thus not only able to reach power savings but also to allow the data center to apply demand response management.

6 Conclusion

In this paper, a methodology was described that allows the data center to improve its energy efficiency by taking into account the IT hardware (servers) and the infrastructure devices (UPS, cooling) when finding VM allocations. Summarized, the methodology described in this paper can be applied (1) to save power by applying a minimum energy state while keeping up flexibility for future load boosts; (2) apply demand response management; (3) be able to immediately react to emergencies (for example cooling failures) or other incidents; (4) data center internal regulation of power consumption to prevent penalties (especially in co-location data centers). The methodology is also able to find fast transitions from the current allocation state to specific power states, enabling the data center to apply demand response management. Our future research will create more detailed thermal models for different cooling strategies (free cooling, chillers,

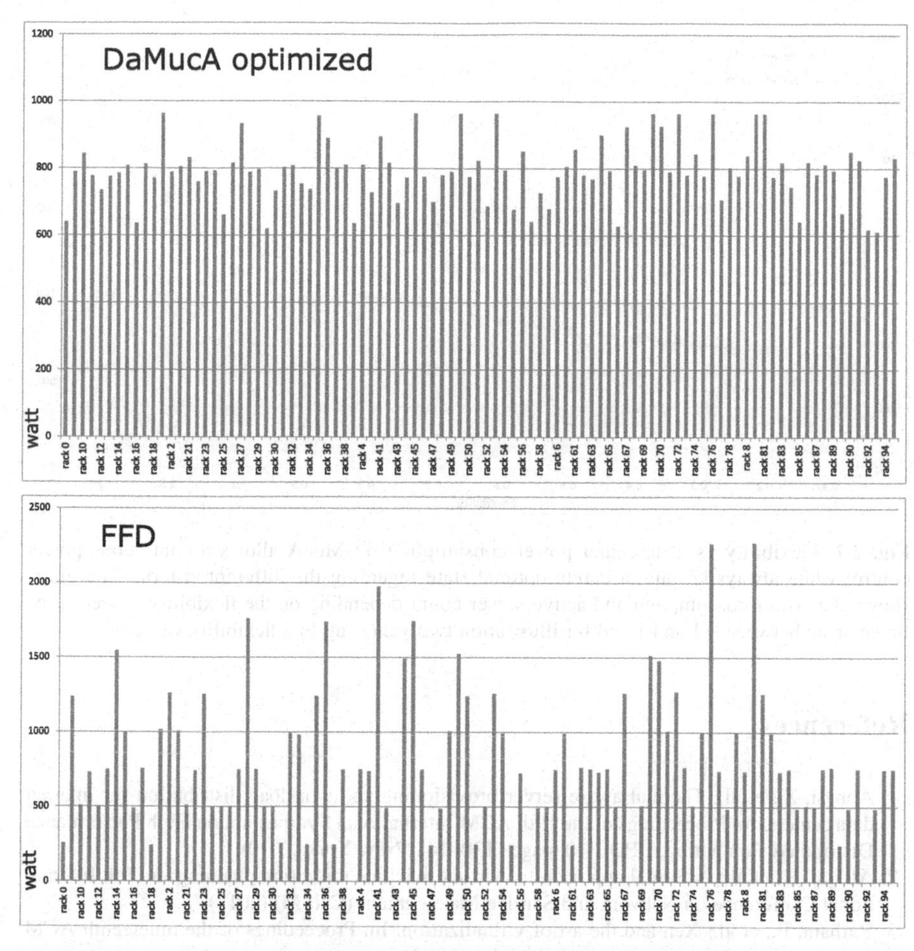

Fig. 2.6 Detailed comparison of each rack's average power consumption under a solution found with DaMucA and FFD. It can be seen that the FFD solution contains racks with no power consumption (all servers off) and some with power peaks. In contrast, the DaMucA solution almost reaches an equilibrium state regarding rack power, enhancing flexibility and cooling efficiency

etc.) and an improved method to retrieve optimal parallel migration plans using the network model. We are also working on evolutionary algorithms to find an allocation state near the optimum while still completing in realistic time frames. Since the problem to solve is very complex and it is generally hard to determine the "real" optimum, a competitive analysis is planned for the evaluation of the algorithm. It is also planned to integrate a load forecasting method from [12] into the methodology that is used to predict VM application load, thus allowing the methodology to act proactive.

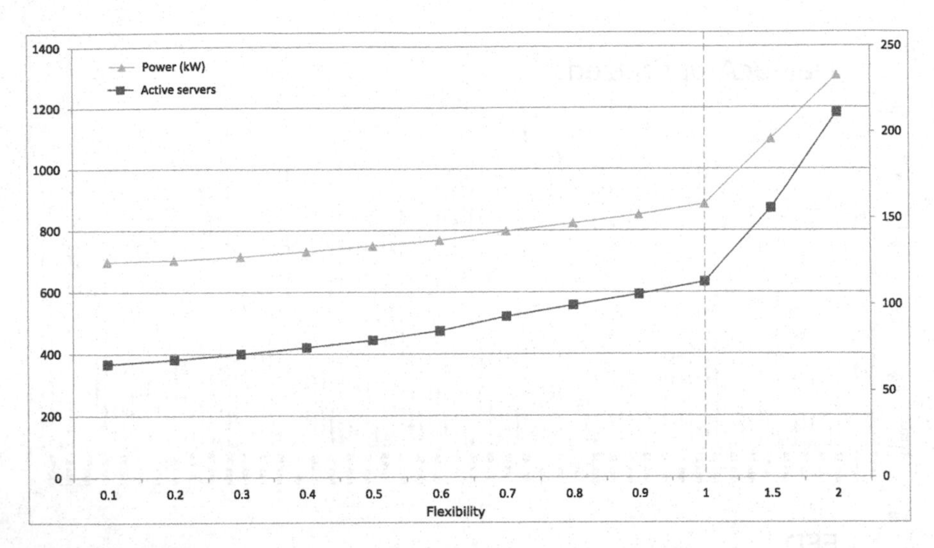

Fig. 2.7 Flexibility vs. data center power consumption: DaMucA allows a continuous power control while always keeping a Pareto-optimal state regarding the different criteria. The graph shows the power consumption and active server count depending on the flexibility grade on the linear scale between 0.1 and 1 and for illustration two values up to a flexibility value of 2

References

1. Abbasi, Z., et al.: Thermal aware server provisioning and workload distribution for internet data centers. In: Proceedings of the 19th ACM International Symposium on High Performance Distributed Computing, HPDC'10, pages 130–141, New York (2010).
2. Al-Haj, S., et al.: A formal approach for virtual machine migration planning. In: 9th International Conference on Network and Service Management (CNSM) (2013).
3. Barham, P., et al.: Xen and the art of virtualization. In: Proceedings of the nineteenth ACM symposium on Operating systems principles (2003).
4. Berl, A., et al.: Integrating data centres into demand-response management: A local case study. In: Industrial Electronics Society, IECON 2013 – 39th Annual Conference of the IEEE, pages 4762–4767, Nov (2013).
5. Carli, T., et al.: A packing problem approach to energy-aware load distribution in clouds. In: arXiv:1403.0493 (2014).
6. Chen, H., et al.: The data center as a grid load stabilizer. In: Proceedings of the Asia and South Pacific Design Automation Conference (ASP-DAC) (2014).
7. Choi, J., et al.: Modeling and managing thermal profiles of rack-mounted servers with thermostat. In: Proceedings of HPCA, IEEE, pages 205–215 (2007).
8. Cochran, R., et al.: Consistent runtime thermal prediction and control through workload phase detection. In: Proceedings of the 47th Design Automation Conference, DAC'10, pages 62–67, New York, NY, USA (2010).
9. Dalvandi, A., et al.: Time-aware vm-placement and routing with bandwidth guarantees in green cloud data centers. In: IEEE 5th International Conference on Cloud Computing Technology and Science (CloudCom), vol. 1, pages 212–217 (2013).
10. Epstein, L., et al.: Bin packing with general cost structures. In: Mathematical Programming, Volume 132, Issue 1–2, pages 355–391 (2012).

11. Ghamkhari, M., et al.: Optimal integration of renewable energy resources in data centers with behind-the-meter renewable generator. In: IEEE International Conference on Communications (ICC) (2012).
12. Hoyer, M., et al.: Proactive dynamic resource management in virtualized data centers. In: Proceedings of the 2nd International Conference on Energy-Efficient Computing and Networking (2011).
13. Janacek, S., et al.: Data center smart grid integration considering renewable energies and waste heat usage. In: 2nd International Workshop on Energy-Efficient Data Centres (2013).
14. Janacek, S., et al.: Modeling and approaching a cost transparent, specific data center power consumption. In: 2012 International Conference on Energy Aware Computing, pages 68–73 (2012).
15. Janacek, S., et al.: Expansion of Data Center's Energetic Degrees of Freedom to Employ Green Energy Sources. In: EnviroInfo 2014 – 28th International Conference on Informatics for Environmental Protection, pp. 445–452, BIS-Verlag, Oldenburg (2014).
16. Jonas, M., et al.: A transient model for data center thermal prediction. In: 2012 International Green Computing Conference (IGCC), pages 1–10 (2012).
17. Mohsenian-Rad, H., et al.: Coordination of cloud computing and smart power grids. In: Proc. Of IEEE Smart Grid Communications Conference, Gaithersburg, Maryland (2010).
18. Moore, J., et al.: Lowcost thermal mapping of data centers. In: Proceedings of the Workshop on Tackling Computer Systems Problems with Machine Learning Techniques (SysML) (2006).
19. Moore, J., et al.: Making scheduling "cool": Temperature-aware workload placement in data centers. In: Proceedings of the Annual Conference on USENIX Annual Technical Conference, ATEC'05, pages 5–5, Berkeley, CA, USA (2005).
20. Mukherjee, T., et al.: Spatio-temporal thermal-aware job scheduling to minimize energy consumption in virtualized heterogeneous data centers. In: Comput. Netw. (2009).
21. Pakbaznia, E., et al.: Minimizing data center cooling and server power costs. In: Proceedings of the 14th ACM/IEEE international symposium on Low power electronics and design, ISLPED'09, pages 145–150, New York (2009).
22. Parolini, L., et al.: Reducing data center energy consumption via coordinated cooling and load management. In: Proceedings of the 2008 Conference on Power Aware Computing and Systems, HotPower'08, pages 14–14, Berkeley, CA, USA (2008).
23. Patel, C., et al.: Thermal considerations in cooling large scale high compute density data centers. In: Thermal and Thermomechanical Phenomena in Electronic Systems, The Eighth Intersociety Conference on ITHERM 2002, pages 767–776 (2002).
24. Pedram, M., et al.: Power and performance modeling in a virtualized server system. In: Proceedings of the 2010 39th International Conference on Parallel Processing Workshops (2010).
25. Ren, C., et al.: Carbon-aware energy capacity planning for datacenters. In: Modeling, Analysis Simulation of Computer and Telecommunication Systems (MASCOTS), IEEE 20th International Symposium on, pages 391–400 (2012).
26. Sharma, R., et al.: Balance of power: dynamic thermal management for internet data centers. In: Internet Computing, IEEE, 9(1): 42–49 (2005).
27. Tang, Q., et al.: Thermal-aware task scheduling to minimize energy usage of blade server based datacenters. In: Dependable, Autonomic and Secure Computing, 2nd IEEE International Symposium on, pages 195–202 (2006).
28. Tang, Q., et al.: Sensor-based fast thermal evaluation model for energy efficient high-performance datacenters. In: Proceedings – 4th International Conference on Intelligent Sensing and Information Processing (2006).
29. Wang, R., et al.: Datacenters as controllable load resources in the electricity market. In: Proceedings of the 2013 I.E. 33rd International Conference on Distributed Computing Systems, ICDCS'13, pages 176–185, Washington, DC, USA (2013).
30. Wu, W., et al.: Efficient power modeling and software thermal sensing for runtime temperature monitoring. In: ACM Trans. Des. Autom. Electron. Syst., 12(3):25:1–25:29 (2008).

31. Yeo, S., et al.: Simware: A holistic warehouse-scale computer simulator. Computer, 45 (9):48–55 (2012).
32. Zhang, Y., et al.: Greenware: Greening cloud-scale data centers to maximize the use of renewable energy. In: F. Kon and A.-M. Kermarrec, editors, Middleware 2011, volume 7049 of Lecture Notes in Computer Science, pages 143–164. Springer Berlin Heidelberg (2011).
33. Zhou, R., et al.: Modeling and control for cooling management of data centers with hot aisle containment. In: ASME Conference Proceedings, 2011(54907):739–746 (2011).

Chapter 3
A Data Center Simulation Framework Based on an Ontological Foundation

Ammar Memari, Jan Vornberger, Jorge Marx Gómez, and Wolfgang Nebel

Abstract The IT-for-Green project aims at developing the next generation of Corporate Environmental Management Information Systems (CEMIS). Green IT being one important aspect of this, the IT-for-Green project seeks to support the analysis of a given data center situation and to support the simulation of alternative architectures to find ways to increase energy efficiency. To facilitate this, we develop a data center simulation framework, designed to be part of a larger CEMIS platform. This is achieved through a focus on flexibility, high interoperability and open standards. Flexibility is especially achieved by building the framework on top of an underlying data center ontology. This ontological approach allows us to derive many components of the framework from a single source, maintaining the ability to quickly adapt to future requirements.

Keywords Simulation framework • Machine learning • Data center simulation • Energy modeling

1 Introduction

Worldwide spending on data center infrastructure is projected to surpass $126.2 billion in 2015 [1], and in line with this, energy demand of Information and Communication Technology (ICT) infrastructure is rising continuously and almost doubling between 2000 and 2006 [2, 3]. While server hardware has generally become more energy-efficient, it has increased even more in performance, resulting in a net increase in energy demand. This has reached a point where energy costs are now a dominant factor in the total cost of ownership [4].

This development has spurred new research into more energy-efficient data center architectures. One approach in this research is to build some type of model of the data center, which can then be used to quickly iterate on the design and try to optimize both individual components (depending on how fine-grained the model is) as well as the interplay between them. But besides optimizing the architecture, such

A. Memari (✉) • J. Vornberger • J. Marx Gómez • W. Nebel
Carl von Ossietzky University of Oldenburg, Oldenburg, Germany
e-mail: ammar.memari@uni-oldenburg.de; jan.vornberger@uni-oldenburg.de; jorge.marx.gomez@uni-oldenburg.de; wolfgang.nebel@uni-oldenburg.de

J. Marx Gómez et al. (eds.), *Advances and New Trends in Environmental and Energy Informatics*, Progress in IS, DOI 10.1007/978-3-319-23455-7_3

a model can also be used to get a better understanding of which application workloads are using the most resources. Creating accountability in this way can create a feedback loop that helps to drive efficiency improvements on the software side.

The research we describe in this paper was conducted as part of the IT-for-Green project. This project aims at developing the next generation of Corporate Environmental Management Information Systems (CEMIS) and also addresses aspects of Green IT in its context. For those businesses which operate their own data center, the IT-for-Green project develops tools to help with analyzing their current data center situation and allow simulating the effects of changes and upgrades on the infrastructure to be able to find ways to increase energy efficiency.

This paper builds upon, and extends the work conducted in [5]. A newer version of the framework is described in the present paper which provides extra functionality and a completely new frontend and a revamped ontology.

In the remainder of this paper we describe a simulation framework that we have designed to meet these goals. The following section will explain the approach we took. Section 3 will then detail the complete simulation framework. We evaluate the framework in Sect. 4 based on the initial goals.

2 Materials and Methods

Since our framework is part of a larger CEMIS solution, one of our main design requirements is high interoperability and flexibility. We have therefore taken great care in choosing appropriate open standards where applicable, and a flexible architecture to facilitate integration with other systems. This also led us to the decision of utilizing an ontological approach as the foundation of the framework. The main goal here is to have as few hardcoded components in the framework as possible, and derive everything from a single source instead. Evolving the framework is then a matter of updating this source and having the updates propagate through the rest of the system.

An ontology is a capable tool for reaching this goal. We can capture all the components of a data center and their possible interactions in the form of a rich ontology. From this we can then derive many different tools and other representations needed for a complete data center simulation. By choosing a standard ontology format, Web Ontology Language (OWL), we also take our goal of compatibility and interoperability with existing tools and systems into account. Among other things, we derive a toolbox for a graphical editor from such an ontology. This editor can then be used to model an existing data center to prepare for a simulation. One goal here is to provide a reasonably fast and user-friendly way of creating such a model, also taking into account that the user may want to import inventory data from other sources.

Finally, we have decided to target the modeling language Modelica to be used in performing the actual simulation. It is an object-oriented, declarative, and multidomain modeling language, and by leveraging it we can reuse many existing modeling tools. The next sections will describe each of these steps in more detail and how they tie together to form the full simulation framework.

3 Related Work

Research related to the present approach finds its root in different disciplines. Traditionally, Data Center Infrastructure Management (DCIM) tools aim at modeling the static state of a data center: location of the racks and their occupying IT equipment. Maximum power demand of each device is stored and the sum is compared to a given maximum that the rack can provide. Not only is this power balance maintained by the tool, but also the weight and size of contained equipment against the respective capacity of the rack. DCIM tools provide topology simulation of data centers that supports design-time decisions and is based on peak power demand values. Some of these tools provide calculation of Key Performance Indicators (KPIs) such as the Power Usage Efficiency (PUE) based on real-time monitoring different sensors throughout the data center. Our present research has a different focus in which it provides a more holistic approach to capacity planning by allowing custom rules to be embedded into the metamodel. These rules can be used to assess capacity as well as many other different uses as will be exemplified later in this paper. Calculation of PUE and other (custom) KPIs is possible in our case without the need for hardware power demand sensors since the power demand will be simulated per device based on its utilization and relation to other devices. Other rules are useful for simulation power failures and assessing redundancy. Our framework integrates well with existing DCIM solutions due to the multi-faceted nature of the metamodel ontology as will be explained later.

The other discipline which had influenced our research is artificial intelligence. A properly-modeled data center can be "understood" and reasoned-about by machines. To create a metamodel for a data center, we have studied a number of relevant ontologies. Ontologies from three main categories were investigated: Top-level ontologies are ones that provide definitions for general-purpose terms, and act as a foundation for more specific domain ontologies [7]. Mid-level ones come directly underneath and are a bit more domain-specific. Low-level ontologies are the domain-specific ones defining terms and relations of a certain narrow domain. For example, [8] have built their low-level "Green IT" domain ontology complying with the mid-level "observation and measurement" ontology of [9], which in turn is built in conformance to the Descriptive Ontology for Linguistic and Cognitive Engineering (DOLCE) foundational ontology [10]. We share the motivation with the authors of [8]; especially to the point that no public data describing the energetic behavior of devices are available, and when such data are empirically generated, they need to have a common form in order to have them integrated into tools or used for benchmarking. Details of this literature study can be found in [6].

Simulation of data center power was addressed by the CloudSim toolkit [11]. It provides the capabilities for modeling and simulating data centers with focus on Virtual Machines (VMs) allocation policies as a main method for optimizing power demand. It provides at its core implementation of various policies and allows the user to implement custom ones. CloudSim is written in Java and exposes an Application Programming Interface (API), but does not provide any graphical interface for designing the data center or running the simulation. This programming-only interface requires a power-user to handle and diminishes readability of the design. Moreover, the toolkit does not provide a reasoning mechanism for recommending design best practices or recognizing shortcomings. Our approach can interface with CloudSim at the stage of "code generating" where instead of transforming the Platform-Independent Model (PIM) into a Platform-Specific Model (PSM) as Modelica code; it is transformed into Java code compatible with CloudSim. This feature is a direct benefit of following a model-driven approach allowing different PSMs to be generated for different platforms.

On the other side one can find general-purpose simulation frameworks that feature a graphical interface but are not customized for data center simulations such as the Kepler Project [12], the OpenModelica connections editor (OMEdit) [13], and Simulink [14]. These tools can be customized with components for simulating a data center, but would still lack the semantic layer that mainly provides integration and reasonability.

4 Simulation Framework

As the previous section already touched on, and as visible in Fig. 3.1, the workflow of the simulation framework looks as follows: Domain experts create the metamodel ontology (we provide an initial attempt at that, but it can be extended further). From that ontology a toolbox for the graphical data center editor is generated. The editor is used by a user of the framework to create a model of a data center he or she wishes to study. This step is supported by both syntactic and semantic checks of the model, in addition to an assessment of best-practices enabled by the underlying ontology. This model is then converted into a Modelica output using a Modelica component library as an additional input, and then used to run the actual simulations. Output of the simulation can be visualized directly from the editor.

4.1 Data Center Ontology

Authors in [10] characterize foundational top-level ontologies to be ones that (1): have a large scope, (2): can be highly reusable in different modeling scenarios, (3): are philosophically and conceptually well-founded, and (4): are semantically

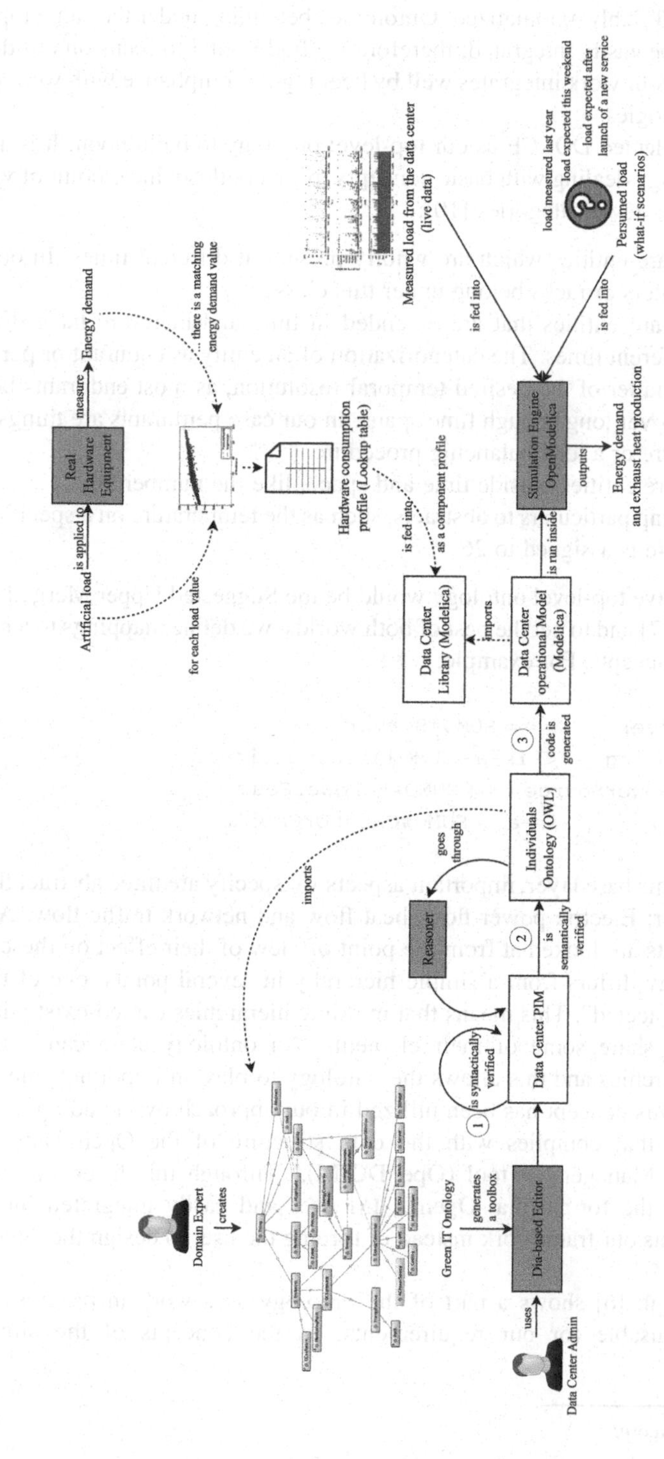

Fig. 3.1 The main workflow used in the framework

transparent and richly axiomatized. Ontologies belonging under the same top-level ontology can be easily integrated; therefore we find it vital to focus on our domain and make sure our work integrates well by keeping its compliance with well-known top-level ontologies.

We have selected DOLCE as our top-level ontology to build upon. It is a fairly abstract ontology dealing with basic concepts from a philosophical point of view. It defines four top level categories [10]:

- Endurants are entities which are wholly present at different times. In our case servers, routers or racks belong under this class.
- Perdurants are entities that are extended in time and therefore have different parts at different times. The categorization of an entity as endurant or perdurant is often a matter of the desired temporal resolution, as most endurants become perdurants over long enough time spans. In our case perdurants are things like a power failure or a load balancing procedure.
- Abstracts are entities outside time and space, like the number 26.
- Qualities map particulars to abstracts, such as the temperature on a specific point of a hot aisle is assigned to 26.

An alternative top-level ontology would be the Suggested Upper Merged Ontology (SUMO) [7] and to get the best of both worlds, we define mappings to a number of SUMO's concepts. For example:

```
GreenIT#Server        === SUMO#Server
GreenIT#Cooling      <-is-a- SUMO#AirConditioner
GreenIT#ACPowerSource === SUMO#ACPowerSource
GreenIT#Rack          -is-a-> SUMO#ChestOrCabinet
```

On top of this base layer, important aspects we specify are three abstract flows in the data center: Electric power flow, heat flow and network traffic flow. All data center elements are looked at from the point of view of their effect on these flows.

An ontology differs from a simple hierarchy in several points, one of them is being "multi-faceted". This means that multiple hierarchies can co-exist within an ontology and share some of their elements. An ontology class can belong to multiple hierarchies and this allows the ontology to play an important role in data integration. This concept has been utilized in our approach by including a facet in our ontology that complies with the data structure of the Open Data Center Infrastructure Management tool (OpenDCIM).[1] Through this facet, data can be exchanged in the format that OpenDCIM uses and easily integrated into other systems such as our framework instead of forcing the user to design the data center twice.

The paper in [6] shows a part of the ontology as a work in progress that is nevertheless usable for our requirements, as the concepts of the simulation

[1] http://opendcim.org/.

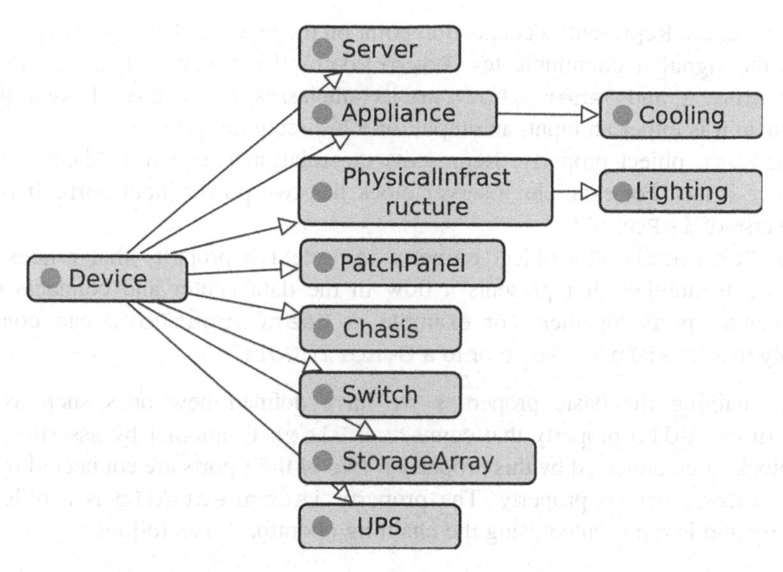

Fig. 3.2 OpenDCIM facet of the Green IT ontology

framework is derived from it. Figure 3.2 represents the facet used for integrating the data model of OpenDCIM. As visible in this figure, and as discussed in [6], the `Server` class for example falls under both the `EnergyConverter` and the `Device` classes. Further facets can be easily added to the ontology in order to support data models of other tools and standards.

4.2 Toolbox Ontology and Generation

One more facet of the ontology is responsible for communicating the data to the graphical editor. We found it suitable for the end user to be able to graphically design the data center, and for that we have prepared a graphical editor that allows each data center component to be represented as a block which has multiple ports. Arcs connecting the ports represent the different flows and relationships within the data center.

We chose to implement this facet as an independent ontology, namely the Toolbox Ontology, which `imports` the Green IT Ontology. This ontology contains the `Block` and `Port` classes that represent the aforementioned concepts in addition to two basic `Object Properties` that represent the relations as follows:

- `Block` class: Represents a graphical box in the editor. These boxes can represent either `Physical` data center components such as the `UPS`, or `Virtual` ones such as a virtual `MonitoringPoint`, a time series loader, or an arithmetic operation. Each `Block` has multiple `Ports`

- `Port` class: Represents a connection point on the box, and has a type depending on the signal it communicates. `PowerPort`, `HeatPort`, `NetworkPort`, `LoadPort`, and `GenericPort` are its subclasses. A `Port` can have a direction so it is either an input, an output, or a bidirectional port.
- `hasPort` object property: Represents the relation between a `Block` and a `Port` it has. For example, a server block has two power input ports. It is the inverse of `isPortOf`.
- `hasConnectionTo` object property: A reflexive property that relates one `Port` to another. It represents a flow in the data center and connects only matching ports together. For example, a `HeatOutputPort` can connect only to a `HeatInputPort` or to a `GenericPort`.

By chaining the basic properties we have defined new ones such as the `isConnectedTo` property that connects a `Block` to another by asserting that two blocks are connected by this property if two of their ports are connected by the `hasConnectionTo` property. The property `isConnectedTo` is a reflexive property and is represented using the chaining operator (o) as follows:

```
hasPort o hasConnectionTo o isPortOf
```

Creating this composite property facilitated creating rules for usage later in assessing best practices. The Semantic Web Rules Language (SWRL), being the proposed rules language for the semantic web, provides a practical way to define rules and logic within the ontology. Through an SWRL rule we have defined the property `isPoweredBy` and its inverse `powers` as:

```
PowerInputPort(?port1),PowerOutputPort(?port2),
hasConnectionTo(?port1,?port2),hasPort(?block1,?port1),
isPortOf(?port2,?block2)->isPoweredBy(?block1,?block2)
```

The rule states that if a block has a power input port which is connected to a power output port of another block, then the first block is powered by the second.

Building upon this rule, we have defined rules for assessing power redundancy attributes of the data center components. A `Data Property` is defined first under the name `hasPowerRedundancy` that can take the values 0 for no power redundancy when the device is connected directly to an interruptible power source, 1 when the device is connected to an Uninterruptible Power Supply (`UPS`) which is connected to a power source, and 2 when the device is connected to two different `UPSs`. The property is assigned the values of 1 and 2 based on the following SWRL rules:

```
ACPowerSource(?source),UPS(?powblock1),
isPoweredBy(?block,?powblock1),isPoweredBy(?powblock,?source)  ->
hasPowerRedundancy(?block,1)
ACPowerSource(?source1),UPS(?powblock1),
```

```
ACPowerSource(?source2),UPS(?powblock2),
isPoweredBy(?block,?powblock1),
isPoweredBy(?block,?powblock2),
isPoweredBy(?powblock1,?source2),
isPoweredBy(?powblock2,?source2),
DifferentFrom(?powblock1, ?powblock2)
->hasPowerRedundancy(?block, 2)
```

Based on this ontology, we then generate a toolbox in the next step. The goal here is to provide building blocks, which can be assembled in a graphical interface to build a model of the data center. We chose in our first two versions of the framework to extend the graphical editor Visual Understanding Environment (VUE) [15], which is provided with the generated toolbox and enables the user to build the data center model by dragging and dropping the provided components onto a canvas and specifying relevant relations (e.g. heat exchange, electricity exchange). In version 3 however, we have switched to the Dia diagram editor as it provides a plugin architecture, and a better concept of a toolbox.

The generation task is performed by the backend that runs on an application server and provides web services to be consumed by the developed editor plugin as their client. The client application needs to specify ontology to be used, the root class, and other parameters. The server resolves subclasses through a reasoner (HermiT [16] in our case) and returns the tools as Extensible Markup Language (XML) in the form of Dia *Shapes*. These shapes are then integrated into Dia by the client plugin and are ready to use for designing the data center.

4.3 Design and Verification

After generating a toolbox and integrating it into the graphical editor, the user can start designing the data center. Each tool can be dragged to the canvas to create an instant of it. Physical component such as servers and cooling systems are color-coded differently from the virtual ones such as load time series and arithmetic components. Ports are also color-coded based on their types. These color codes were defined right in the ontology.

Figure 3.3 shows the design of an example data center. It contains two UPSs each powering two Power Distribution Units (PDUs) that in turn power the servers. Generated heat from all components is used as input for the model of the air conditioner. Virtual components (white background) are used for calculating the sums of generated heat in addition to other calculations. They also serve the purpose of loading the *Load* and *Outside Temperature* time series. Arrows used on the relations are a convention only since `hasConnectionTo` is a symmetric relation.

As the toolbox that was used to build the data center model is based on our initial ontology, we can leverage the ontology again to run a number of consistency checks in the next step. These are both syntactic checks (e.g. every server needs to have a

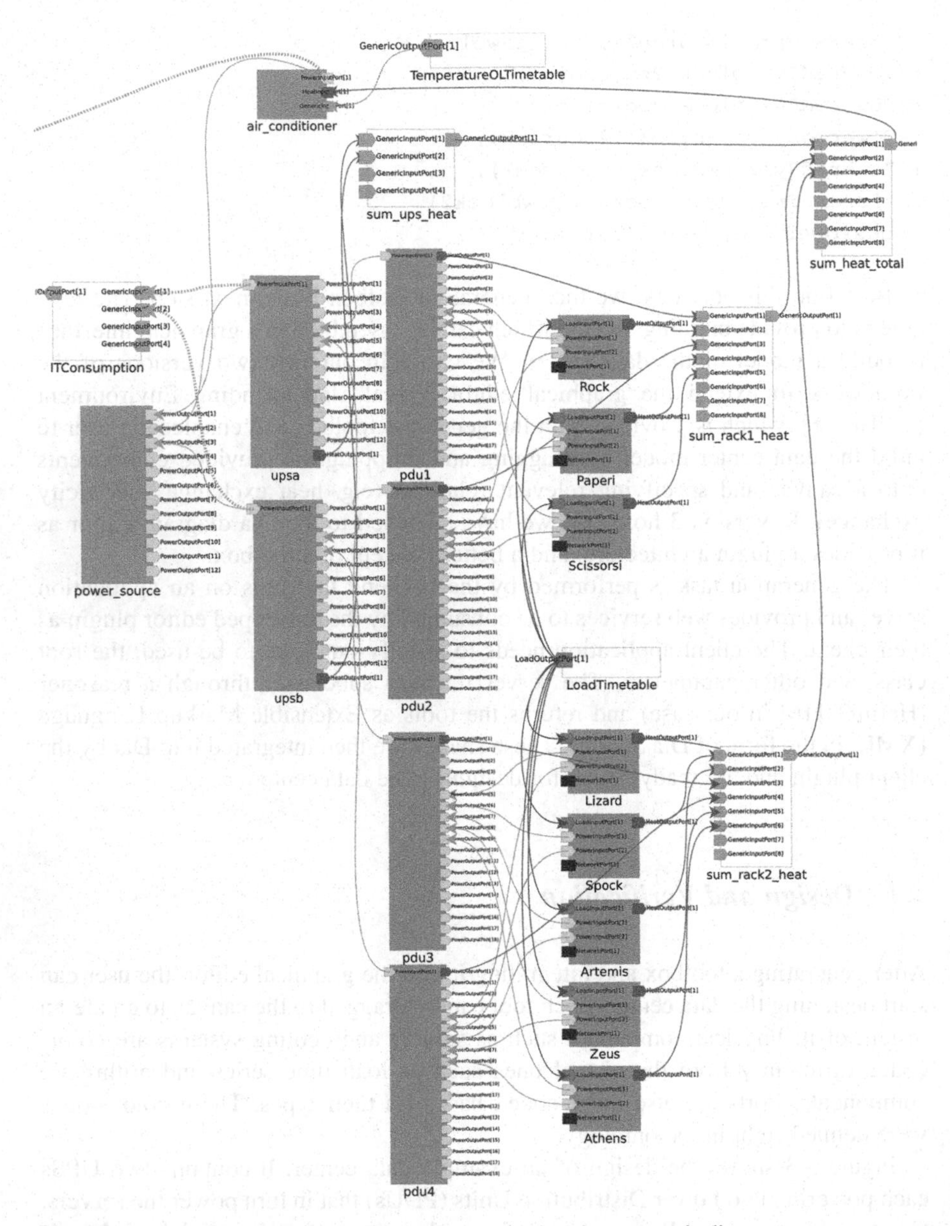

Fig. 3.3 An example small data center as designed using the proposed editor

label) as well as more complicated semantic checks. An example for the latter would be the requirements, specified in the ontology, that every port is connected to a suitable other port.

Semantic verification starts by transforming the design elements into instances of the ontology classes and the links into object properties. These *individuals* are collected into a new ontology that imports the Toolbox Ontology and we call it the Individuals Ontology.

The semantic reasoner HermiT [16] is used here once more against the Individuals Ontology to verify these requirements. Any errors that are found are transformed into a visual representation, and presented back to the user for correction. Similarly, checks on best practices (which are not design semantic errors) can be run by executing a Description Logic (DL) query against the generated Individuals Ontology. For example the query: `hasPowerRedundancy value 2` returns all the components which have a power redundancy of 2 according to the rules mentioned earlier and stored in the ontology. The editor represents the results of this query as visual annotation on the matching elements and provides the user thereby with a clue on which components might need extra power redundancy.

4.4 Operational Simulation Model

After ending up with an error-free data center design, the workflow goes further into generating an operational model out of the design. This model is written in Modelica language and captures only operational attributes of the design. To follow best-practices and keep intra-component mathematical models apart from inter-component relations, the model is generated as two linked parts: Modelica package as a library that contains components' mathematical models, and the operational data center design which contains the instantiation of these models and the relations among these instances; it refers to component models from the former.

4.4.1 Data Center Modelica Package

This package contains all component models and serves as a library. It is generated from the metamodel ontology. Object-oriented quality of Modelica is utilized in this package to define generic types of component models, and then parameterize them with training data into final models of concrete components.

An example of a generic server model is one that relies on a lookup table of load and power demand to predict demand value for a given load. The lookup table set in this generic model would be assigned values later in the concrete components. Another generic model of a server could be formulated by measuring its power demand at the idle point when zero load is applied, and then again when it is 100 % loaded. Under the assumption that between these points power demand relates to load in a linear fashion, we get a simple model of a server which requires minimal training but provides lower prognosis accuracy.

Concrete component models are defined as extensions of the generic models. For example, the server Dell PowerEdge R710 is modeled using the `ServerLookup` as shown in Fig. 3.4. Training data concerning the Dell PowerEdge R710 are used

```
block DellPowerEdgeR710
extends ServerLookup(lookupTable(tableOnFile = true,
fileName = "./profiles/DellPowerEdgeR710.txt",
tableName ="DellPowerEdgeR710"));
end DellPowerEdgeR710;
```

Fig. 3.4 An excerpt from the data center package showing the model of the concrete server "DellPowerEdgeR710" as an extension of a generic model

for populating the lookup table. These models of concrete components are instantiated in the operational data center design explained next.

4.4.2 Operational Data Center Design

This design is a translation of the data center XML design into the Modelica language. It captures, however, only operational properties of components and relations ignoring information about their physical location, layout, or graphical representation properties. Instances of the components are related to their corresponding classes contained in the data center package where all intra-component information resides.

4.4.3 Simulation and Visualization

Simulation tools that run Modelica models, such as the OpenModelica Shell, are able to run the operational model and return simulation results. The design can be run with live load measurements as input stream, outputting power demand and exhaust heat production at each component and measurement point. It can be run on historical or presumed load data as well with the aim of examining different what-if scenarios.

The developed editor is able to communicate with an OpenModelica server or start one if needed. Through a menu item the editor simulates the operational model on the started OpenModelica server and saves the results in a Comma Separated Values (CSV) file. After running a successful simulation, specific ports in the design can be selected and then visualized. For this the editor utilizes OMPlot application which comes bundled with the OpenModelica suite.

4.4.4 Hardware Profiling

In building the data center Modelica package we face the challenge, that data sheets for data center components are often not detailed enough to allow for the construction of an accurate simulation model of this component. To address this challenge, we develop tools to utilize to quickly profile a given component (i.e. measuring power demand in different load situations) and then perform regression analysis to derive an approximate model to be used in Modelica.

One major category of components we look at this way contains the servers themselves. They typically exhibit different energy demand depending on their current work load. A number of researchers have looked at creating energy demand models for servers [4, 17–24]. Some of the simpler models assume that the Central Processing Unit (CPU) is the most important factor and postulate a roughly linear relationship between CPU utilization and power demand. Authors of [17] have shown that even such a simple model can work fairly well and be within 10 % mean accuracy. Of course this works best if indeed the CPU has the biggest demand for energy. In combination with memory, the CPU did indeed dominate total power in the research done in [18]. An additional beneficial factor for a CPU-only model might be the fact that activities of the CPU seem to be well correlated with the power demand of the rest of the system, as authors of [19] were able to show. Further successful applications of a simple CPU-based linear model can be seen in [18, 20–22]. On the other hand, authors of [4] claim that the CPU no longer dominates platform power in modern servers and expect this trend to continue, as processors are adopting more and more energy-efficiency techniques. This is supported by the findings in [23], which reports that power demand of a server can vary by a factor of two while CPU utilization levels stay constant. Authors of [24] go as far as suggesting that accurate power modeling is only possible if the specific workload running on the server is taken into account and report errors as high as 50 % in application-oblivious power models. Because of these different approaches to server models, we have striven to keep the ontology and simulation framework flexible enough to support different kinds of models. At the same time we have implemented a simple model to be used as a starting point.

Our current approach to profiling servers and other load-sensitive components is therefore to instrumentalize the component temporarily with power measuring equipment and at the same time record the system status (for example CPU utilization) via the Simple Network Management Protocol (SNMP). This workflow can fairly easily be carried out inside a production data center, as it requires no software agents on the servers and therefore poses minimal risk to the production work load. On the collected data we run regression analysis to build models for the data center Modelica package. For a better accuracy of the models, we try to generate them based on further parameters besides the CPU utilization. The aim of this is to produce the model as the polynomial:

$$P = x_1.C + x_2.M + x_3.I + x_4.D + x_5$$

Coefficients vector (x) of this polynomial is estimated from the measured data by calculating multiple correlation. The method conducted by [25] is widely used in this regard, and is very well supported by the various tools. The approach used for estimating and testing the models starts by loading the measured data as a matrix containing power demand (P), CPU utilization (C), used memory bytes (M), Internet Protocol (IP) datagrams sent and received (I), and disk read and written bytes (D) measurements at specific timestamps. After loading and preparing the matrix, further variables are generated. These include the square of each measured variable so that correlation is calculated later for the squares and not only the linear variables. The next step is to apply the method of [25] to calculate the coefficients then compose the model. Calculated power demand is then added to the matrix as an extra column, and the difference to the measured demand is visualized.

5 Results and Discussion

5.1 Flexibility: Example Scenario

The sought-after flexibility of the framework is defined in terms of adaptivity to changes in data center hardware components and their interrelations. This attribute is measured by how smoothly these changes are incorporated and streamlined. To discuss this, we propose an example of an emerging technology in data center power supply and study the effect of its incorporation into the framework. The example technology is summarized in using Direct Current (DC) power in the data center instead of the Alternating Current (AC) power [26]. This is motivated by the energy savings that result from avoiding AC to DC and DC to AC conversions at the UPS, at the PDU, and at the front end of the Power Supply Unit (PSU) on servers. Evaluating this trend is out of scope for this article, but we will focus instead on outcomes of its adoption, and ability of the framework to absorb them.

Researchers supporting this idea suggest that one AC to DC conversion takes place at the UPS. All power thereafter is DC power. For the framework this would mean representing a new class of UPS that does not convert DC power coming out of its battery back to AC power, but rather delivers it as it is as DC power output. A new class of PDUs for distributing DC power needs also to be represented, and a new PSU class which would result in a new DC server class is required. These new classes of equipment come accompanied with new restrictions on their power inputs and outputs.

Representing the new classes and relations within the framework requires modifying the ontology representing the metamodel. This can be achieved simply by adding an attribute to the `PowerPort` classes that determines if the input or output is an AC or a DC. This attribute is then inherited automatically by subclasses `PowerInputPort` and `PowerOutputPort`. Setting its default value to AC releases the designer from the burden of changing its value in existing components. Restrictions on the `hasConnectionTo` property must be updated to include restricting to the same type of current.

New component classes that use DC are then added normally into the ontology with one difference from conventional components that they use DC `PowerPorts`. Mathematical models of these new components are inserted into the ontology advocating thereby its role as the central metamodel. In a following step they will be automatically extracted from the ontology and collected into the data center Modelica package as mentioned earlier.

This is all it takes for the new components to be fully incorporated into the framework. Changes in the ontology are reflected automatically on the toolbox which will start listing the new components making them available for the designer. Semantic verification would check the domain and range of `hasConnectionTo` relations and detect inconsistencies without any additional effort. Generation of the operational model would translate the verified design including the new components into Modelica code and link them properly to their mathematical models. Flexibility is additionally incarnated in the ability of choosing different mathematical models to represent the same component. This allows seamless update of the component models through the ontology without having to edit the data center design itself.

5.2 The Framework in Real-Life Operation

During the lifetime of the project a training day was scheduled for introducing the project partners to the framework. On this day trainees were required to model the server room of the Business Informatics department, run the required verifications, run a simulation, try to recognize energy saving potentials, and then simulate different scenarios that lead to improved efficiency. The server room contained fifteen servers in one rack, three switches, one air conditioner, and two UPSs.

Before starting this day we had monitored the server room devices in operation for 5 months through our monitoring software and hardware in order to collect real utilization and energy demand time series, and to create energy models of the devices using our machine learning-based tools. These data were provided respectively as CSV files and lookup tables in the Modelica format. Additionally, a weather data file was provided containing outside temperature on the course of the 5 months.

Trainees could independently model the server room from the provided description by generating a toolbox and using it in a drag-and-drop fashion. They were able to use arcs for connecting the matching types of ports on the created components, and to utilize with little help the *virtual components* for calculating the PUE. After that and through a menu item in the editor, they have run the syntax and semantic verifications and could detect and correct some errors in the design. By clicking another menu item they have verified that the servers do have a degree of power redundancy since each was powered by two different UPSs. Following the given order of steps, they have generated the Modelica code and have run the simulation. After that the trainees were free to plot the simulation results for the different ports they choose. Some have plotted the PUE and the outside temperature and were able to detect a correlation. With a total power demand between 4.000 and 5.638 kW, and an IT power demand between 1.246 and 1.536 kW, the calculated PUE ranged from 3.1 to 3.8. Others have plotted the input and output power of the UPSs and have concluded that they were underutilized and therefore operating in a low-efficiency point. The measured power demand before one UPS (demand of the UPS itself together with its connected devices) had an average of 1.990 kW while the UPS itself had a power demand that averaged at 1.267 kW contributing more than 60 % of the total power demand. An energy saving potential was identified at this point, and there were different suggestions for optimization. One suggestion was to keep one UPS and connect the ports that were originally connected to the first UPS directly to the power source. Trainees were able to test this scenario by simply rearranging the power connections in the model and deleting the other UPS, and got the following results: PUE was significantly improved and ranged between 2.15 and 2.63 with a total power demand between 2.726 and 3.907 kW. The utilization of the remaining UPS was not influenced with this change and it kept operating with low efficiency. Another scenario was suggested by the trainees: to connect all devices to one UPS trying to increase its utilization. Again with simple steps the new scenario was designed, and the results were as follows: PUE between 2.35 and 3.14, total power demand between 3.331 and 4.280 kW, power demand before the UPS was 3.039 kW in average, and demand of the UPS itself was averaged at 1.615 kW.

The best scenario for the PUE and the total power demand of the server room (under the given utilization values) is clearly the second one. Through the "integration" tool trainees were able to estimate total *energy* demand for each of the scenarios along the 5 months period. These were respectively: 18.196, 12.564, and 14.230 MWh. Potential savings over the actual state (first scenario) by implementing the second scenario is 5.632 MWh over 5 months, which amounts to about 13.500 MWh over a year. For a kWh price of €0.15, a total of €2025 could have been saved per year. Other (potentially better) scenarios can be easily examined using the framework and the effect on PUE, total consumption, and power redundancy of devices can be simply assessed.

5.3 Further Features

We define framework interoperability as the ability of the framework to work in conjunction with other systems. This feature is achieved in our case mainly through relying on open standards like XML, OWL, Scalable Vector Graphics (SVG), Modelica, and Extensible Stylesheet Language Transformations (XSLT) which enables interoperability out of the box. Additionally, interoperability is approached by keeping the metamodel ontology aligned with widely-used top-level ontologies like DOLCE [10], and SUMO [7] as mentioned in Sect. 4.1 and detailed in [6]. This alignment allows interoperation on the data level with other systems that utilize these top-level ontologies or their descendants, and the number of such systems is not to be underestimated. For example, ontology alignment allows our framework to seamlessly integrate data collected from monitored data centers around the globe through alignment with the ontology that is created in [8] to serve the purpose of such data curation. This data serve parameterize mathematical models of components with training data without having to go through the training phase. This interoperability on the data level is achieved by aligning the two ontologies, which is a relatively simple task having in mind that both correspond to DOLCE.

Ability of the ontology to include different facets allows importing and exporting data from and to various systems. An example was discussed before for the case of OpenDCIM where a facet that matches its data structure was included in the ontology. Adding new facets would allow the framework to integrate into further tool chains through communicating data with new systems.

6 Conclusion

In this chapter we demonstrated a framework for data center modeling. The framework is composed of an ontology as a metamodel, and a workflow of transformations leading the designer through the design process from the toolbox through a fully operational model to the simulation and visualization. All stages of the design throughout the workflow are governed by and accorded with the metamodel ontology, which was built with interoperability in mind. Additionally, we have included tools for hardware profiling to provide basic component energy demand models. The framework was discussed later from the flexibility and interoperability points of view. The studies were based on actual example cases where these features exhibit a high demand. Having a central, interoperable, (re-) usable, and comprehensive metamodel proved useful under these circumstances and conferred flexibility and interoperability on the framework as a whole. Additionally, sticking to open standards in representing the metamodel, transformations, and the different stages of the design grants the framework additional interoperability, prevents a vendor lock-in, and allows a wide array of tools to access and manipulate data and processes. An implementation on a real-life situation was discussed in the end including real devices and data center circumstances on the one hand and a real user group of varying experience on the other.

Acknowledgement This work is part of the project IT-for-Green (Next Generation CEMIS for Environmental, Energy and Resource Management). The IT-for-Green project is funded by the European regional development fund (grant number W/A III 80119242).

References

1. Gartner, Inc.: Press Release: Gartner Says Worldwide Data Center Hardware Spending on Pace to Reach $99 Billion in 2011. (2011).
2. Koomey, J. G.: Estimating total power consumption by servers in the U.S. and the world. (2007).

3. EPA, US Environmental Protection Agency: Report to congress on server and data center energy efficiency: Public law 109–431, (2008).
4. Barroso, L. A., Hölzle, U.: The case for energy-proportional computing. IEEE Comput., 40 (12), 33–37 (2007).
5. Memari, A., Vornberger, J., Nebel, W., Marx Gómez, J.: A Data Center Simulation Framework Based on an Ontological Foundation. In Marx Gómez, J., Sonnenschein, M., Vogel, U., Rapp, B., Giesen, N. (eds.) Proceedings of EnviroInfo 2014 – 28th International Conference on Informatics for Environmental Protection, Oldenburg, Germany (2014).
6. Memari, A.: Angewandtes Semantisches Metamodell von Rechenzentren für Green IT. In 5. BUIS-Tage: IT-gestütztes Ressourcen- und Energiemanagement, Oldenburg, Germany (2013).
7. Niles, I., Pease, A.: Towards a standard upper ontology. In Proceedings of the international conference on Formal Ontology in Information Systems-Volume 2001, pp. 2–9. ACM (2001).
8. Germain-Renaud, C., Furst, F., Jouvin, M., Kassel, G., Nauroy, J., Philippon, G.: The Green Computing Observatory: A Data Curation Approach for Green IT. In IEEE Ninth International Conference on Dependable, Autonomic and Secure Computing (DASC), pp. 798–799. IEEE (2011).
9. Kuhn, W.: A Functional Ontology of Observation and Measurement. In Proceedings of the 3rd International Conference on GeoSpatial Semantics. pp. 26–43. Springer, Berlin Heidelberg (2009).
10. Borgo, S., Masolo, C.: Foundational Choices in DOLCE. In Staab, S., Studer, R. (eds.): Handbook on Ontologies. pp. 361–381. Springer, Berlin Heidelberg (2009).
11. Calheiros, R. N., Ranjan, R., Beloglazov, A., De Rose, C. A., Buyya, R.: CloudSim: a toolkit for modeling and simulation of cloud computing environments and evaluation of resource provisioning algorithms. Software: Practice and Experience, 41(1), 23–50 (2011).
12. Altintas, I., Berkley, C., Jaeger, E., Jones, M., Ludascher, B., Mock, S.: Kepler: an extensible system for design and execution of scientific workflows. In Proceedings of the 16th International Conference on Scientific and Statistical Database Management. pp. 423–424. IEEE (2004).
13. Asghar, S. A., Tariq, S., Torabzadeh-Tari, M., Fritzson, P., Pop, A., Sjölund, M., Vasaiely, P., Schamai W.: An open source modelica graphic editor integrated with electronic notebooks and interactive simulation. (2011).
14. The MathWorks, Inc.: MATLAB and SIMULINK 2012b, Natick, Massachusetts, United States (2012).
15. Kumar, A., Saigal, R.: Visual Understanding Environment. In Proceedings of the 5th ACM/IEEE-CS joint conference on Digital libraries. pp. 413–413. ACM (2005).
16. Shearer, R., Motik, B., Horrocks, I.: HermiT: A highly-efficient OWL reasoner. In Proceedings of the 5th International Workshop on OWL: Experiences and Directions (OWLED 2008). pp. 26–27. (2008).
17. Rivoire, S., Ranganathan, P., Kozyrakis, C.: A comparison of high-level full-system power models. In Proceedings of the 2008 conference on Power aware computing and systems. pp. 3–3. USENIX Association, San Diego, California (2008).
18. Fan, X., Weber, W. D., Barroso, L. A.: Power provisioning for a warehouse-sized computer. ACM SIGARCH Comput. Archit. News, 35(2), 13–23 (2007).
19. Bircher, W. L., John, L. K.: Complete system power estimation: A trickle-down approach based on performance events. In IEEE International Symposium on Performance Analysis of Systems & Software (ISPASS). pp. 158–168. IEEE (2007).
20. Pelley, S., Meisner, D., Wenisch, T. F., VanGilder, J. W.: Understanding and abstracting total data center power. In Workshop on Energy-Efficient Design. (2009).
21. Bircher, W. L., Valluri, M., Law, J., John, L. K.: Runtime identification of microprocessor energy saving opportunities. In Proceedings of the 2005 International Symposium on Low Power Electronics and Design (ISLPED'05). pp. 275–280. IEEE (2005).

22. Pedram, M., Hwang, I.: Power and performance modeling in a virtualized server system. In 39th International Conference on Parallel Processing Workshops (ICPPW), pp. 520–526. IEEE (2010).
23. Dhiman, G., Mihic, K., Rosing, T.: A system for online power prediction in virtualized environments using gaussian mixture models. In 47th ACM/IEEE Design Automation Conference (DAC), pp. 807–812. IEEE (2010).
24. Koller, R., Verma, A., Neogi, A.: WattApp: an application aware power meter for shared data centers. In Proceeding of the 7th international conference on autonomic computing, pp. 31–40. ACM (2010).
25. Pearson, K.: Note on regression and inheritance in the case of two parents. Proc. R. Soc. Lond., 58(347–352), 240–242 (1895).
26. Garling, C.: AC/DC Battle Returns to Rock Data-Center World. http://www.wired.com/2011/12/ac-dc-power-data-center/ (2011). Accessed 26 September 2014.

Chapter 4
The Contexto Framework: Leveraging Energy Awareness in the Development of Context-Aware Applications

Maximilian Schirmer, Sven Bertel, and Jonas Pencke

Abstract We introduce a new context classification and recognition framework for the development and deployment of mobile, context-aware applications. The framework is complemented by an energy calculator that specifically assists mobile developers in estimating the energy footprint of context-aware applications during the development process with the framework. The framework abstracts from the raw context information gathering, allows for sensor fusion, enables the prediction of custom and higher-level contexts, and provides for context sharing.

Keywords Energy awareness • Context-aware frameworks and applications • Ubiquitous computing • Mobile sensing

1 Introduction and Motivation

The evolution of mobile devices and the general availability of information sources that describe the situation and environment (i.e., the context) of mobile users offer new opportunities for innovative applications [1]. By constantly monitoring the contexts in which mobile users are situated, applications obtain a potential to adapt their behaviour to current contexts more intelligently than previously possible, and without user intervention. However, such mobile context awareness comes at a price: Novel challenges of the mobile environment and specific constraints of mobile devices and their use (e.g., limited battery life, a comparably small screen size, dependence on network infrastructure) can severely impact the acceptance of mobile context-based approaches. In addition, adequate developer support for the realisation of context-aware applications is currently lacking. Consequently, most application developers are on their own when realising the sensing and interpreting of context information, or the sharing of context. With the increasing interest in, and a growing market for, context-aware applications, developers are more and more in charge of carefully designing context-aware applications. They need to be

M. Schirmer (✉) • S. Bertel • J. Pencke
Usability Research Group, Bauhaus-Universität Weimar, Weimar, Germany
e-mail: max@schirmers.de; sven.bertel@uni-weimar.de; jonas.pencke@uni-weimar.de

J. Marx Gómez et al. (eds.), *Advances and New Trends in Environmental and Energy Informatics*, Progress in IS, DOI 10.1007/978-3-319-23455-7_4

able to competently address issues such as privacy [2], availability, precision of context recognition, or energy requirements.

In this contribution (which is an extended version of [3]), we address the energy-related implications of developers' choices of sensing components, processing algorithms, and granularity or temporal frequency of sensing. We specifically aim at developer energy awareness and present *Contexto*, an energy-aware framework for offline context classification and recognition on mobile devices. The framework provides a layered, component-based architecture that can easily be extended, modified, or customised. It follows established software engineering patterns to provide high learnability and a low threshold for beginners. Within the framework, the energy requirements for all used components on a specific device are always made transparent, and information about energy requirements can be used early in the design process with the help of the framework's energy calculator, and at runtime.

The following section will introduce the main concept and energy model of *Contexto*. Section 3 will address the software architecture and implementation details. Section 4 will give an overview of related work. Section 5 will provide an outlook on future work.

2 Energy Model for the Contexto Framework

Contexto aims at providing energy awareness for developers of context-aware applications. In order to do so, it requires an energy model with information about the individual energy demands of the relevant components of the framework. We hope that insights into the specific energy footprints of alternative implementations of context recognition will lead to more energy-efficient applications that, in turn, will be more widely accepted by their users. At its core, the framework employs an energy model for a number of smartphone devices. The model provides information on the individual power consumption and required energy for all of a device's sensors in relation to a chosen sampling interval. We conducted extensive measurement experiments with a software-based remaining capacity approach to build energy models for the Apple iPhone 4, 4S, and 5. In essence, we read the remaining battery capacity in mAh, as provided by the *IOPowerSources* part of the *IOKit* framework (please see [4] for a detailed description of the technical measurement setup). As we showed previously in [5], this software-based, model-based approach is well-suited for the total demand measurement required for building the energy model. Information about the energy demand of the current selection of sensors is made available to developers at runtime, directly within the framework.

In contrast to most of the classic context platforms and toolkits that rely on a distributed or client/server-based architecture, *Contexto* is completely self-contained and provides true offline context classification and recognition. All steps of the processing pipeline (data acquisition, pre-processing, context classification, context prediction) take place directly on the mobile device, and there is no external service or context platform required. Our tests of the framework have shown that recent smartphone models provide the required resources (e.g., CPU speed and RAM size) for all of the involved processing steps. We think that offline approaches are superior to online approaches because they allow for higher levels of privacy (all gathered data remain on the device), reduce the amount of energy required for context recognition (UMTS and WiFi hardware requires most of the energy on a smartphone [6]), and do not rely on any kind of network infrastructure.

We conducted an extensive series of experiments on all three types of devices. On each device, the energy demand of about 20 hardware and software sensors (the exact numbers vary, depending on specific device capabilities and available sensors) as well as of 10 features was measured with a sampling frequency of 1 Hz over the course of 10 min—a timeframe that typically corresponds to the average amount of time that a user spends in a mobile application at a time over the course of the day. Over the course of this paper, we will refer to each test of a given sensor or feature on a given device as a *trial*. For every sensor/feature, we repeated each trial three times on every device. Before the tests, we configured all devices by performing the following steps (steps 5 and 6 were repeated before every trial):

1. Force a battery calibration—also known as capacity relearning—by applying a full charge, discharge, and charge
2. Remove the SIM card
3. Deactivate influencing services and tasks, like Bluetooth, WLAN, notifications, push, and background updates
4. Decrease screen brightness to minimum
5. Terminate all active applications
6. Restart the device to free up/release unused memory

The first step is particularly important as charges and discharges influence the estimation of the remaining amount of energy of the phone's power management components. Varying conditions and timeframes under which the battery charges and discharges contribute to a deviation in the battery's true capacity compared to the readout. Battery recalibration counteracts this effect. The record of a trial gives information about the maximum battery capacity (in mAh), the starting battery capacity (in mAh), current battery voltage (in mV), and current battery capacity (in mAh). From this data, we calculated the following characteristics for the energy model:

- *Drained rated capacity*: $C_{N\_DRAIN} = C_{N\_START} - C_{N\_END}$ (subtract the rated capacity remaining at the end of a test trial from the rated capacity at the start)
- *Electrical work*: $W = C_{N\_DRAIN} {}^* U_{AVG}$ (obtain watt-hours by multiplying the drained rated capacity with the average voltage applied over the discharge interval, then convert to watt-seconds, a unit that is equivalent to Joule, in both units and meaning: 1 Wh = 3600 Ws = 3600 J)

Table 4.1 Device-specific energy demand overview of hardware sensors

Component	Energy demand (J, over 10 min)		
	iPhone 4	iPhone 4S	iPhone 5
Sensors			
Accelerometer	29.15	25.13	20.42
Camera	217.02	317.00	293.89
GPS	143.29	143.56	180.44
Gyroscope	61.67	49.53	34.26
Heading (compass)	69.23	70.34	37.41
Features (measurement/computed)			
Camera + GPS	543.31/ 549.63	662.45/ 657.00	638.49/ 646.16
Accelerometer + Orientation	237.68/ 240.86	247.28/ 241.93	202.97/ 205.79
Framework (local/remote)			
Baseline	211.86/ 268.18	245.12/ 318.92	181.89/ 249.30
Battery + Carrier + Contacts + Date + Reminder	302.45/ 380.18	358.31/ 486.87	307.44/ 399.47
Camera + GPS + Heading + NetworkInfo	671.80/ 750.35	846.49/ 977.99	694.60/ 808.45

- Real energy demand: $E_{REAL} = abs(E_{BASE} - E_{ALL})$ (subtract the energy baseline of the device from the energy demand of the device while using a specific component)

We also evaluated if our initial argument for local (offline) processing and classification directly on the device had a positive energy impact, compared to a remote solution where gathered sensor data was sent to a server using WiFi. Table 4.1 gives a short overview of some of the most interesting findings. In the table, we compare the energy demand (expressed in Joule for a 10-min trial, all values are averages across three trials) of selected sensors and features across our tested devices. Furthermore, we used the measured energy demand of individual sensors to approximate the energy demand of our features. Finally, we evaluated the difference between keeping all data local vs. sending data to a remote server for processing.

Our gathered data indicates that most of the sensors in the iPhone 5 (the most recent device among our test devices) show a lower energy demand than the older sensor components in the iPhone 4 and iPhone 4S. There is an exception with the GPS sensor of the iPhone 5. As that sensor's performance has drastically improved since the iPhone 4S, we suppose that, for the iPhone 5, Apple opted for high accuracy and low latency over low energy demand.

We were also able to show that the sum of the individual energy demands of sensors very well approximates the energy demand of features using these sensors. In most cases, the computed energy demand overshoots the measured energy demand; this is probably due to internal optimisations of the underlying iOS sensor APIs. However, the highest difference amounts to about 2 % (for the Camera + GPS feature); this leads to our conclusion that we can estimate the energy demand of a feature with the following equation:

$$E_{REAL} \approx E_{COMP} = \left(\sum\nolimits_{i=1}^{n} E_{SENSOR_i}\right) + E_{DEVICE_{BASELINE}} \tag{4.1}$$

Which can then be further generalised to:

$$\begin{aligned} E_{REAL} &\approx E_{COMP}(Sources, Device) \\ &= \left(\sum_{Sensor \in Sensors(Sources)} E(Sensor) \right) + E(Device) \end{aligned} \tag{4.2}$$

Where

$$Sensors(Sources) = \{Sensor | \exists_{Source \in Sources} : SourceUsesSensor\} \tag{4.3}$$

The "SourceUsesSensor" relation enumerates all sensors used by a feature. If a feature uses another feature as a source, the sensors that the latter feature depends on are enumerated recursively. The set returned by the "Sensors" function counts the energy demand of each sensor only once.

Finally, our evaluation clearly shows that sending sensor data over a WiFi connection requires a considerable amount of energy; and that this amount cannot be compensated for with the energy saved via outsourcing the processing and classification algorithms. We are still looking into this issue, since we assume that very complex processing algorithms and classifiers running on the smartphone may change this circumstance.

3 Implementation

Following the description of the *Contexto* framework's energy model in the previous section, we will now focus on the framework itself. The *Contexto* framework follows a layered architecture, based on the proposed architectures in [7, 8]. As shown in Fig. 4.1, it separates context usage from context detection. In the *Context Detection* layers, we set a *Managing* layer that contains context models and persistence functionality on top of classifiers and features of the *Thinking* layer. The basic *Sensing* layer contains all sensors that are used as data sources.

Our prototype implementation is currently available as an iOS framework, but the concept can be adapted to other smartphone operating systems with little effort. The sandbox design of iOS currently prohibits the implementation of a service architecture for providing context data to other applications, but this could easily be

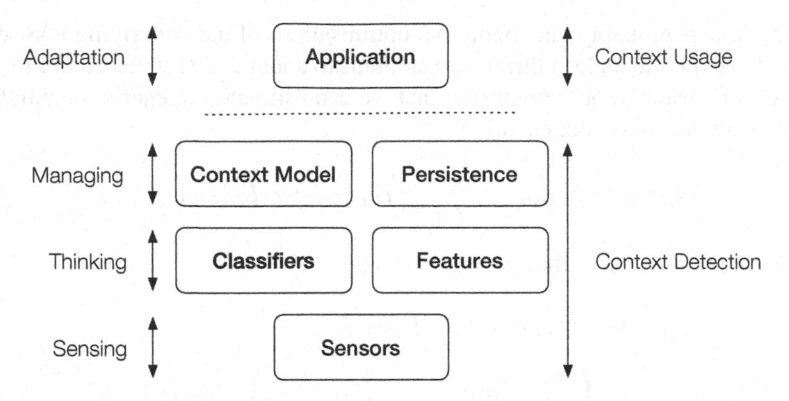

Fig. 4.1 Layer architecture of *Contexto*

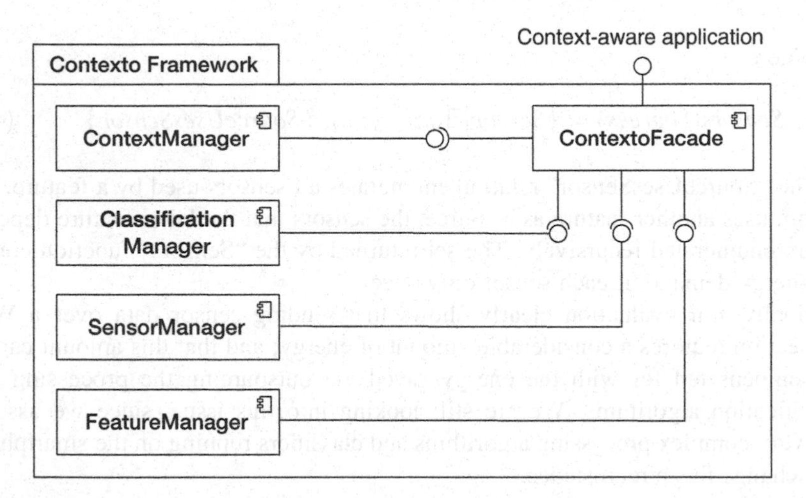

Fig. 4.2 Overview of client interactions with the *Contexto* façade

overcome on other platforms. The framework implementation was tailored to support easy extensibility, learnability, customisability, and maintainability by paying particular attention to accepted design principles (Separation of Concerns, Single Responsibility, Interface Segregation, and Dependency Inversion).

This means in particular that developers familiar with common iOS frameworks should have a fast learning experience and quickly make efficient use of *Contexto*. All components can easily be extended or replaced. In our prototype implementation, we currently provide a naïve Bayes classifier; additional classifiers can be added by overriding our well-documented *JPClassifier* class. All key components follow this pattern of extensibility.

Instead of exposing developers to a set of classes and their APIs, the *Contexto* framework provides a single interface, called the *ContextoFacade*. The unified entry point simplifies the framework usage for developers by hiding complex subsystems and their interactions (cf. Fig. 4.2).

The facade decouples the application code that uses the framework from the internal protocols and implementations. It reduces direct dependencies between the application code and the inner workings of the framework. Besides that, it consistently retains the same API while manager classes remain exchangeable with custom-tailored implementations by developers. The functionality of the *ContextoFacade* includes allocating framework instances, configuring the framework before it is used, accessing sensors, features, and classifiers provided by the framework; registering framework and custom features for the classification process, announcing contexts the context-aware application is interested in, registering observers interested in context changes, registering an oracle that answers label requests, initiating context classification manually, training the classification model, handling application background mode, and computing the average energy demand of the running framework.

The *Contexto* framework requires an initial configuration before it starts the context classification and model training. At the minimum, developers must select features for the classification and training process, announce desired contexts, register as observer to receive notifications on context changes, and define a part of the application that answers label requests. The *Contexto* framework configures the remaining options automatically by selecting default components or values, such as a classification sampling frequency or a classifier. As a real world example, a possible configuration for a messaging application is shown in the following listing (Fig. 4.3). This messaging application aims to avoid inappropriate interruptions (e.g., through notifications and alerts) by considering the context the application user is currently in. It acquires a simplified context—such as, whether or not a person is currently in a meeting—and decides to postpone the delivery of messages or reject incoming calls if necessary.

The few lines of code in Fig. 4.3 are sufficient to cover the entire configuration of the framework and setup context acquisition for the application. First, a framework instance is allocated. The second line tells the framework to use a default classifier

```
Contexto *contexto = [Contexto sharedInstance];
[contexto configureWithClassifier:nil samplingFrequency:@(60.0f)];

[contexto addFrameworkFeatureWithType:ContextoFeatureTypeAppointments];
[contexto addFrameworkFeatureWithType:ContextoFeatureTypeBattery];
[contexto addFrameworkFeatureWithType:ContextoFeatureTypeConnectivity];
[contexto addFrameworkFeatureWithType:ContextoFeatureTypeDayTime];
[contexto addFrameworkFeatureWithType:ContextoFeatureTypeDuringWeek];
[contexto addFrameworkFeatureWithType:ContextoFeatureTypeIsMoving];
[contexto addCustomFeature:[[CustomFeatureNoiseLevel alloc] init]];
[contexto addCustomFeature:[[CustomFeatureIsOfficialHoliday alloc]
init]];

[contexto registerContexts:[@"IsInAMeeting", @"IsNotInAMeeting"]];

[contexto registerObserverForContextChanges:self];
[contexto registerContextProvider:self withConfidenceThreshold:@(0.65)];
```

Fig. 4.3 Example *Contexto* configuration in a context-aware application

for all context instances and to periodically initiate the classification process every minute. Developers can use a different framework classifier, or their own implementation, by passing an appropriate instance instead. The framework allows developers to retrieve existing classifiers by type, like *ContextoClassifierTypeNaiveBayes*, using the *frameworkClassifierWithType:* method. The next code block registers existing framework features and afterwards custom features provided by the developer. The sources of a feature—sensors or other features—are automatically instantiated and registered by the framework. The selection of features is a complex task and completely unique to the usage scenario. Quality features are crucial for an accurate prediction result. Line number 13 declares the two contexts the application is interested in. Finally, the application registers itself as observer for context changes and as oracle providing labels for the active learning part. The *self* keyword represents the part of the context-aware application that implements the required context change and context request protocol methods.

Having successfully configured the *Contexto* framework, developers can use the *computeAverageEnergyDemand:* method of the facade to compute the expected energy demand for context acquisition. Through the registered features, the framework knows which underlying sensors are sampled, and how frequently. Using this knowledge, it accumulates the energy demand of the lowest units and adds a fixed energy demand for the baseline of the framework. The result of the computation gives developers an idea of the additional average energy demand that their application needs.

The implementation of the *Contexto* framework is backed up by over 200 logic unit tests to ensure a high degree of code quality. The test suite pays particular attention to verifying manager and data storage classes, as well as to static factories. According to our gcov analysis, the test coverage of default manager and data storage implementations is above 80 %. Along with the test cases, the *Contexto* framework provides comprehensive documentation of all of the important external and internal classes. The documentation is available as HTML and as a fully indexed *Xcode* documentation set, giving developers the possibility to integrate the documentation as part of the auto-completion functionality of *Xcode*.

4 Contexto Energy Demand Calculator

The *Energy Demand Calculator* (see Fig. 4.4) is a tool for developers to estimate additional energy demands of selected information sources (i.e., sensors and features) for context acquisition. The calculator complements our *Contexto* framework and makes use of the previously described energy model. The tool gives developers the ability to determine an approximated energy demand early in the development process of a context-aware application with our framework. The calculator allows developers to try out different combinations of information sources and visualises their impact on the energy demand of the application. Such information allows

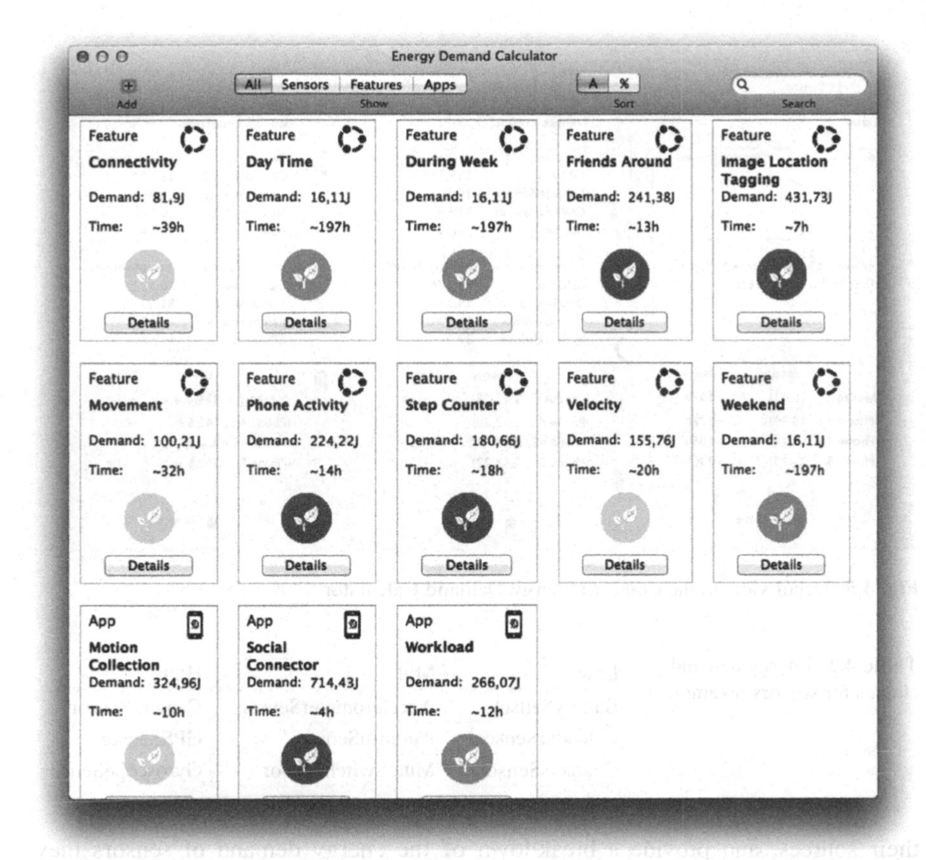

Fig. 4.4 Graphical user interface of the *Contexto* Energy Demand Calculator. Colours *green* (dark grey), *yellow* (light grey), and *red* (medium grey) denote components' energy demands

optimising an application's energy footprint and increases the awareness for major energy consumers.

The previously introduced energy model serves as the basis for all estimations. In order to compute the energy demand of an application, the calculator searches recursively for related sensors, generates a unique sensor pool by removing duplicates.

The construction kit of the calculator provides a *sensor*, *feature*, and *application* component to interact with. This allows developers to rebuild their context-aware application and simulate possible configurations. The *application* component represents the real-world application that is yet to be developed by holding a set of assigned feature components. A *feature* component is composed of necessary *sensor* and *feature* components, just as is the case in the framework. As lowest-level component, *sensors* provide information about their average and device-specific energy demand. Based on this information, *feature* and *application* components accumulate their own energy demand, show the overall energy demand of

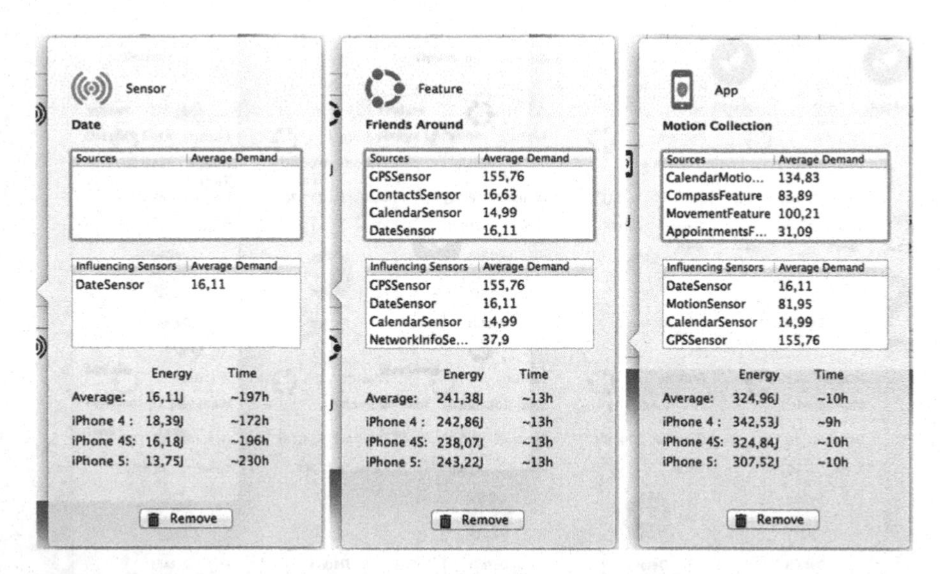

Fig. 4.5 Detail view in the *Contexto* Energy Demand Calculator

Table 4.2 Energy demand classes for sensors (examples)

Low	Mid	High
BatterySensor	AccelerometerSensor	CameraSensor
CalendarSensor	AudioInSensor	GPSSensor
ContactsSensor	MuteSwitchSensor	GyroscopeSensor

their sources, and provide a breakdown of the energy demand of sensors they depend on. Figure 4.5 shows detail views of all three component types. In addition, the calculator roughly estimates the time the component would need to completely drain a charged battery. This estimate helps developers to get a better sense of the actual impact of their application.

For better orientation, the calculator introduces three general energy demand classes: *Low*, *Mid*, and *High*. Class membership is defined in terms of varying energy thresholds for sensors, features, and applications. Sensors belonging to the *Low* energy class require barely more energy than an idle application. Sensors in the *Mid* class have a noticeable impact, but are still far from the *High* energy demand class. The average energy demand of all sensors is approximately 40 J in 10 min. The classification uses this as a reference value and assigns sensors with an above-average energy demand into the *High* energy demand class. The remaining range is divided into two identical parts. Sensors belonging to the *Low* energy demand class do not exceed an energy demand of 19 J. The *Mid* class ranges from 20 to 39 J. Table 4.2 provides an examples for the energy demand classes of sensors within the *Contexto* framework. The graphical user interface of the Contexto Energy Demand Calculator makes intuitive use of these three demand levels by mapping them to a traffic light colour scheme: *green* components possess low energy demands, *yellow* components possess mid-level energy demands, and *red* components possess high energy demands.

We envision that app developers will be using the *Contexto* framework and the calculator in order to assess individual sensor combination's energy demands. This requires an integration into the regular app development process. Developers would benefit from the energy demand estimation and could then adjust two important factors that influence the energy demand: choice of sensors in context-aware mobile applications, and the sampling rate of those sensors.

5 Evaluation

Evaluating *Contexto* is a twofold approach, consisting of a technical evaluation and a user study with human participants. The technical evaluation aims to show that our energy model's prediction of energy demand for components of the *Contexto* framework actually works and that it is possible to estimate the energy demand by measuring only the use of these components. We have been able to show this in Sect. 2 of this contribution.

The second part, a between-subject design user study, will focus on the long term evaluation of the framework, with its users (i.e., mobile application developers) and in daily use. There should be a control group in which each participant develops a context-aware application without the energy-aware methods of *Contexto*, and a second group in which developers may use these methods. After a development period of at least a week, all resulting mobile applications will be analysed regarding their energy demands. This procedure will allow us to evaluate the actual impact of *Contexto* on the energy-awareness of its users; it will require significant funding, since mobile developers will have to be paid to work with our framework for days or weeks.

6 Related Work

The research presented here is embedded in a broad range of mobile, pervasive, and ubiquitous computing activities. Within these communities, there has been active research on context models, context recognition, and context-aware applications and devices. Concept and implementation of *Contexto* highly benefit from this previous research. Various frameworks for context classification and recognition exist.

Context Toolkit [9] supports developers in rapid prototyping of context-aware applications. The framework relies on a distributed architecture utilising five key components: *context widgets*, *interpreters*, *aggregators*, *services*, and *discoverers*. A distributed infrastructure is responsible for hosting the components, acquiring recognised contexts and delivering them to interested applications. *Context widgets* provide access to context information while hiding the details of the acquisition. *Interpreters* raise the level of abstraction by transforming low-level context information from single or multiple sources to higher-level context information. *Aggregators* group logically related context information together to make them easy accessible to interested applications. *Services* are actuators that applications trigger to execute predefined actions. *Discoverers* provide an overview of environmental capabilities allowing developers to find existing components, like *context widgets*, *interpreters*, *aggregators* or *services*. Applications can either query certain components in regard to their information, or register themselves as subscribers to changes. The distributed architecture enables the sharing of context information to multiple applications. The framework components abstract the acquisition of sensor data and allow the creation of higher-level context information through sensor fusion. New contexts are addable by creating appropriate components with the aid of existing components. To access and use existing components, the application can use a network API or register itself as subscriber.

The *Hydrogen Context-Framework* [10] is based on a centralised architecture comprising *Adaptor*, *Management*, and *Application* layers. The *Adaptor* layer gathers sensor information, enriches them with additional contextual information, and passes the result to the *Management* layer. This layer accommodates a centralised *ContextServer* that provides and stores contexts from the previous layer and manages context exchange with surrounding devices using peer-to-peer communication. The *Application* layer encapsulates applications that are subscribed for context changes, applications can use asynchronous or synchronous methods to retrieve these context changes. This approach distinguishes between remote and local contexts. The remote context is information provided by other devices, whereas the local context describes the context knowledge of the local device. The centralised design—without using a distributed server—makes this approach more robust against network disconnections and allows multiple applications to retrieve contexts per subscriptions, but heavily depends on other remote devices providing contexts.

CoBra [11] is a broker-centric architecture to support prototyping of context-aware systems. The approach introduces an intelligent server agent called *Context Broker* to help on the following central aspects: context modelling, context reasoning, knowledge sharing, and user privacy protection. The server entity is outsourced to a resource-rich stationary device employing four modular components. The *Context Knowledge Base* uses ontologies—defined with OWL—to provide a model describing contextual information and metadata. The *Context Reasoning Engine* acts as interpreter and reasoner, transforming acquired sensor data into contextual information by applying aggregation and domain heuristics. In addition, this component is responsible for detecting and resolving inconsistent context knowledge stored in the shared context model. The *Context Acquisition Module* provides general sensing capabilities and hides the low-level acquisition of information sources by providing reusable procedures. Finally, the *Privacy Management Module* takes care of establishing individual privacy policies on information sharing. Using a policy language, users define which personal information is revealed to the Context Broker. The broker centric design in combination with the ontology-based model enables multiple applications to make use of the shared context model, the inferred context changes, and enrich the context knowledge.

The *Sentient* framework [12] provides an approach to develop context-aware applications using a *Sentient Object Model*. The model consists of autonomous freely combinable *Sentient Objects*. A sentient object assimilates three components: *sensory capture* component, *context hierarchy* component, and *inference engine* component. The *sensory capture* component allows the acquisition of information gathered by sensors and fusion of higher-level information by combining multiple sensor information. In the *context hierarchy*, facts, such as discrete data, about individual context fragments are deposited. Representing the intelligent part of the *sentient object*, the *inference engine* evaluates incoming sensor information against the contextual deposited data. Based on predefined conditional rules, the inference engine decides which result is actuated. Each *sentient object* provides important characteristics including *Sentience*, *Autonomy*, and *Pro-activeness*. *Sentience* describes the possibility that sentient objects can perceive the environmental state via sensors. *Autonomy* stands for the independent operability without the necessity of human interaction. Pro-activeness means the ability to act in anticipation of upcoming conditions. Nesting *sentient objects* and context fragments result in an overall context hierarchy. The *context hierarchy* encapsulates knowledge about relevant sensors, possible actions and dependencies for each context. This approach tries to solve three main requirements for creating context-aware applications: each context-aware application needs a set of sensors for capturing context data, a set of rules that decide on the behaviour according to context, and a set of actuators for generating responses.

CASP—Context-Aware Service Platform [13] is a context gathering framework uses a client–server architecture with two components—a *Client-side Sensory API* and a *Server-side Sensory API,* both connected through TCP. The *Client-side Sensory API* provides a uniform interface for the server to access sensor data on clients. In addition, it translates sensed raw data supplied by a sensor abstraction subcomponent into a platform-understandable format. After transmitting the data to the *Server-side Sensory API*, subcomponents behind the *Sensory API* of the server start the further processing. The *Ontology Manager* subcomponent maps the sensory information to object instances of the OWL model to determine context data. The reasoning utilises a rule-based inference engine. In contrast to other approaches, *CASP* provides a *Context Service Execution Engine* subcomponent that executes services relevant to context domains, instead of returning context data to clients. Besides thinking, the server overtakes the acting part that is usually in the responsibility of clients. Each client has the ability to create new services via a *Service Creation API*. The client and server, as well as subcomponents of the framework, communicate via XML messages. *CASP* distinguishes between active and passive sensors. A passive sensor provides data when queried, whereas an active sensor periodically reports its sensed information to the server. New sensors are added by using the sensory API on the client. The abstraction of physical sensors, sensor data acquisition and sensor data dissemination separate concerns in *CASP*.

From a recent perspective, the sharp division between context acquisition and context usage is an inalienable requirement for building frameworks that facilitate the development of context-aware applications, as applied by the *Hydrogen* and *CoBra* approach. This concept dissolves former dependencies by strictly decoupling the generalizable context acquisition part from the application-specific adaptation. The clear assignment of responsibilities allows the framework to provide reusable functionality, and applications to focus on appropriate adaptations optimally suited to their use-cases. However, it appears reasonable to outsource the *Service* concept of the *Context Toolkit* to assist developers with one aspect of possible adaptations. A stand-alone framework could provide trigger-able actuators that execute predefined actions on the mobile operating system, like reducing the volume, opening a browser, locking the display, activating airplane mode. Even though the relocation into an additional framework would preserve the separation of concerns and keep the frameworks as slim as possible, it requires further examination and was not subject of the work presented in this paper.

Former approaches (*Context Toolkit*, *CoBra,* and *CASP*) use a distributed architecture, in contrast to the centralized approach that we introduce in *Contexto*. The evolution of mobile devices allowed shifting from this type to centralised on-device architectures, as partly used by the *Hydrogen* framework. It is valuable for multiple reasons: First, modern mobile devices supersede the supplementary installation of sensors in the environment and dissolve the limited availability of information sources by a plethora of hardware and software sensors. Second, the continuous improvement of processing capabilities provides enough resources to reason about context entirely on the device—rather on resource-rich remote servers. Besides robustness against discontinuities in network connections and increased operational reliability, it significantly reduces the network expenses by removing energy-hungry network transmission. Though an important aspect of mobile applications, none of the approaches consider the aspect of energy-awareness. The only framework using at least a centralised architecture (*Hydrogen*) depends on other mobile devices nearby, building ad-hoc networks to share context knowledge in a cooperative manner. Besides issues like missing applicability to personal trained models and privacy concerns, the continuous polling for other devices, establishing connections, answering incoming inquiries, and transferring information to other devices, consume the battery power at a very high rate.

The approaches differ in the models used to represent context knowledge and reason about context. Newer systems move towards either rule-based or ontology-based models. The former are popular because they are easy to implement for simple scenarios, whereas the latter simplify sharing and reusing. The *Contexto* framework takes another path. It uses an active supervised learning approach that—in comparison to others—allows the construction of personal context models considering individual characteristics and habits of users. This dissolves the static predefinition of contexts and the disparity in definition when created by experts or an impersonal group vs. the user itself. The construction of custom-tailored models within an active supervised learning process not only compensates issues of other models, but rather confronts further modern mobile requirements. A personal model provides significantly improved recognition rates—even with a short training period—and is capable of handling shifting user behavior by retraining the model. Active training handles falsely predicted contexts and unconfident situations.

Contexto is also related to research in the field of energy-aware software engineering and development: [14] presents the concept of energy labels for Android applications, a simple mechanism that easily allows end-users to assess the energy demand of their apps. *PowerTutor* [15] was one of the first energy models that were used for a mobile application. Numerous ongoing research regarding the measurement of energy demand on smartphones also exists (e.g., Refs. [5, 16]).

Furthermore, *Contexto* is related to the field of green and sustainable software engineering. This community has established a lot of concepts and models for building sustainable and resource-friendly software (e.g., Refs. [17–20]). There is even an agile approach to green software engineering [21].

7 Future Work

Contexto aims to provide energy awareness to developers of context-aware applications. In the future, we would like to make the framework itself aware of energy requirements. Using an energy budget system, developers will then specify a desired energy footprint, and the framework will make sure that the allocated budget is respected. This raises questions of the relation between energy demand and user requirements such as accuracy, precision, availability, or actuality of sensor data and context recognition. These parameters greatly influence the user acceptance of context-aware applications, which we will investigate further with the help of the framework. On a closer time horizon, we will conduct a user study with developers, to see if our design goals regarding the ease of use of the framework have been met.

It also seems worthwhile to explore whether the usability of the Contexto Energy Demand Calculator may benefit from offering a more hierarchical view of components. For example, if energy demands of a certain app are rated as *red* (corresponding to high energy demands), the user may open a detail inspection window of that app in order to investigate respective energy demands of features and sensors employed by it. In this vein, an iterative, structured exploration of energy demands may be provided, offering information about causality of energy demand ratings for composite components. Such an extension of the calculator by structure inspection facilities is similar to approaches which we chose for CONTEX(T)PLORER [22] to provide users with structured insights on the flow of information within component networks inside ubiquitous environments.

Another interesting aspect for further investigations is the provisioning of pre-defined context templates for specific domains. The breakdown of context domains into relevant sensors and features, as well as optimal settings like sampling frequency, may be possible, but tough for complex contexts. However, the benefit of context templates would be the support of novice developers by automatically selecting high quality features with corresponding sensors. Experts would still be able to fine-tune features as desired, but may gain helpful insights from templates. The optimal selection of features for a given context would improve the recognition rate and have a positive impact on context-aware application users.

The feasibility of framework-integrated actuators is worth examining. Actuators could provide default functionality or services that are triggered based on sensed contexts or requested on demand by the application developer. A further review of the four categories of context-aware application behaviour—*proximate selection*, *automatic contextual reconfiguration*, *contextual information and commands*, and *context-triggered actions*—may highlight specific examples that can be integrated into a separated framework. At a first glance, it might be possible to change settings of the mobile device like audio volume or brightness, or request and then show additional information, if hooks into the context-aware application are provided.

A smaller but no less interesting point of investigation relates to user interruption by the active learning. As implemented, developers can decide whether the active learning mechanism is turned on or off. But, it is not possible to define a time interval that must be at least expired before a subsequent query can be sent to the oracle. It should be possible to not only provide some meaningful settings for the active learning setup, but rather introduce an intelligent interruption algorithm. With the availability of sensors, features and the currently predicted context, the framework may be able to compute a disturbance value describing the attitude of the user towards an interruption. Moreover, the acuteness and actuality of the query must be taken into account. If the user answers a query after prolonged absence from the device, the validity of the label could be out-dated or at least questionable. Based on the value of the disturbance attribute, a decision whether a moment in time is appropriate to interrupt the oracle or not could be made. In addition, helpful information regarding labels could be inferred directly from the user interface. For example, it should be possible to observe the state of user interface controls, like buttons, sliders or personal settings to automatically determine label information without querying the user. A good example is a navigation application that allows the determination of different movement types by just evaluating the tapped button illustrating by foot, by car or by rail.

At the moment, the framework uses a simple periodic-triggered context classification. It would be beneficial if the framework could adapt the context sampling frequency in a more intelligent way. For example, the sampling frequency could be decreased automatically, if a certain amount of time has passed without recognising a context change. As a consequence, the time until processing the next classification could be doubled. If a context change occurs afterwards, the sampling frequency could be reset to a default or in between value. A more intelligent approach would be the observation of or subscription to information changes of sensors or features, such as device orientation, battery status, and network conditions. The classification process would be triggered only if the observer recognises a change. This approach would eliminate the need for a fixed timer that unintelligently starts a classification in a periodic manner. Moreover, fine-tuned time parameters, including when to increase or decrease the sampling frequency, when to reset it, or how to determine added or removed time interval or under which conditions it should be done, would no longer be necessary. The advantage of this approach is that fewer features need to be observed constantly. But, determining the features most significant to a given context beforehand, is not trivial.

A problem with various context-aware applications are reoccurring fast context changes leading to a flickering user interface. It is clear that context changes should be used defensively and only if a certain threshold is reached to constrain the problem as much as possible. Another possibility is to allow users to disable user interface changes for a certain time period. But it could be worthwhile to evaluate reasonable thresholds in-depth or to develop a mechanism that validates and intercepts context changes if they are all too frequent. In addition, a conflict management component could be introduced and encapsulated into the inferring logic. The predicted context may not match the context expected by the user, therefore options must be available to control the decision making as well as user interface adaptations. In the case of an incorrect prediction, the user could improve subsequent predictions by adding examples to correct the "false" prediction.

The integration of new classifiers and corresponding pre-processing steps seems more a diligent but routine piece of work. Including classifiers optimised for active online learning may be worthwhile. In addition, services providing trained context models or ontology-based models could be consulted under a stable Internet connection if the confidence of a predicted context does not exceed a pre-defined threshold. It is unsure if unambiguous conditions for merging an external context model can be defined, and how such a merging should take place. At least, this approach could be used to eliminate the cold start problem of the framework. At first start, a restricted basic model or simple examples could be loaded, which can be used for generating an initial model and then be adapted based on the personal preferences of the application user later on.

References

1. W. Clark, D. W. Cearley, and A. Litan. (2012, 8/5/2014). *Context-Aware Computing and Social Media Are Transforming the User Experience*. Available: http://www.gartner.com/doc/1916115/contextaware-computing-social-media-transforming
2. V. Bellotti and A. Sellen, "Design for privacy in ubiquitous computing environments," in *ECSCW'93*, New York, NY, USA (1993).
3. M. Schirmer, S. Bertel, and J. Pencke, "Contexto: Leveraging Energy Awareness in the Development of Context-Aware Applications," in *Proceedings of the 28th Conference on Environmental Informatics – Informatics for Environmental Protection, Sustainable Development and Risk Management*, Oldenburg, Germany (2014), pp. 753–758.

4. M. Schirmer and H. Höpfner, "Software-based Energy Requirement Measurement for Smartphones," in *First Workshop for the Development of Energy-aware Software (EEbS 2012)*, Braunschweig, Germany (2012).
5. H. Höpfner and M. Schirmer, "On Measuring Smartphones' Software Energy Requirements," in *ICSOFT 2012*, Rome, Italy (2012).
6. M. Schirmer and H. Höpfner, "SenST: Approaches for Reducing the Energy Consumption of Smartphone-Based Context Recognition," in *CONTEXT'11*, Karlsruhe, Germany (2011).
7. M. Baldauf, S. Dustdar, and F. Rosenberg, "A survey on context-aware systems," *Int. J. Ad Hoc Ubiquitous Comput.*, vol. 2 (2006).
8. S. Loke, "The Structure And Elements Of Context-Aware Pervasive Systems," in *Context-Aware Pervasive Systems: Architecture for a New Breed of Applications*, ed Boca Raton, FL, USA: Auerbach Publications (2007), p. 25.
9. A. K. Dey, G. D. Abowd, and D. Salber, "A conceptual framework and a toolkit for supporting the rapid prototyping of context-aware applications," *Hum.-Comput. Interact.*, vol. 16 (2001).
10. T. Hofer, W. Schwinger, M. Pichler, G. Leonhartsberger, J. Altmann, and W. Retschitzegger, "Context-awareness on mobile devices – the hydrogen approach," in *System Sciences 2003* (2003).
11. H. Chen, T. Finin, and A. Joshi, "A Context Broker for Building Smart Meeting Rooms," in *AAAI Symposium on Knowledge Representation and Ontology for Autonomous Systems Symposium, 2004 AAAI Spring Symposium* (2004), pp. 53–60.
12. G. Biegel and V. Cahill, "A framework for developing mobile, context-aware applications," presented at the Pervasive Computing and Communications – PerCom 2004 (2004).
13. A. Devaraju, S. Hoh, and M. Hartley, "A Context Gathering Framework for Context-Aware Mobile Solutions," presented at the Mobility '07 (2007).
14. C. Wilke, S. Richly, G. Püschel, C. Piechnick, S. Götz, and U. Aßmannn, "Energy Labels for Mobile Applications," in *First Workshop for the Development of Energy-aware Software (EEbS 2012)*, Braunschweig, Germany (2012).
15. L. Zhang, B. Tiwana, Z. Qian, Z. Wang, R. P. Dick, Z. M. Mao, *et al.*, "Accurate online power estimation and automatic battery behavior based power model generation for smartphones," in *Proceedings of the eighth IEEE/ACM/IFIP international conference on Hardware/software codesign and system synthesis*, Scottsdale, Arizona, USA (2010).
16. A. Pathak, Y. C. Hu, and M. Zhang, "Where is the energy spent inside my app?: fine grained energy accounting on smartphones with Eprof," in *Proceedings of the 7th ACM european conference on Computer Systems*, Bern, Switzerland (2012).
17. S. Agarwal, A. Nath, and D. Chowdhury, "Sustainable Approaches and Good Practices in Green Software Engineering," *International Journal of Research & Reviews in Computer Science*, vol. 3 (2012).
18. B. Penzenstadler, "What does Sustainability mean in and for Software Engineering?," in *Proceedings of the 1st International Conference on ICT for Sustainability (ICT4S)* (2013).
19. S. Naumann, M. Dick, E. Kern, and T. Johann, "The greensoft model: A reference model for green and sustainable software and its engineering," *Sustainable Computing: Informatics and Systems*, vol. 1, pp. 294–304 (2011).
20. J. Taina, "How green is your software?," in *Software Business*, ed: Springer (2010), pp. 151–162.
21. M. Dick, J. Drangmeister, E. Kern, and S. Naumann, "Green software engineering with agile methods," in *Green and Sustainable Software (GREENS), 2013 2nd International Workshop on* (2013), pp. 78–85.
22. M. Schirmer, S. Bertel, and R. L. Wilkening, "CONTEX(T)PLORER: A Mobile Inspector for Ubiquitous Environments," in *Proceedings of Mensch und Computer 2014 Conference*, Munich, Germany (2014), pp. 195–204.

Chapter 5
Refactorings for Energy-Efficiency

Marion Gottschalk, Jan Jelschen, and Andreas Winter

Abstract Energy-efficiency is an important topic in information and communication technology and has to be considered for mobile devices, in particular. On the one hand, the environment should be protected by consuming less energy. On the other hand, users are also interested in more functionality of their mobile devices on hardware and software side, and at the same time, longer battery durations are expected. Hence, the energy consumption for mobile devices should be more efficient. This paper shows an approach to save energy on application level on mobile devices. This approach includes the definition, detection, and restructuring of energy-inefficient code parts within apps. Energy savings are validated by different software-based energy measurement techniques.

Keywords Energy-efficient programming • Refactoring • Static analysis • Graph transformations

1 Motivation

Mobile devices and the mobile communication network are big energy consumers of information and communication technologies (ICT) in Germany [1] and their rising sales volume in the last years shows their significance in ICT [2]. Meanwhile, the energy consumption of mobile devices and the mobile network amounts to 12.9 PJ in Germany per year. This corresponds to the complete energy production of the atomic power plant Emsland [3] within 4 months. This amount shows that energy saving for mobile devices is also important to reduce the energy consumption in Germany, and hence, to protect the environment. In addition, the scope of mobile devices is increasing, due to new requirements, more powerful processors, and a

M. Gottschalk (✉)
Division Energy, Architecture Engineering and Interoperability, OFFIS e.V, 26111 Oldenburg, Germany
e-mail: gottschalk@offis.de

J. Jelschen • A. Winter
Department of Computing Science, Software Engineering, Carl von Ossietzky University, 26111 Oldenburg, Germany
e-mail: jelschen@se.uni-oldenburg.de; winter@se.uni-oldenburg.de

J. Marx Gómez et al. (eds.), *Advances and New Trends in Environmental and Energy Informatics*, Progress in IS, DOI 10.1007/978-3-319-23455-7_5

broad variety of different apps provided on mobile devices. All these functionalities have an influence on battery duration and battery lifetime.[1] This paper focuses on supporting apps' developers to save energy on application level and does not consider the operating system (OS), e.g. selecting the most energy efficient network connection.

An approach for saving energy on mobile devices on application level is described. Therefore, software evolution techniques, such as reverse engineering and reengineering, are used. *Reverse engineering* describes approaches to create abstract views of software systems to enable efficient code analyses. *Software reengineering* is a process to improve the software quality by reconstituting it in a new form without changing its functionality [4]. In this work, the energy consumption instead of the internal software quality of existing Android apps [5] is changed by code analysis and transformation. Like general reengineering activities, these energy improvements do not change apps' external behavior. The process of code analysis and transformation is called *refactoring* in this work. The approach starts with identifying energy-inefficient code parts called *energy code smells*. For these, a reengineering process [6] is described and validated by different, software-based energy measurement techniques [7]. Aim of this work is the collection and validation of energy code smells and corresponding refactorings which are presented in form of a catalog.

This paper is structured as follows: Firstly, the reengineering and the evaluation processes are described in Sect. 2 to show how energy can be saved by refactoring. Secondly, in Sect. 3 validated energy code smells are described, in which the energy code smells *Data Transfer*, *Backlight*, *Statement Change*, *Third Party Advertisement,* and *Binding Resources Too Early* and their restructurings are presented in more detail. Also, energy code smells, which were not validated yet, are listed at the end of Sect. 3. Thirdly, measurement results of all validated energy code smells are summarized in Sect. 4. Finally, Sect. 5 concludes this paper with a summary and an outlook.

2 Basic Techniques

Energy saving on mobile devices by refactoring needs two fundamental steps. Firstly, a *reengineering process* must be defined to perform the bad smell detection and the code restructuring. Secondly, the benefit (energy saving) for the refactoring must be shown by an *evaluation* which includes measurement techniques, used smartphones, apps, and settings during the evaluation process.

---

[1] This paper extends Gottschalk et al. [36] by discussing further energy smells and presenting deeper evaluations. Also, the validation process (energy measurement techniques) is described in more detail.

2.1 Reengineering Process

The reengineering process to generate more energy-efficient code is depicted in Fig. 5.1. It describes the platform-independent process from an energy-inefficient app to a more efficient one. In addition, the process is declared for Java code because the implementation of restructurings described later is realized for Android apps, which are written in Java. The process starts with reverse engineering the Java code of an app. In Step 1, the code is parsed into an abstract representation which conforms to a Java meta model [8] and it is stored into a central repository to allow efficient analyses [5]. In the Steps 2 and 3, the refactoring by analyzing and restructuring the abstracted code is done and saved. The analysis is done by static code analysis which has proved itself in software evolution, to identify energy code smells. The restructuring changes the abstract view without changing the intended functionality. In the last Step 4, the abstract view is unparsed back into Java code (Android app) which can be compiled and executed.

The abstract view of app code is provided by TGraphs [9] which can represent each piece of code by nodes and edges according a given metamodel representing the corresponding programming language. TGraphs are directed graphs, whose nodes and edges are typed, attributed, and ordered. Thanks to the SOAMIG project [8], the tooling for the Java programming language is available and used for this work. This abstract view allows performing efficient reverse engineering and transformations by static analyses and graph transformations, using GReQL (Graph Repository Querying Language) [8] and the JGraLab API [10]. GReQL can be used to extract static information and dependencies in source code represented as TGraphs. The JGraLab API implements the GReQL approach and provides means for graph manipulation. An example of the analysis and restructuring by JGraLab is shown in Sect. 3, explaining the reengineering process in more detail.

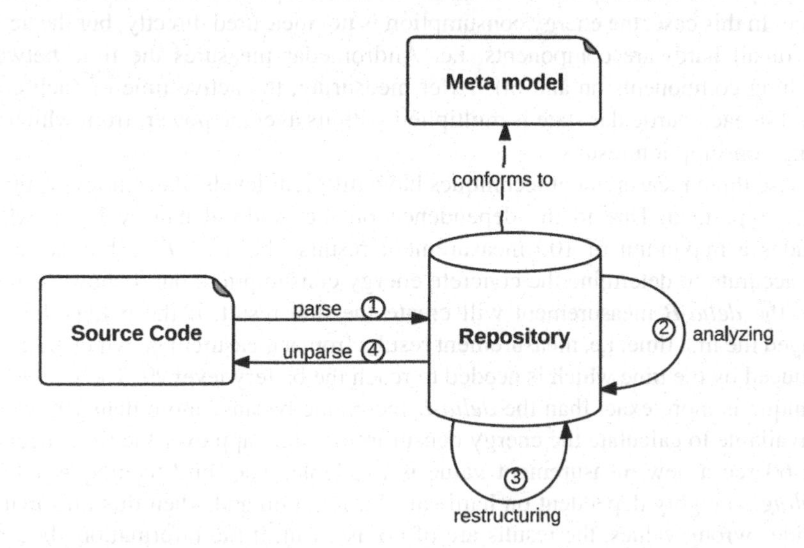

Fig. 5.1 Reengineering process [6]

2.2 Evaluation Process

After the refactoring, the energy improvement of refactored apps must be proven, which is done by different software-based *energy measurement techniques*. Also, the evaluation process is described in more detail to make the described energy refactorings repeatable. Thus, the used *hard-* and *software* are introduced, and the *measurement procedure* involving the duration and the mobile devices' settings during the energy measurement are demonstrated.

2.2.1 Energy Measurement Techniques

The approach uses three software-based energy measurement techniques: *file-based*, *delta-B*, and *energy profiles* [7]. These techniques are realized by the Android app Andromedar [11] which is used for validating the energy consumption of apps before and after refactoring. The *delta-B* technique calculates the energy consumption using the battery level, which can be retrieved with the Android BatteryManager API [12]. Android provides 100 battery levels. If the battery level changes, Andromedar calculates a new value of the current energy consumption by using the BatteryManager API. The *file-based* technique uses an internal system file which is implemented by some mobile device vendors, only. It stores information about the current battery discharge and battery voltage which are updated every 60 sec by the OS. This information is used by Andromedar to calculate the energy consumption. *Energy Profiling* uses an internal XML file, which is created by the vendor or someone who has detailed information about the mobile devices' hardware components. This XML file gives information on the average power consumption of each hardware component built into the mobile device. In this case, the energy consumption is not measured directly, but the active time of all hardware components, i.e. Andromedar measures the time between switching components on and off. After measuring, the active time of each component in each particular state is multiplied with its average power, from which the energy consumption results.

These three measurement techniques have different levels of accuracy, which is briefly explained. Due to the dependence on the Android battery level, which provides a maximum of 100 measurement results, the *delta-B* technique is not very accurate to determine the concrete energy consumption, but it shows a trend. Also, the *delta-B* measurement will create the first result, if the battery level is changed the first time, i.e. measurement results from a measurement which takes 2 h is reduced by the time which is needed to reach the battery level 99. The *system-file* technique is more exact than the *delta-B* technique because more than 100 values are available to calculate the energy consumption of an app over the time, because each 60 sec a new measurement value is available. The third technique, *energy profiling*, is highly dependent on hardware information and when this information provides wrong values, the results are of no use. But, if the information about the average power of each component is reasonably good, this technique will provide the best results. However, the *energy profiling* shows only differences in the measurement results after restructuring, when the active time of components is changed. To sum up, the *energy profiling* provides the best measurement results, when the hardware information is correct. Otherwise, the *system-file* and *delta-B* techniques provide good results to give a statement about the energy consumption of an app.

2.2.2 Hardware

For validation, two mobile devices are used: a HTC One X [13] and a Samsung Galaxy S4 [14]. The HTC One X is used for the validation of all energy refactorings and the Samsung Galaxy S4 is only used for validating the *Backlight* energy refactoring (cf. Sect. 3). Both devices use the Android OS which is necessary to apply the reengineering process with the Java meta-model and the energy measurements by Andromedar. The HTC One X uses Android 4.1.1 with HTC Sense 4+. It has a 4.7 in. HD, super LCD 2 screen with a resolution of 1280×720 pixels. The Samsung Galaxy S4 uses Android 4.2.2 with TouchWiz. The screen is a 5 in. Full HD Super AMOLED screen with a resolution of 1920×1080 pixels. This information is listed to give an overview about considered components during the energy measurements in Sect. 3.

2.2.3 Applications

For validation, two Android apps are chosen: *GpsPrint* [15] and *TreeGenerator*. GpsPrint is a free Android app with about 1642 LOC. It locates the position of a mobile device and looks up the street address on the internet. TreeGenerator is a self-written Android app for this validation with about 345 LOC. Each second it displays another name of a tree, its type, and a picture of it. Both apps display advertisements at the bottom of the screen. GpsPrint is used for analyzing and validating the energy refactorings: *Third-Party Advertisement* and *Binding Resources Too Early* (paragraph on Sect. 3). TreeGenerator is used for the refactorings: *Third-Party Advertisement*, *Statement Change*, *Backlight* and *Data Transfer* (paragraph on Sect. 3).

2.2.4 Measurement Procedure

The evaluation process for each energy-refactoring is the same; hence, a short overview about the process is given. For each energy-refactoring, ten energy measurements are done before and after the refactoring and an average is created to compare the results. The energy measurement includes the usage of all three techniques (*delta-B*, *system-file*, and *energy profiling*), so that each energy measurement contains three measurement results. To guarantee correct, comparable, and reproducible results, the mobile devices' settings should be the same during the different energy measurements. This includes *screen*, *notification*, *application*, *wireless and network*, and *gesture* settings.

- **Screen:** The screen is permanently on, but on lowest brightness, because apps only run when the screen is active. The lowest brightness is chosen to reduce and to normalize the energy consumption of the screen. Auto rotation of the screen is turned off, because the rotation sometimes interrupts the GPS connection. This was observed by first measurements with GpsPrint.
- **Notification:** Automatic notifications are turned off, such as email client, play store, calendar, etc. This allows comparable energy measurements which are not interrupted by individual notifications, such as emails or updates.
- **Application:** All background apps, such as HTC Services, Google Services, etc. are set off. Only the tested app is running, so that the energy consumption is not affected by further apps. Also, the power saver of the OS is turned off.
- **Wireless and network:** The modules mobile data, blue-tooth, GPS, NFC, and Wi-Fi are set to off, if they are not needed.
- **Gesture:** Additional gestures, such as the three finger gestures, are turned off. It should reduce the possibility to execute unwanted actions during the energy measurements, e.g. to switch between different screen views.
- **Miscellaneous:** The HTC does not have a SIM card installed, with the result that it seeks for it, permanently. Additional measurements have been performed to check the influence of the missing SIM card, and have shown that the influence is very small [16]. Also, no further apps are installed apart from the standard apps, the app under test, and Andromedar.

3 Energy Refactoring Catalog

The energy refactoring catalog includes a list of five energy code smells which are validated: *Data Transfer*, *Backlight*, *Statement Change*, *Third-Party Advertising*, and *Binding Resources Too Early* [16]. In this paper, these energy code smells are described in detail. The aim is to demonstrate possible areas of wasting energy by bad programming and strategies to improve apps' energy behavior. Further energy code smells are presented to show more possibilities for improving energy behavior of apps by applying refactorings.

The detailed description of the energy code smells follows a template which was developed in [16]. The template includes: a name, a definition, a motivation, constraints, an example, an analysis, a restructuring, and an evaluation. Thus, a consistent presentation of the energy code smells is ensured. A part of the template is derived from the description of code smells by Fowler et al. [17]. For the first described energy code smell *Data Transfer*, the example, analysis, and restructure part of the template is completely described. For the further energy code smells, these parts are shortened and can be looked up in [16].

3.1 Energy Code Smell: Data Transfer

Description Data Transfer refers to loading data from a server via a network connection instead of reading pre-fetched data from the app's storage [18].

Motivation Many apps use data, such as images, videos, and sound effects. Programmers can decide whether this data is stored on the apps' storage or on an external server. The measurements by [18] show that the usage of an external server needs more energy than the local one. Additionally, Android apps offer a further possibility to access data. Data is stored on an external server and is loaded into the apps' cache during runtime. As long as the cache is not cleared, the data does not have to be loaded again from the server. Hence, three variants exist: (1) data is loaded from a server (every time), (2) data is stored on the mobile device, and (3) data is stored in the cache. These variants are considered in the following description and evaluation.

Constraints The variants (2) and (3) request storage on the mobile devices while an app is installed. With large amounts of data, users will face issues when installing many data-intensive apps. Variant (1) needs less storage on mobile devices, but generates a lot of data traffic to load all data each time. If the data access of apps should be changed, many changes must be done. At first, the data must be collected to upload them to a server (for variant (1) and (3)) or to download and integrate them into an app (for variant (2)). For all variants, different Android APIs exist that must be known to detect this energy code smell.

Example Example source code (TreeGenerator) showing the "Data Transfer" bad smell is depicted in Fig. 5.2. On the left side it shows, the code for loading data from server (1), and on the right side data is read from the memory card (2). The code on the left side loads data from the server in line 8 and 11 using the `image` method. If `applicationCache` on the left side is set to `true`, the usage of app's cache will be possible (3). The code on the right side loads the image files from disk in line 8 and 12 using the method `getDrawable(file)`, which contains the path to the image.

Analysis Apps must be detected which use the `image` method of the `AQuery` API to know which part within the app must be changed. To this end, a GReQL query is used. It seeks for a class (`AQuery`) and an access (`image()`) vertex within the TGraph which is created by parsing the app's code (cf. Fig. 5.1). These vertices must be a part of the same block and called in a specific order.

Figure 5.3 shows the GReQL query (cf. Sect. 3) which is used to detect this energy code smell. The query consists of three parts: *from*, *with*, and *report*. The

```
1  @Override
2  public void run() {
3      i = (int) (Math.random() * 51);
4      AQuery aq = new AQuery(pic);
5      //tree list
6      if(i == 1) {
7          value.setText("Nikko-Tanne");
8          aq.id(pic).image("http://
               mgottschalk.eu/img/bilder/
               nikko.jpg", false, false, 200,
                R.drawable.ic_launcher);
9      } else if(i == 2){
10         value.setText("Riesen-Tanne");
11         aq.id(pic).image("http://
               mgottschalk.eu/img/bilder/
               riesentanne.jpg", false, false
               , 200, R.drawable.ic_launcher)
               ;
12     }
13     [...]
14 }
```

```
1  @Override
2  public void run() {
3      i = (int) (Math.random() * 51);
4      Resources res = getResources();
5      //tree list
6      if(i == 1) {
7          value.setText("Nikko-Tanne");
8          Drawable picture = res.getDrawable
               (R.drawable.nikko);
9          pic.setImageDrawable(picture);
10     } else if(i == 2){
11         value.setText("Riesen-Tanne");
12         Drawable picture = res.getDrawable
               (R.drawable.riesentanne);
13         pic.setImageDrawable(picture);
14     }
15     [...]
16 }
```

Fig. 5.2 Example for Data Transfer [16]

```
1  from cache : V{frontend.java.Literal},  block : V{frontend.java.Block}, image : V{
       frontend.java.Access}, aquery : V{frontend.java.Class}
2  with cache.value = "false" and aquery.name = "AQuery" and
       cache <--{frontend.java.HasOperand} ... block ->frontend.java.BlockContainsStatement
       ... aquery  and image <--{frontend.java.HasOperand} ... block
3  report cache
4  end
```

Fig. 5.3 Query for Data Transfer [16]

from-part declares the search space of the query. In this case, vertices and edges are defined which are needed to detect the energy code smell, e.g. the vertex `cache` of the type `Literal` must exist to detect the attribute *applicationCache* of the `image` method (cf. Fig. 5.2). The second part is the *with*-part, which includes the content of the query. Firstly, the known values of the vertices are checked (in line 2: *cache.value* = *"false"* *and* *aquery.name* = *"AQuery"*). Secondly, the link between these vertices is checked by *cache <--{frontend. java.HasOperand} ... block*. This means, that the vertex `cache` is an operand of a further vertex which is connected via further edges with the `block` vertex. The last part is the *report*-part, which defines a list, containing an element for each tuple of the search space satisfying the constraints in the *with*-part. Those elements will consist of the value of the cache variable corresponding to the respective tuple.

Restructuring If the query detects such a code part, the restructuring will start. Due to the complexity (saving images locally, loading each image) of the changes, this restructuring must be done manually from variant (1) to (2). The restructuring from variant (1) to (3) can be done automatically by the JGraLab API. For detection, nearly the same query can be used, only a further vertex for the boolean *applicationCache* must be added which must be detected and changed to *true*.

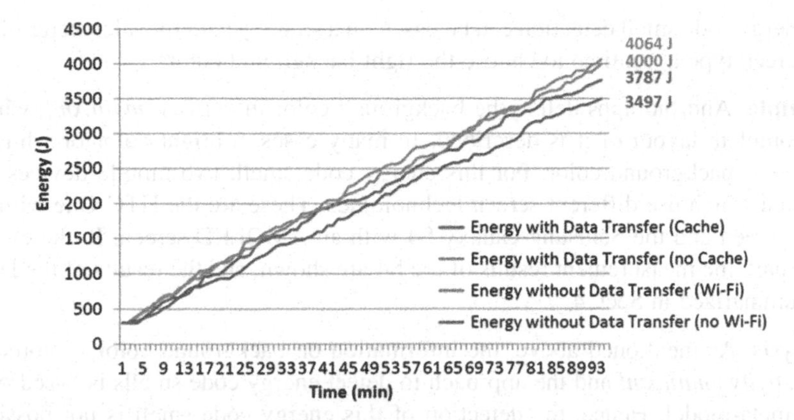

Fig. 5.4 Delta-B measurement for Data Transfer [16]

Evaluation Figure 5.4 shows the energy consumption of the TreeGenerator app in all variants. Variant (1) (data is loaded from server) consumes the most energy with 4064 J in 2 h. The difference to variant (2) (data is stored locally) with a switched-off Wi-Fi module amounts to 567 J. If the Wi-Fi module is not switched off, the difference is very small, which shows that the data traffic does not have a significant influence on the switched on Wi-Fi module. Variant (3) (data is stored in the app's cache) also saves energy, which amounts to 277 J in 2 h. The results of the other two energy measurement techniques (*file based* and *energy profile*) are summarized in Sect. 4. They show the same trend as the *delta-B* technique used here.

3.2 Energy Code Smell: Backlight

Description Backlight refers to the background color of an app. For different screen technologies (e.g. Super LCD and Super AMOLED) the energy consumption can vary for different background colors [19].

Motivation In recent years, LCD (Liquid Crystal Display) screens have been replaced by Super AMOLED (Active-Matrix Organic Light Emitting Diode) screens in the smartphone area, because AMOLED screens offer a better quality and are more energy-efficient [19]. The main advantage of the AMOLED screen is that each pixel can be controlled to display colors or to switch a part of the screen off or on. I.e., if black elements are used in an app, pixels can be switched off and energy is saved. LCD screens cannot control individual pixels, and needs the backlight to display colors, so the backlight is also on, when apps run.

Constraints The motivation also contains the constraint that several devices use different screen technologies, on which they consume different energy amount for several colors. Thus, apps must be adapted according to hardware information. So, the energy code smell detection can be based on a strategy pattern which determines the screen type at runtime to choose the right background colors.

Example Android apps define the background color in *activity_main.xml*, where the complete layout of it is described. In many cases, a bright color or white is chosen as background color. For this energy code smell, two mobile devices are checked which use different screen technologies. These are the HTC One with an LCD screen and the Samsung Galaxy S4 with an AMOLED screen. In the evaluation part, the measurement results of the S4 are shown, and the results of the HTC are summarized in Sect. 4.

Analysis As mentioned above, the information on background color is stored in the *activity_main.xml* and the approach to detect energy code smells is based on a Java meta-model. Hence, the detection of this energy code smell is not possible with the existing tooling, the Java TGraph approach. An additional meta model for the XML part is needed to depict the XML file in a TGraph and to run graph queries on it. So, the detection and also the restructuring part are done manually to show energy savings.

Restructuring If this energy code smell exists, a new color type must be stored and assigned to the attribute *@color* in *activity_main.xml*. After changing the color, the app can be built and executed on mobile devices.

Evaluation The validation of this energy code smell shows significant differences in the energy consumption of the S4. These are illustrated in Fig. 5.5. The app TreeGenerator with the white background consumes 5883 J more energy within approx. 2 h, representing an energy saving of approx. 60 %. This conforms to the statement of Chen et al. [19] who describe the energy saving of AMOLED screens by using black backgrounds. This saving can be explained through the switched-off pixels on the screen to depict black. The measurement results for the HTC show considerably lesser savings when a white background is used (approx. 3 %, cf. Fig. 5.9).

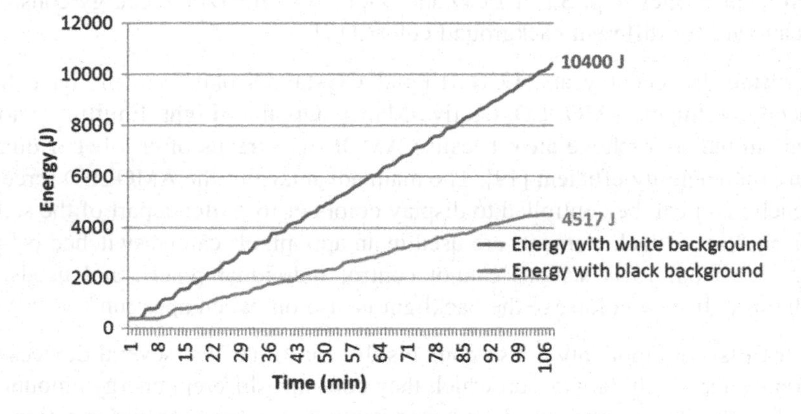

Fig. 5.5 Delta-B measurement for Backlight (S4) [16]

3.3 Energy Code Smell: Statement Change

Description Statement Change describes alternative programming statements, such as Java's `if` and `switch`, which can be substituted with each other, because they have the same functionality, but potential different energy consumption [20].

Motivation During the 2nd EASED Workshop (Energy-Aware Software-Engineering Development) [20] some ideas about energy-aware programming were discussed. One idea is the exchange of *if*- and *switch*-statements within apps. For Android apps, this is checked in this work. `Switch`-statements are used for better code maintenance, i.e. a sequence of `switch`-statements is easier to read than a sequence of nested `if`-statements [21].

Constraints *Switch* and *if* are used to make decisions, but they do not work completely identical, e.g. after a match with a *switch*-statement, all other assignments are executed when no break is used (called fall-through [21]). The bodies of *if*-statements are only executed, when the condition is *true*, and hence, no break is needed. Furthermore, *if* can be used for comparisons with all data types and *switch* only with `integer`, `enumerations`, and `Strings` (since Java 1.7). Android SDK 18.0 works with Java 1.6, and hence, a comparison with `Strings` is not allowed. These differences limit the usage of *switch* within Android apps. Also, *switch* works only with constant operators [21].

Example If *if*-statements are replaced by *switch*-statement, the app's code will need more lines of code to implement the same functionality. This has two main reasons. First, each case in the *switch* statement has to be terminated with a `break` to prevent fall-through [21]. Second, if `Strings` are used in *if*-statements, the comparison is more complex in Android apps, e.g. `enumerations` are used. A more detailed example with a code excerpt is given in [16].

Analysis This Energy Refactoring can be detected by a querying for *if*-statements with more than three *else*-statements. Three *else*-statements should be the minimum when *if*-statements are replaced through *switch*-statements because the effort is too high when each simple *if*-statement is identified and must be checked by a programmer. Furthermore, the restructuring should have an influence on the energy consumption, and hence, changes should not be too small.

Restructuring An automatic restructuring is very difficult because many changes must be done, e.g. the `String` comparison has to be done with `enumeration`. Furthermore, the nodes for the *if*-statement must be changed and shifted into a new method when the condition is not constant (from `String` to `enumeration`). Hence, it is easier to have the change made manually by a programmer, who decides whether a manually restructuring is sensible or not.

Evaluation The measurement results in Fig. 5.6 show that no major difference exists concerning the energy consumption between *if*- and *switch*-statements. For the energy measurement, the app TreeGenerator and the HTC are used. The app

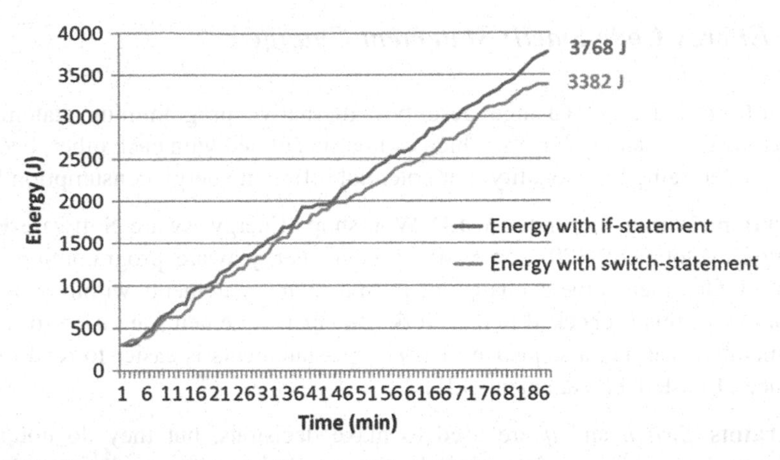

Fig. 5.6 Delta-B measurement for Statement Change [16]

with the *if*-statement consumes a little more energy than the *switch*-statement. The difference amounts 386 J within approx. 90 min. Due to the high effort of the restructuring, it seems not sensible to perform the restructuring in this case. But, it has to be checked how this energy code smell acts in other use cases, e.g. if `integer` are used to compare something in the *if*- or *switch*-condition.

3.4 Energy Code Smell: Third-Party Advertisement

Description Third-party advertisements are integrated code parts within apps which display advertisements during operation. Thereby, advertisements do not have an influence on apps' functionality, but might consume energy through 3G or Wi-Fi connections [22].

Motivation Tests have shown that approx. 65 % of energy consumption result from additional functionalities, such as advertisements, wakelocks, and location pinpoints, which are not important for an app's actual functionality [22]. Many popular, free apps, like AngryBirds [23] and Leo [24], include advertisements. Due to the frequent usage of the apps (AngryBird has been installed about 100,000,000 times [23]), the total energy savings by removing this energy code smell are higher than for other apps with less downloads. Advertisements consume much energy through the use of 3G or Wi-Fi connections which update advertisements every few seconds. For example, after installing AngryBirds, a 3G or Wi-Fi connection is not required for the game, but for advertisements' updates, only. Hence, energy is saved for this app when the communication with unrequired components stops [22].

Constraints The main functionality of apps is not changed, but its financial arrangement, inasmuch as advertisements are not requested and not displayed anymore. Programmers and app vendors often use advertisements to finance the development of apps, and hence, advertisements cannot be deleted without any consequences. Due to this restructuring, the GNU General Public License (GPL) [25] or other licenses in general may be violated. Hence, the energy code smell is checked in this work, but it may not be legally permitted without the programmer's and vendor's consent.

Example Advertisements can be included in apps through special APIs. One API is AdMob [26] which is used in the app GpsPrint. AdMob allows different views of advertisements, such as banners or interstitials. With an object of the class *AdView* and the method *loadAd(new AdRequest)*, advertisements are defined and loaded from a server. To allow the usage of advertisements in apps, the *manifest.xml* should be changed to set the correct permissions.

Analysis This energy code smell can be detected by seeking for third-party APIs and their methods' calls to display advertisements. Therefore, further information about apps is needed to know which APIs could be used by programmers. This depends on the OS on which apps run. If the OS is known, third-party APIs and their method calls can be detected by querying special imports, initializations, and methods. However, various advertisement APIs have to be known, as well.

Restructuring Code parts which are responsible for third-party advertisements can be removed without any influence on apps' main functionality. But it is important to remove all code parts of third-party APIs, otherwise the source code might contain errors, e.g. when APIs' imports are deleted but not all method calls.

Evaluation Removing advertisements within apps saves energy and has no influence on apps' main functionality. For the measurement results in Fig. 5.7, the app GpsPrint and the HTC are used. It illustrates a saving of approx. 20 % within 95 min, when advertisements are removed.

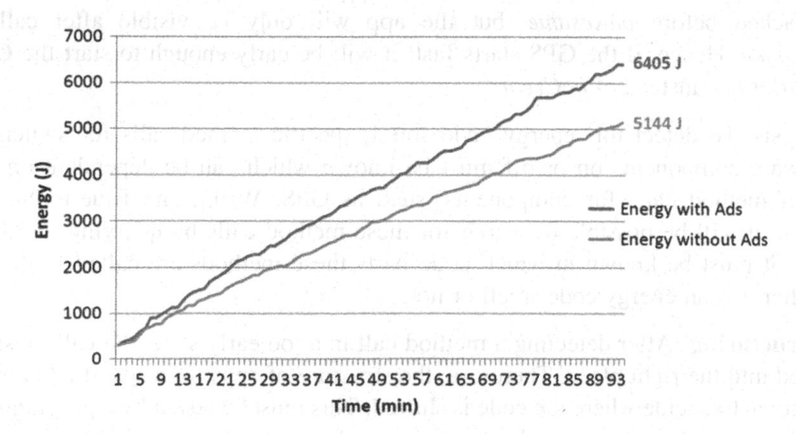

Fig. 5.7 Delta-B measurement for Third-Party Advertisements [16]

3.5 Energy Code Smell: Binding Resources Too Early

Description Binding resources too early refers to hardware components, such as Wi-Fi and GPS, which are switched on by apps at an early stage when they are not yet needed by the app or user [27].

Motivation The energy consumption of hardware components can be very high [18], hence, it is important to reduce the runtime of these components, e.g. for GPS-sensors. One option would be to start these components only when they are really needed. Carroll and Heiser [18] have demonstrated through several measurements that the highest energy savings can be realized by shutting down unused components. Hence, the restructuring tries to reduce the runtime of components to produce maximum energy savings.

Constraints Binding resources too early can only be detected and removed when the programmer knows the app-specific structure with regard to the energy efficiency, i.e. programmer knows which method is called in which app state. This information is sometimes given by the vendor, e.g. the Android life cycle [28]. Furthermore, the programmer must know which notifications or hardware states belongs to the method call, in order not to forget to shift these code parts to prevent errors after refactoring. Hence, manual decisions are necessary to execute a successful refactoring and platform-specific knowledge is also necessary.

Example The Android life cycle [28] defines different states which can be reached by events of an app, such as the events *onCreate*, *onResume*, *onPause*, *onDestroy*, etc. Programmers have to decide which resources are started or stopped in which state to realize the desired app behavior. For example, the GPS can be started by the event *onCreate* or *onResume*. In both cases, the GPS signal is available for users when the app is visible. The event *onCreate* is reached before *onResume*, but the app will only be visible after calling *onResume*. Hence, if the GPS starts fast, it will be early enough to start the GPS by *onResume* instead of *onCreate*.

Analysis To detect this energy code smell, specific method calls for switching hardware components on or off must be known which can be dependent on the OS. If method calls for components, such as GPS, Wi-Fi, and Blue-tooth, are known, it will be possible to search for these method calls by querying. Furthermore, it must be known in which code parts these methods are called to decide whether it is an energy code smell or not.

Restructuring After detecting a method call in a too early state, the call must be shifted into the right place. Therefore, the structure of apps for a specific OS must be known to decide where the code is shifted. This must be tested by a programmer who has semantic program understanding in order not to change apps functionality. In this regard, it should be considered that all code parts are shifted which are connected with the component method call, such as comments.

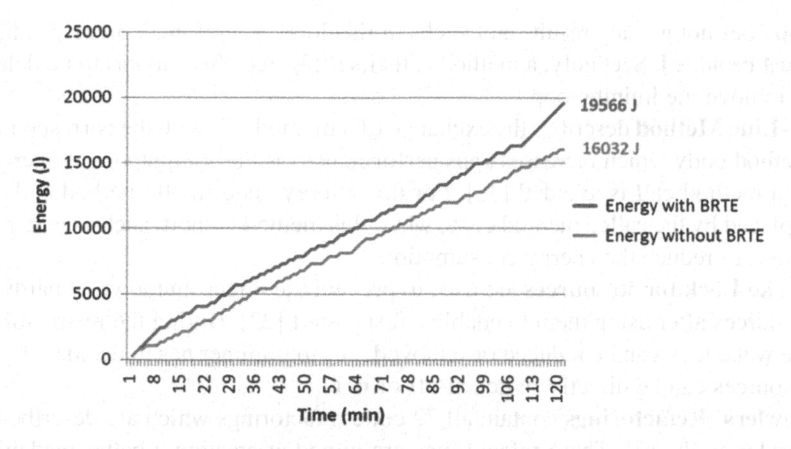

Fig. 5.8 Delta-B measurement for Binding Resources Too Early [16]

Evaluation To validate this energy code smell, the app GpsPrint is started and stopped every 5 sec within 2 h by an additional app, which explains the high energy consumption in comparison to the other energy measurements. For the measurements, the HTC is used. In Fig. 5.8, the measurement results show that the app consumes 3534 J more when the GPS is started by *onCreate* instead of *onResume*. Hence, the energy saving is about 18 %.

3.6 *Further Energy Code Smells*

The following energy code smells are explained briefly to give an overview on further areas for saving energy. More detailed description of the first eight can be found in [16]. The last energy code smell, *Cloud Computing,* is described in [29].

- **Using expensive Resources** are the possibility to swap energy expensive hardware resources for apps against energy-efficient alternatives [30], e.g. using the signal of cell towers instead of GPS to localize mobile devices.
- **Dead Code** describes code parts which are never reached during apps' runtime. But it is also loaded into memory, and thus, it consumes energy [31]. Simply removing these code parts would lead to save energy on application level without changing apps' functionality.
- **Replace Sorting Algorithm** means to replace one sorting algorithm with another more energy-efficient sorting algorithm [32]. In Höpfner et al. [32], a substitution of sorting algorithms in an app on a special hardware shows that the energy consumption varies between algorithms. If the most energy-efficient sorting algorithm for hardware is known, algorithms can be replaced by this.
- **Loop Bug** describes a program behavior wherein apps are repeating the same activity over and over again without reaching intended results [27]. Two kinds of this energy code smell can exist: Firstly, a server is not acquirable, so that the

app does not get any results and sticks in this loop. Therefore, a break condition must be added. Secondly, a method calls itself; hence, this call has to be deleted to remove the infinite loop.

- **In-Line Method** describes the exchange of a method call with the corresponding method body which increases apps performance, as the computational overhead of a method call is avoided [33]. For this energy code smell, method calls are replaced by the called method code, when this method is short, such as getter and setter, to reduce the energy consumption.
- **Wake Lock for Resources** are used to prevent the direct shutdown of hardware resources after using them to enable a fast restart [22]. During the restructuring, the wake lock can be reduced or removed. A programmer has to decide whether resources can be directly switched-off or not.
- **Fowlers' Refactorings** contain all 72 code refactorings which are described by Fowler et al. [17]. These refactorings are aimed at creating a better readability and maintainability for source code than before. In addition, the impact on energy consumption of these refactorings can be identified to provide further information for programmers.
- **Design Patterns** are defined by Gamma et al. [34] and describe solutions for special problems that can be applied again and again within several apps [35]. As with Fowlers' Refactorings, the energy consumption of the design pattern and an alternative implementation has to be measured to decide for an energy-efficient solution.
- **Cloud Computing** describes the usage of external services via internet to run energy-intensive apps instead of running them on mobile devices. Energy is saved on mobile devices, and allows further optimizations on servers [29].

4 Energy Measurements

For most energy code smells described in Sect. 3, the three energy measurement techniques (Sect. 2) were applied. For each use case and technique, ten measurements, each lasting 2 h, were performed and averaged; hence about 400 measurements were done. The energy code smells that are described in Sect. 3.6 have not been validated, except for *Cloud Computing,* which is described in another work [29], hence the results are shown there. In Fig. 5.9, the measurement results with and without energy code smells are depicted. In addition, the difference between the energy consumption before and after refactoring is shown. If the difference is negative, the energy consumption rose after refactoring, i.e. the original implementation was more energy-efficient. The *file-based* measurement technique is not available for the Samsung Galaxy S4. Energy code smells which were validated for different apps or smartphones, *Third-Party Advertisement* and *Backlight*, are marked by a suffix.

Energy Code Smell	Measurement with Energy Code Smell			Measurement without Energy Code Smell			Difference (in %)		
	File Based	Energy Profiling	Delta-B	File Based	Energy Profiling	Delta-B	File Based	Energy Profiling	Delta-B
Third-Party Advertisement "GpsPrint"	6628 J	268 J	6405 J	5272 J	300 J	5144 J	20.5	-10.7	19.7
Third-Party Advertisement "TreeGenerator"	5310 J	308 J	4814 J	3681 J	269 J	3413 J	30.7	12.7	29.1
Binding Resources Too Early	11903 J	981 J	19566 J	11276 J	706 J	16032 J	5.3	28.0	18.1
Statement Change	3965 J	274 J	3768 J	3823 J	297 J	3382 J	3.5	-7.7	10.2
Backlight HTC	3681 J	269 J	3413 J	3796 J	272 J	3768 J	-3.0	-1.1	-2.9
Backlight S4	---	26152 J	10400 J	---	19184 J	4517 J	---	26.6	56.6
Data Transfer	4232 J	289 J	4064 J	3864 J	265 J	3497 J	8.6	8.3	14.0

Fig. 5.9 Energy measurement results [16]

The results are explained using *Data Transfer* as example. All three energy measurement techniques show an energy saving after restructuring. The *delta-B* measurement results are described in Sect. 3.1, its energy saving amounts to 14 %. The other two measurements, *file-based* and *energy profiling*, display the same trend. The difference of the file-based measurement after refactoring amounts to 368 J, which is an energy saving of about 8.6 %. The energy profiling shows a different of 24 J, which is an energy saving of about 8.3 %. To sum up, removing the energy code smell *Data Transfer* reduces the energy consumption by decreasing the data traffic through using devices' storage or apps' cache.

As can be seen, energy profiling has a major problem, which was detected during validation. After a firmware update, the internal XML files used for realizing the *energy profiling* was also updated and the measurement results differ dramatically from the other two techniques [16]. This indicates that the vendor provides an internal XML file for the latest mobile devices, which are transferred to all devices; independent of the actual hardware (e.g. the new internal XML file has more processor states for the same mobile device after updating). In [16] the new power profile is adapted to the old one and the results of the energy consumption are manually calculated. With the adapted profile, the energy consumption for *Third-Party Advertisement "GpsPrint"* amounts 4461 J before refactoring and 4426 J after refactoring, a difference of 1 % [16]. This small difference results from the active time of the Wi-Fi component because in both calculations (before and after refactoring) the Wi-Fi is switched on to search for address of the current smartphone location. I.e., if the advertisements are deleted, the Wi-Fi component is still switched on. Hence, the technique *energy profiling* calculates the same energy consumption for both measurements because the active time of the components is the same and the different data traffic (before refactoring the address and advertisements are loaded, after refactoring only the address is loaded) does not have an influence on its calculation.

The energy code smell *Backlight* is checked on two smartphones, the HTC and the S4. For the S4, the energy saving is up to 56.6 %, when a black background is used in apps. In contrast, the HTC consumes more energy when a black background is used; the difference is up to 3 %. This result from the different screen technologies, the HTC uses an LCD screen and the S4 uses an AMOLED screen.

These results show that it is sensible to consider energy refactorings to save energy on mobile devices. The presented energy refactorings are validated for Android apps, but the definition of energy code smells can be transferred to other platforms. But, the analyses and restructurings must be changed. Moreover, the energy code smells must be validated for other platforms to confirm their savings.

5 Conclusion

The definition and validation of energy code smells present an approach to save energy on mobile devices. The energy code smells are described by a template to get a uniform definition in the energy refactoring catalog. The template includes: a name, a definition, a motivation, constraints, an example, an analysis, a restructuring, and an evaluation. The detection and restructuring of energy code smells are done with the graph query language GReQL [8] and the JGraLab API [10]. For validation, software-based measurement techniques were used [7].

All measurement results show that it is reasonable to perform energy refactorings to reduce energy consumption on mobile devices. The presented energy code smells are validated for Android apps. However, the definition of the code smells can also be applied on other platforms, such as Windows Phone and iOS, but has to respect hardware and operating system specifics. These results show that saving energy by refactoring can support programmers to improve their existing apps and avoid energy code smells while developing new apps. Also, apps' users can be supported by providing free apps with advertisements and the same apps for a fee without advertisements, because programmers can use these energy refactorings to change their existing free apps without great effort.

Next steps could be to extend the reengineering process to be applicable Windows Phone and iOS apps. Thereupon, the existing energy code smells could be validated for these platforms. Furthermore, hardware-based energy measurement techniques could be used to confirm the software-based measurement results. These would help to extend and to confirm the energy refactoring catalog in the mobile devices' area.

References

1. L. Stobbe, N. Nisse, M. Proske, A. Middendorf, B. Schlomann, M. Friedewald, P. Georgieff, and T. Leimbach, "Abschätzung des Energiebedarfs der weiteren Entwicklung der Informationsgesellschaft," 2008.
2. Statista, "Absatzprognosen für 2013: Smartphones verkaufen sich am besten," 2013. [Online]. Available: http://de.statista.com/themen/647/itk-branche/infografik/711/prognosen-zum-weltweiten-absatz-von-itk-geraeten/
3. RWE Power AG, "Kernkraftwerk Emsland," 2013. [Online]. Available: https://www.rwe.com/web/cms/de/16646/rwe-power-ag/standorte/kernkraft/kkw-emsland/
4. J. Chikofsky and J. H. Cross, "Reverse Engineering and Design Recovery: A Taxonomy," *IEEE*, 1990.
5. Android Developers, "Android, the world's most popular mobile platform," 2013. [Online]. Available: http://developer.android.com/about/index.html
6. M. Gottschalk, J. Josefiok, J. Jelschen, and A. Winter, "Removing Energy Code Smells with Reengineering Services," *Lect. Notes informatics, GI*, 2012.
7. M. Josefiok, M. Schröder, and A. Winter, "An Energy Abstraction Layer for Mobile Computing Devices," *Oldenbg. Lect. Notes Softw. Eng.*, vol. 5, 2013.
8. A. Fuhr, A. Winter, U. Erdmenger, T. Horn, U. Kaiser, V. Riediger, and W. Teppe, "Model-Driven Software-Migration – Process Model, Tool Support and Application," in *Migrating Legacy Applications: Challenges in Service Oriented Architecture and Cloud Computing Environment*, Hershey, PA: IGI Global, 2012.
9. J. Ebert, V. Riediger, and A. Winter, "Graph Technology in Reverse Engineering, The TGraph Approach," in *10th Workshop Software Reengineering*, Bonn: GI, 2008, pp. 67–81.
10. T. Horn and J. Ebert, "The GReTL Transformation Language," in *Theory and Practice of Model Transformations – 4th International Conference*, 2011.
11. M. Schröder, "Erfassung des Energieverbrauchs von Android Apps," Oldenburg University, 2013.
12. Android Developers, "Monitoring the Battery Level and Charging State," 2013. [Online]. Available: http://developer.android.com/training/monitoring-device-state/battery-monitoring.html
13. HTC Corporation, "HTC One X," 2013. [Online]. Available: http://www.htc.com/uk/smartphones/htc-one/
14. Samsung, "Samsung Galaxy S4," 2013. [Online]. Available: http://galaxys4.samsung.de/technik
15. Robotmafia.org, "GpsPrint," 2012. [Online]. Available: https://play.google.com/store/apps/details?id=com.tyfon.gpsprint&hl=en
16. M. Gottschalk, "Energy Refactorings," Oldenburg University. Masters' Thesis, 2013.
17. M. Fowler, K. Beck, W. Brant, W. Opdyke, and D. Roberts, *Refactoring: Improving the Design of Existing Code*. Addison-Wesley, 2002, p. 431.
18. A. Carroll and G. Heiser, "An Analysis of Power Consumption in a Smartphone," *USENIX Annu. Tech. Conf.*, 2010.
19. X. Chen, Y. Chen, Z. Ma, and F. Fernandes, "How is Energy Consumed in Smartphone Display Applications?," *ACM*, 2013.
20. C. Bunse, M. Gottschalk, S. Naumann, and A. Winter, "2nd Workshop EASED@BUIS," 2013.
21. C. Ullenboom, *Java ist auch eine Insel*. 2011.
22. A. Pathak, Y. Charlie Hu, and M. Zhang, "Fine Grained Energy Accounting on Smartphones with Eprof," *EuroSys'12*, 2012.
23. Rovio Mobile, "Angry Birds," 2013. [Online]. Available: https://play.google.com/store/apps/details?id=com.rovio.angrybirds
24. Leo GmbH, "LEO Wörterbuch," 2013. [Online]. Available: https://play.google.com/store/apps/details?id=org.leo.android.dict&hl=de

25. "GNU," 2013. [Online]. Available: http://www.gnu.de
26. "Build a great app business with AdMob," 2013. [Online]. Available: http://www.google.com/ads/admob/
27. A. Pathak, Y. Charlie Hu, and M. Zhang, "Bootstrapping Energy Debugging on Smartphones: A First Look at Energy Bugs in Mobile Devices," *ACM*, 2011.
28. Android Developers, "Activity," 2013. [Online]. Available: http://developer.android.com/reference/android/app/Activity.html
29. V. Strokova, S. Sapegin, and A. Winter, "Cloud Computing for Mobile Devices," in *Proceedings of the 28th EnviroInfo 2014 Conference*, 2014.
30. M. Schirmer and H. Höpfner, "Towards Using Location Poly-Hierarchies for Energy-Efficient Continuous Location Determination," in *Proceedings of the 24th GI – Workshop on Foundations of Databases*, 2012.
31. Y. Chen, E. R. Ganser, and O. Koutso, "A C++ Data Model Supporting Reachability Analysis and Dead Code Detection," in *Proceedings 6th European Software Engineering Conference*, 1997.
32. H. Höpfner, C. Bunse, S. Roychoudhury, and E. Mansour, "Choosing the 'best' Sorting Algorithm for optimal Energy Consumption," in *4th International Conference of Software and Data Technologies*, 2009.
33. W. G. P. da Silva and L. Brisolara, "Evaluation of the Impact of Code Refactoring on Embedded Software Efficiency," in *1. Workshop de Sistemas Embarcados*, 2010.
34. E. Gamma, R. Helm, R. E. Johnson, and J. M. Vlissides, "Design Patterns: Abstraction and reuse of object-oriented design," in *ECOOP*, 1993.
35. C. Bunse and S. Stiemer, "On the Energy Consumption of Design Patterns," in *Energy Aware Software-Engineering and Development*, 2013.
36. M. Gottschalk, J. Jelschen, and A. Winter, "Saving Energy on Mobile Devices by Refactoring," in *Proceedings of the 28th EnviroInfo 2014 Conference*, 2014.

Part II
From Smart Grids to Smart Homes

Chapter 6
The 5 % Approach as Building Block of an Energy System Dominated by Renewables

Enno Wieben, Thomas Kumm, Riccardo Treydel, Xin Guo, Elke Hohn, Till Luhmann, Matthias Rohr, and Michael Stadler

Abstract We describe an approach for doubling distribution grid capacity for connecting renewable generators based on curtailing a maximum of 5 % of the yearly energy fed into the grid on a per-generator basis. The paper contains information about the control unit needed for automatic minimum curtailment and the field test that has been set up to validate the approach. Furthermore, topics concerning the operationalization of the 5 % approach using both operational technology and information technology are discussed.

Keywords 5 % approach • Connection capacity • Distribution grid • Dynamic curtailment • Grid controller • Grid curtailment • Grid integration • Photovoltaic generator • Renewables control • Wind turbine

1 Introduction

In order to meet climate challenges the German federal government has issued targets for the electrical energy generation mix. In 2050, the amount of renewable energy sources should account for 80 % of the total electricity generation. For the grid of EWE NETZ, a distribution system operator (DSO) in the northwest of Germany, already for 2020 renewable generation is estimated to reach 150 % of electricity consumption. Today already, installed decentralized power generation capacity in the grid of EWE NETZ is higher than the maximum annual load (in 2013: installed power generation capacity was 5.5 GW while the maximum annual load was about 2.3 GW).

Following the current German legislation [5] the distribution grid has to be laid out such that it can absorb the entire electricity generated from renewable energy sources (RES). The usage of distribution grids for electricity in Germany is more

E. Wieben (✉) • T. Kumm • R. Treydel • X. Guo
EWE NETZ GmbH, Cloppenburger Str. 302, 26133 Oldenburg, Germany
e-mail: enno.wieben@ewe-netz.de

E. Hohn • T. Luhmann • M. Rohr • M. Stadler (✉)
BTC Business Technology Consulting AG, Escherweg 5, 26121 Oldenburg, Germany
e-mail: michael.stadler@btc-ag.com

J. Marx Gómez et al. (eds.), *Advances and New Trends in Environmental and Energy Informatics*, Progress in IS, DOI 10.1007/978-3-319-23455-7_6

and more determined by feed-in distributed energy resources (DER). This leads to situations in which allowable transformer loads or cable loads are exceeded or in which voltage thresholds are violated. Grid operators are only allowed to temporarily throttle renewable generators (grid curtailment) if there are no other options to prevent harm to the power grid infrastructure. Furthermore, they are forced to execute grid expansion after grid curtailment actions have taken place, which additionally provides security for investments into renewables as subsidies are connected to the amount of feed-in.

Assessment of load and system design both follow a worst case approach resulting in the system being dimensioned towards a maximum load. In the case of grids dominated by decentralized power feed-in, this maximum load is given by the cumulated installed generation capacity combined with minimal electricity consumption. Frequency and duration of such load situations are not taken into account in the worst case approach. This typically results in a low number of hours of full grid capacity utilization, since utilization is determined by the feed-in characteristics of connected generators. Figure 6.1 depicts the measured annual feed-in duration curve of a photovoltaic generator. The maximum power feed-in from such a generator is 300 kW. Only in about 20 h per year a feed-in of more than 285 kW (95 % of maximum power feed-in) is reached. During these periods only a small fraction (e.g., 2 %) of the total yearly electric energy is delivered during feed-in peaks. Therefore, it seems plausible, that (partial) curtailment of feed-in from such a generator during a couple of hours per year may have a big effect on the grid load while not substantially reducing the amount of energy available.

It is obvious that the generator only reaches its maximum output for a few hours per year. However, distribution grid dimensioning in Germany is currently adjusted to feed-in situations only occurring a couple of hours per year. As a consequence enormous investments in grid construction are required. Therefore, the following questions concerning system layout of today's distribution grid structures arise:

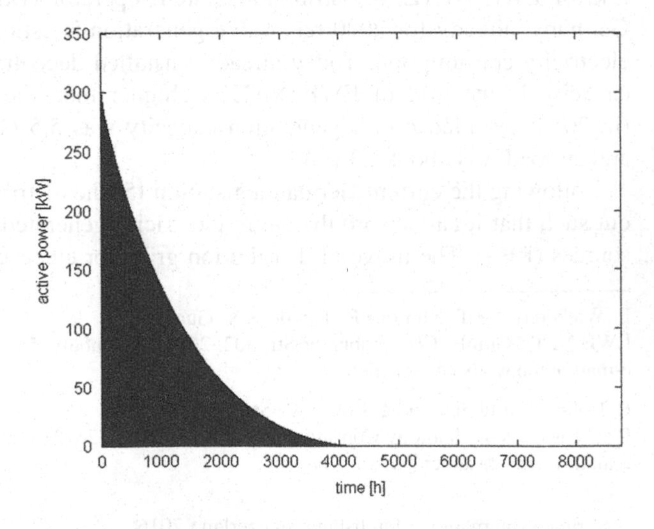

Fig. 6.1 Illustration of a measured annual feed-in duration curve of a photovoltaic generator

- Is it macroeconomically reasonable to plan distribution grids based on rare maximum loads?
- By which percentage can grid connection capacity be augmented when the distribution grid does not have to account for rare maximum loads?
- What is the overall macroeconomic balance when substituting grid expansion by grid capacity extension by means of fine-grained curtailment?

In this chapter, the first two questions are answered to a certain extent. Refining the answers and answering the third question is subject for further research.

In the following we first present dynamic intelligent power feed-in management and the 5 % approach. After mentioning simulations which have yielded the probable effect of the 5 % approach, we describe a field test setting suited for evaluating the 5 % approach in practice. After presenting the first results achieved during the first few months of a 1-year field test, we present the requirements towards a control unit used to implement the 5 % approach. Following this, we give information on the design of the control unit used within the field test.

2 Idea of a Dynamic Intelligent Power Feed-In Management

An early result from considerations about enhancing the load curtailment practice prescribed by the current German legislation [5] is that overall economic efficiency can be increased by 15 % using static settings [7]. In this scenario, called static curtailment, output of generators is reduced by a fixed percentage of their nominal power.

In contrast to this, dynamic intelligent power feed-in management (dynamic curtailment) is a more sensitive approach, reducing feed-in from individual generators only in case of bottlenecks and to a percentage adapted to the prevailing load situation. This significantly improves grid connection capacity. Figure 6.2 depicts the principle of a sensitive, dynamic curtailment and its difference to static curtailment. Sensitive control leads to significantly reduced peak generation times while ensuring a highly efficient behavior, since the control function is only activated if it is necessary due to grid requirements. By applying a sensitive, dynamic curtailment method it is possible to increase the grid connection capacity in a higher amount than by applying static curtailment. Static curtailment is based on the simple grid connection capacity. In other words: an equal amount of required control activities caused by a violation of thresholds (voltage, current) leads to a fewer amount of curtailed energy from DER in case of a sensitive, dynamic curtailment.

The economic impact of curtailment in situations with grid peak loads is substantially lower than the impact of static limits to generation. This is due to electricity wholesale market prices often being very low during peak generation times of renewables because of the overall availability of energy is high during these times.

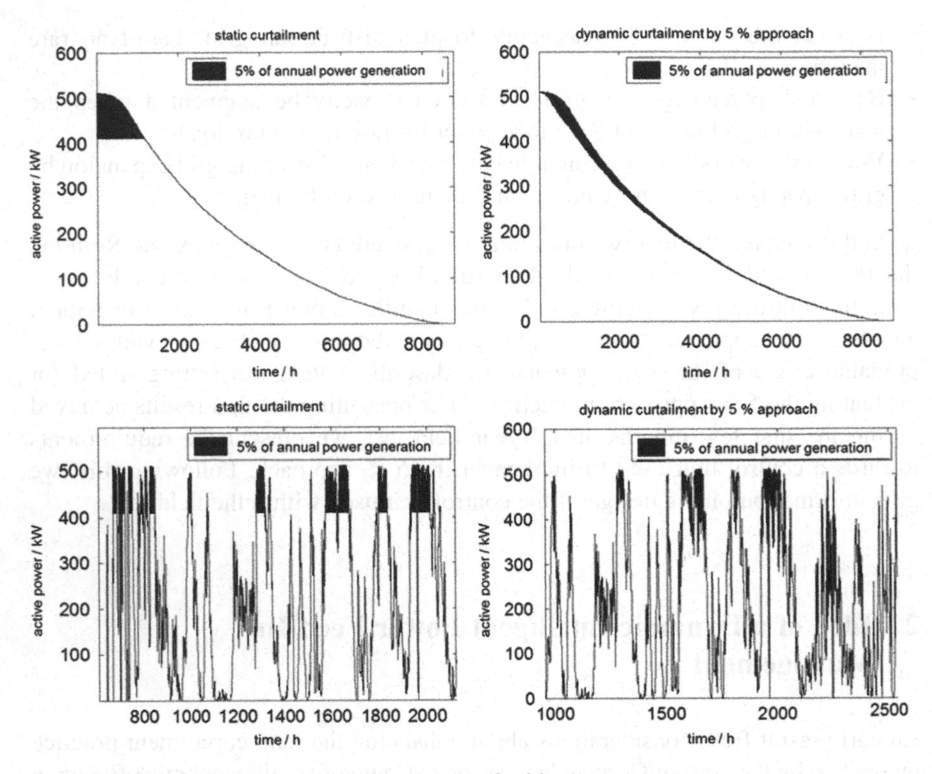

Fig. 6.2 Principle of a dynamic intelligent power feed-in management

3 The 5 % Approach

The main hypothesis of the 5 % approach is that load flow dependent throttling of a low percentage (i.e., less than 5 %) of yearly power feed-in carried out in maximum load situations leads to a drastic increase of grid connection capacity. The approaches' key characteristic is the load flow dependent throttling of generators, since voltage stabilization and equipment usage result from summed up load both from consumption and feed-in. This substantially discriminates the 5 % approach from an overall throttling of generators since both frequency and duration of throttling are minimized by load flow dependent control of generators.

In the sense of smart grids we look at an intelligent system for generation management consisting of the following main components:

- Metrological coverage of all voltage critical and load critical components of the distribution grid.
- Possibility for continuous control of reactive power output of all generators based on ICT.
- Online load flow calculation based upon a grid state identification to continuously monitor all relevant system parameters.

- Continuous identification of sensitivity of monitored system variables towards generator feed-in in order to identify optimal target values (minimum throttling).
- Temporary and well-dosed throttling of relevant generators in case of impending threshold violations of equipment currents or node voltages.

4 Assessment of the Approaches' Potential by Simulation

To assess the potential that can be achieved by the approach described above, simulation experiments were carried out on basis of a model corresponding to a rural type grid as controlled by the distribution system operator EWE NETZ. The model's characteristics were as follows:

- The model included 110–20 kV transformer as connection between the high voltage grid and the medium voltage grid (underground cables). It also included clustered low voltage type grids.
- A typical DER distribution of biomass, photovoltaics, and wind in grids operated by EWE NETZ was assumed.
- A steady state power flow calculation based on a yearly time series (15 min resolution) delivered currents and voltages.
- Time series of measured feed-in for DER (photovoltaics, wind) are used to model feed-in of DER.
- Consumption loads for households were modelled based on a load model devised by RWTH Aachen [3] and measured loads for industry.

Based on the medium voltage grid model, different simulation scenarios were evaluated, using a scenario with 100 % feed-in as reference. To determine 100 % maximum feed-in for the modelled grid (100 % scenario), all generation capacities were iteratively increased until minimum allowed voltage stability and maximum allowed load of transformers and cables were reached. For this grid the first bottleneck was the 110–20 kV transformer. In the subsequent comparison scenarios, installed generation capacity was further increased stepwise from 100 % to 325 %. Whenever system parameters exceeded tolerance limits, feed-in was reduced, the amount of necessary reduction having been calculated by an optimization algorithm.

Simulation results of two different scenarios are shown in Fig. 6.3. In one scenario it is assumed that annual energy production of wind power plants is minor (weak wind year). Due to practical conditions, this means about 1500 full load hours of the wind power plant, while in case of a strong wind year more than 2000 full load hours can be expected. This is covered in another scenario (strong wind year).

In principle, the diagram shows grid connection capacity for feed-in depending on the curtailed generation energy. It is obvious that grid limitations (load, voltage) appear more frequently for a strong wind year than for a weak wind year. Thus, wind power generation has to be reduced more often. That results in a higher

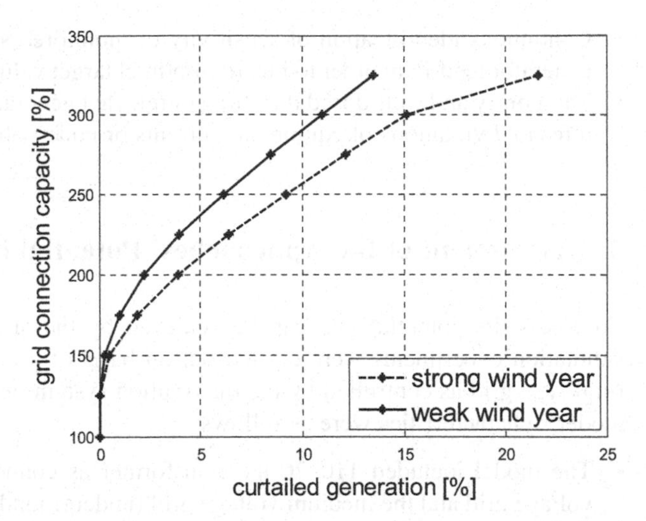

Fig. 6.3 Interdependency between curtailed generation and grid connection capacity during 1 year

amount of curtailed generation in comparison to curtailed generation in a weak wind year. Figure 6.3 depicts the grid connection capacity dependent upon the percentage of curtailed generation during both kinds of wind type years as a result of the simulations. By curtailing power generation by 5 % at most, grid connection capacity can be doubled for both wind type years.

By using the 5 % approach described in this chapter, grid planning can be improved and, especially, grid planning forerun can be increased.

4.1 Validating the 5 % Approach in a Field Test

From the practical point of view, it is necessary to evaluate the 5 % approach under real field conditions. To this end, a field test is carried out to validate the relationship between reduced feed-in and grid connection capacity. The field test is characterized by a power flow dependent scheduling of renewable generators.

In parallel the 5 % approach has to be validated for general distribution grids. To this end, a system study is carried out by an independent scientific institute. Critical parameters are identified by means of a sensitivity analysis. Furthermore, it is studied by which amount distribution grid connection capacity can be increased using the 5 % approach. Note, that during the field test, there is no limitation to the curtailment of generators. Instead, curtailment per generator is regularly evaluated to show that not more than 5 % of its potentially delivered energy has been curtailed.

In the following, we focus on the field test as means of demonstrating how the 5 % approach works in practice and, especially, to answer the question of how the challenges of information technology (e.g., control unit) can be met. Regulatory aspects of the 5 % approach are not part of the field test. However, the results of the field test, which is not yet finished, may serve as an input for changes of the German law on the energy industry whose next amendment is expected in 2015 and is intended to contain requirements to an intelligent dynamic feed-in reduction based on the 5 % approach.

4.2 Field Test Description

Figure 6.4 depicts the selected field test area. The chosen medium voltage grid is characterized by typical feed-in capacity and generators as well as by a representative generation and load combination. Here, low voltage grids do not substantially contribute to the overall power feed-in, so that it can be avoided to include feed-in from generators installed on the low voltage level into the 5 % control. Renewable power generation plants connected to the medium voltage grid have to be controlled by a suitable control unit.

In order to guarantee that equipment usage (electrical current) and grid voltage stability only depend upon measured and controllable feed-in, the switch between switching stations Tettens and Wittmund has to be opened. The field test medium voltage grid is electrically connected to Jever substation that is connected to the high voltage grid.

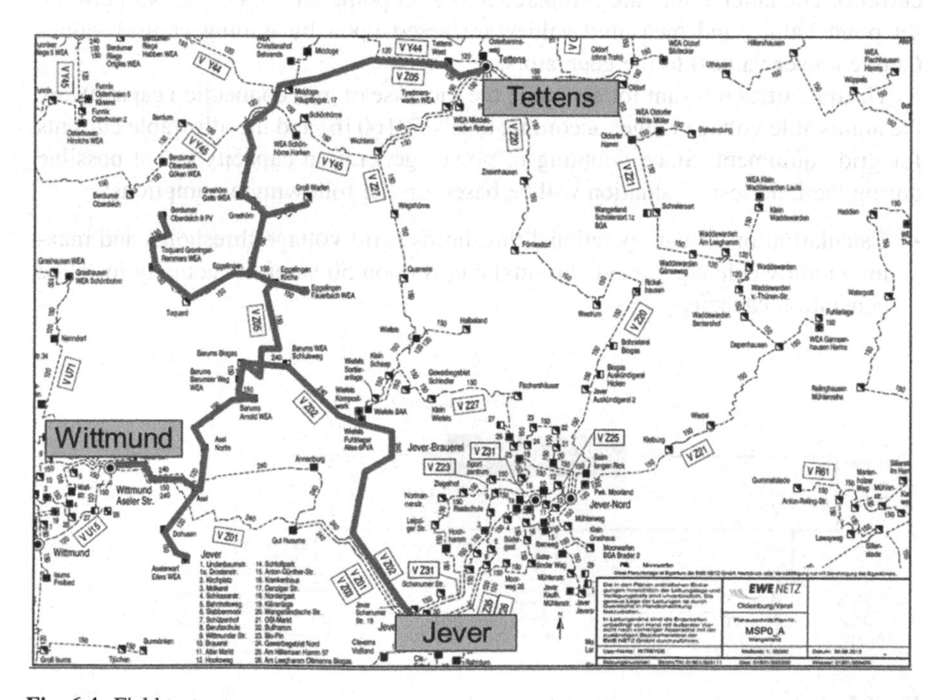

Fig. 6.4 Field test area

The selected grid area contains 11 generators corresponding to a maximum feed-in of 10 MW. Six of these generators are wind turbines, four are photovoltaic generators and one is a combined heat and power plant. During the field test all generators are controlled in their active power behavior. In order to avoid further control activities external to the field test, generators are operated with a constant reactive power ratio. Later on, both active and reactive power control should be dynamically carried out. The following values are continuously measured every 5 sec in order to provide information to a controller performing the task of regulating feed-in from power generators:

- Electrical currents at the medium voltage grid connection point to Jever substation.
- Voltages from substation, switching station, and grid connection points of power generators.
- Power and primary voltage from all low voltage transformer substations.
- Reactive power, active power, and voltage of distributed generators.
- Furthermore, for purpose of validation, wind and radiation measurements are constantly taken.

Figure 6.5 depicts the general system configuration for the field test. The 5 % controller receives measured electric values from the generators (voltage, active power feed-in and reactive power feed-in) and from low voltage and medium voltage measurement equipment at critical points in the distribution grid (voltage, current). The latter values are compared to the set point values. Deviations between set point values and measured values are acted upon by issuing control values (active power values) to the generators.

The quantities relevant for assessing the increase of grid connection capacity are the admissible voltage ranges according to EN 50160 [6] and the allowable currents for grid equipment. Since doubling of power generation capacity is not possible during the field test, evaluation will be based on the following assumptions:

- Calculation of virtual operational thresholds (grid voltage thresholds and maximum allowable equipment currents) based upon 50 % of the actually installed generation capacity.

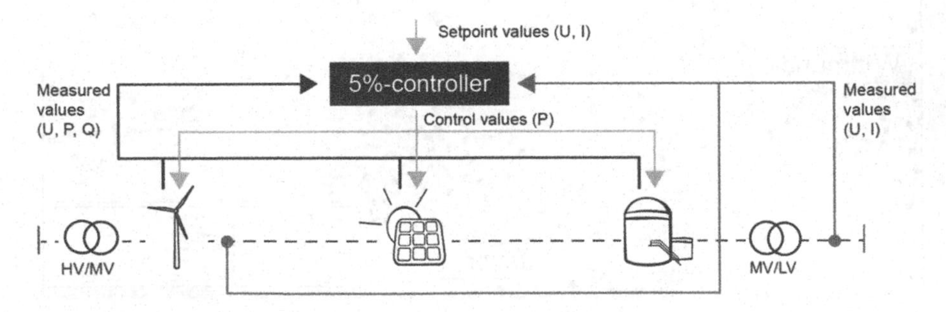

Fig. 6.5 System configuration of the 5 % controller in the field test

- Operation of the field test grid with 100 % of the actually installed generation capacity and control of generators such that the virtual operational thresholds are observed.

In order to validate the 5 % approach, the energy curtailed must not exceed 5 % of possible generation taking into account the weather conditions. To gather statistically adequate evidence, the field test has to run for at least an entire calendar year. Only after this period the ratio between curtailed energy and available energy can be properly calculated.

4.3 Field Test Results

The field test has started in October 2014 and is planned for 1 year. So far, the field test has shown, that the 1-min average of measured values for currents at transformers has rarely been violated and during very short time periods. The following reasons for threshold violations could be identified:

- Generator inertia.
- Weather variability.
- Delay time due to GSM communication.
- The algorithm of distributing control between different generators.

When averaging measurements of current over longer time periods, e.g., 15 min, no threshold violations can be identified at all.

Thus, the control unit performs well (Fig. 6.6). So far, due to the limited field test duration no indication can be given of whether the maximum of 5 % feed-in reduction per generator and year can be met.

Figure 6.7 depicts the curtailment of a wind turbine during an overload situation. The control values issued by the control unit for this particular field test wind turbine lead to the resulting generation limited by the control values. The possible generation of the wind turbine without limitation is shown as reference. The curtailed energy is shown the marked area between the reference and the generation.

So far, the field test showed that control behavior of wind turbines is very good, especially in situations with high power generation. Older turbine types switch off in low power generation situations whereby no sensitive control of power generation is possible in such situations for such turbines.

Until now, during the field test, this behavior did not lead to critical grid situations. Photovoltaic generators exhibit a control behavior similar to wind turbines. However, a required additional electronic control unit causes a delay in reaction time. Combined heat and power generators have limited reaction times due to characteristics of their combustion motor.

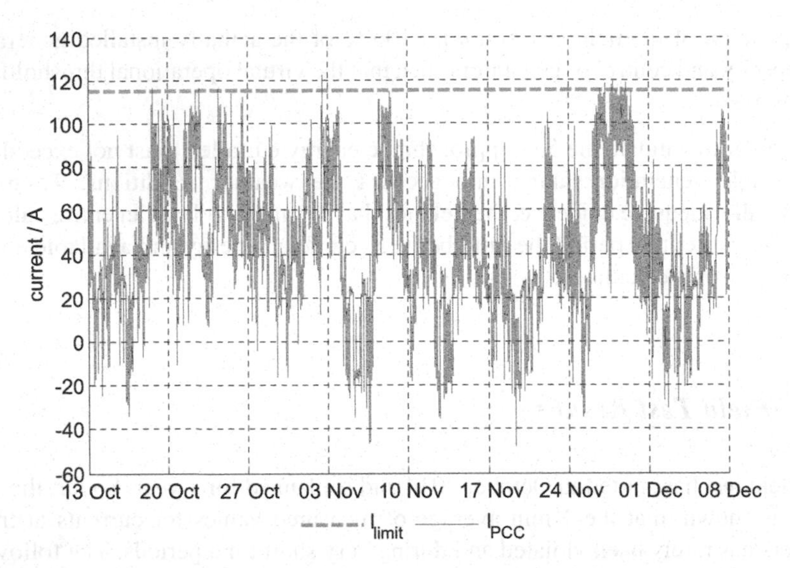

Fig. 6.6 One minute average of electrical current measurements at the substation connection point

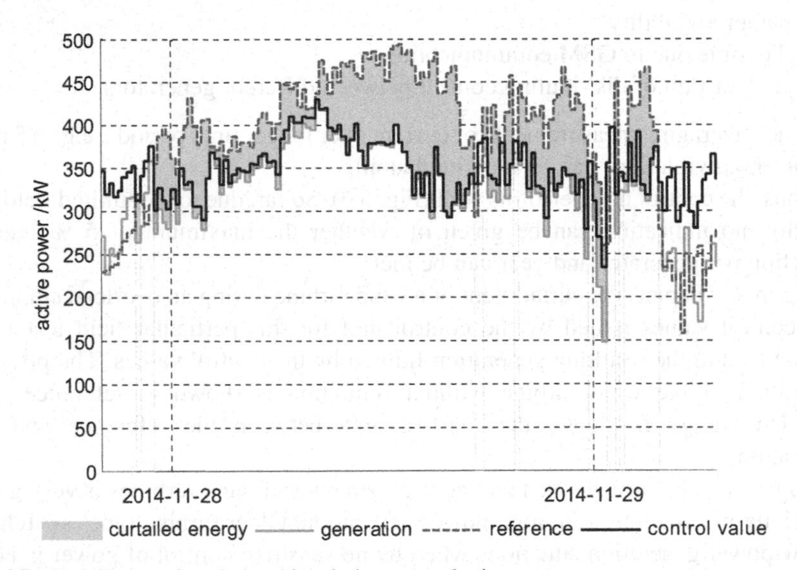

Fig. 6.7 Curtailment of a wind turbine during example time span

During the first 2 months of the field test, mainly wind turbines have been curtailed. This is due to seasonal effects on solar radiation. In order to guarantee the non-discriminatory treatment of generators, a good alignment of curtailment of different generators will have to be focused upon during the remaining field test period.

5 Requirements Toward Control Units Supporting the 5 % Approach

Systems in the energy sector can be distinguished into OT (operational technology) and IT (information technology) [1]. OT is focused on monitoring, supervision, control, and automation; for instance SCADA (supervisory control and data acquisition) systems, automatic control units and sensors are considered as OT systems. Typical non-functional requirements of OT are high availability, 24/7 operation, and redundancy. IT systems provide functions for business, market, documentation, and management that are usually not directly connected to the physical energy system equipment. IT systems are mainly used during office times, typically require less availability than OT systems, and do not typically run on embedded systems. Examples for IT systems are billing systems, geographic information systems (GIS), asset management systems, customer care systems, or energy trading systems. A core element of smart grid architectures (see, e.g. [8]) is to connect IT and OT systems, sometimes called IT/OT convergence.

The 5 % approach can be assigned to the OT domain. We consider it an OT component because it is a non-market mechanism to continuously operate the grid and to deal with exceptional feed-in situations. Thus, control units supporting the 5 % approach must not only meet requirements towards their control capabilities but they also must address problems as reliability under adverse conditions. Grid capacity and safety must be guaranteed even after a communication failure or a breakdown of central system components. Installing a single controller at a central network monitoring center highly dependent on the availability of communication lines and representing a single point of failure would be a risk for grid capacity and grid safety. Placing controllers into decentral substations is one approach for guaranteeing higher availability by decentrality. In the field test we use this approach, however, only a single controller is installed in a single substation. The idea of decentral controllers can be expanded to integrating controllers with emergency network operation software installed in substations.

5.1 *Control Capabilities Required for the Field Test*

Measurements shall take place separately and in a given frequency. Both the necessary frequency and the time span between a threshold violation (characterized by measured values exceeding the desired set point values), detection and triggering of control values will be evaluated during the field test. The control unit continuously outputs control values. As soon as the control unit determines that feed-in reduction can be (partially) taken back without thresholds being violated, feed-in reduction shall (partially) be taken back.

The control strategy shall take into account:

- Safety margins for set point values after threshold violation.
- Avoiding short-term electrical overloading of grid equipment caused by delays in reaction to control signals.
- Grading of control values in order to avoid oscillation.

Since the quality of control has a significant influence on the successful implementation of the 5 % approach, a control quality (deviation from set point values in percent) for voltages and for currents must be guaranteed. Quality of control will be evaluated throughout the field test.

6 Control Unit Design

The control unit supporting the 5 % approach is based upon a product called BTC | GRID Agent. It operates in a continuous loop consisting of three steps:

1. Read measurements and set points.
2. Calculate control values for generators using power flow calculation and taking into account technical limitations.
3. Send control values to generators.

Thus the controller uses a model-based method, a grid model being used for power flow calculations. The main advantage of this method over PID controllers (see [4]) that are not model-based is the reduced number of control actions needed to correct threshold violations due to the higher possible accuracy. This results in faster control process alignment, especially in face of low quality communication links. Generally model-based approaches enhance stability of control due to the higher amount of knowledge of the system under control.

During the field test, the control unit has to evaluate measurements from about 20 measurement points and has to issue about ten control values in each cycle. The frequency of control cycles necessary for ensuring the needed control quality will be evaluated during the field test. Evaluation of the control unit in scenarios for controlling reactive power settings of heterogeneous wind farms had as a result that the control unit can perform multiple control cycles per second. However, this performance will most likely not be reached during the field test for validating the 5 % approach because the size of grid models in this setting is larger than in the mentioned scenarios. Note that, besides control values being sent to generators, event information is sent to distribution management systems or other supervising systems, for instance when a generator does not react to issued control values. Also, parameters, e.g., for transforming set point values, can be modified during run time. They are held in the parameter and curves storage.

The controller's architecture is depicted in Fig. 6.8. The control unit has three main types of modules that are executed in each control cycle: set point modules calculate set point values from set point information (electrical quantities and

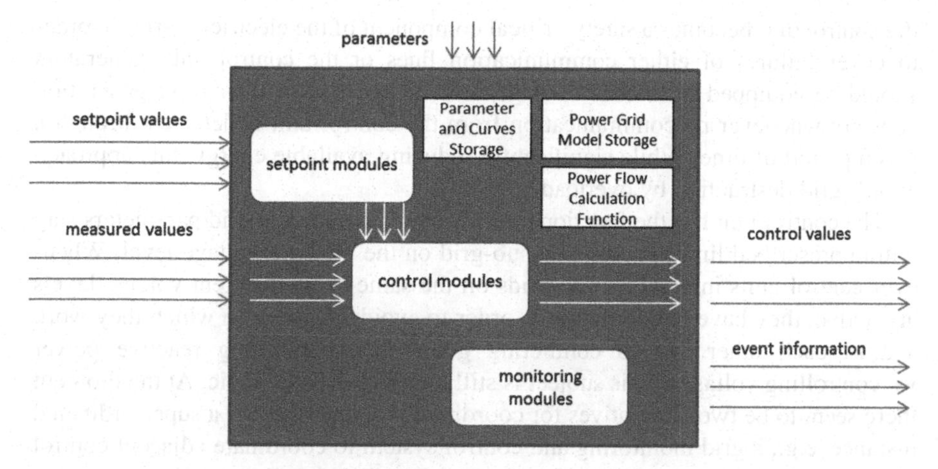

Fig. 6.8 Conceptual architecture of the BTC | GRID Agent

precision information) and ongoing measurements. Control modules calculate control values from set points and measured values. There are several variants of control modules. One variant splits control values to power control values for single generators using power flow calculation. Another variant limits power change rates to acceptable values for generators depending on their operation conditions. Finally, monitoring modules serve the purpose of restricting set point values, e.g., stub currents, to the technical specifications of electrical equipment and generators.

The control unit software has been designed with universal extendable APIs so that it can be adapted to different execution environments (e.g., embedded PCs, matlab, SCADA systems, the simulation framework MOSAIK [10]) by adapters of which some have already been developed.

The power flow calculation function is used by the control modules and relies upon the power grid model stored in the power grid model storage.

6.1 Interplay Between Control Unit and Generators

The controller takes into account whether generators react to the control values issued. If they do not react, they are excluded from the set of controllable generators for a period of time that can be specified. When this time has passed, the generator is again included into the control set.

The control unit also takes into account basic constraints of generators. I.e., some generators do not allow adjusting feed-in below a given threshold. The controller won't issue control values smaller than this threshold value.

Deploying the 5 % approach to a grid means that the connected generation will eventually double, thereby exceeding the physical ability of grid equipment and lines to transport the entire generated current during peak generation times. Thus, the control unit becomes a safety-critical component of the electricity grid. In order to cover failures of either communication lines or the control unit, generators should be equipped to throttle their generation to 50 % of their peak generation capacity whenever no communication from the control unit is detected through a given period of time. While significantly reducing available energy, this approach avoids grid destruction by overload.

The control unit has the function of ensuring that electrical grid parameters stay within prescribed limits for a given sub-grid on the medium voltage level. Whenever control units in adjacent sub-grids on the same or on different voltage levels are active, they have to coordinate in order to avoid situations in which they work against each other due to conflicting goals (e.g. controlling reactive power vs. controlling voltage). This subject is still an open research topic. At the moment there seem to be two alternatives for coordination: either there is a super-ordinated instance, e.g., a grid monitoring and control system to coordinate adjacent control units, or a direct communication between such units is implemented.

7 Related Work

This chapter is based on an article published in [9]. The active network management scheme implemented in the distribution grid on the Orkney Islands, UK, uses controllers to limit power feed-in of renewable generators so that grid expansion can be avoided while increasing the number of renewable generators connected to the grid [2]. The approach used in the Orkney Islands project differs from the 5 % approach in that no limit on yearly feed-in reduction per generator is defined.

8 Conclusion and Further Work

It is the purpose of the described field test to validate the 5 % approach. The next step will be to align the regulatory framework and the 5 % approach and to create solutions allowing distribution grid operators to implement it efficiently, taking into account aspects of integrating OT and IT.

Reliability and maintainability are very important factors for distribution system operation. Their interdependency as well as their dependency upon the degree of centralization of OT and IT and also their influence on costs are important topics that still have to be looked upon in the context of smart grids.

There are a number of other purposes for control in distribution grids. An interesting topic still to be researched is the coexistence of control strategies with different aims and for different grid domains. The cooperation or coordination between control units can either be mediated by distribution management systems or take place directly between controllers.

A possible extension to the 5 % approach could be to combine it with approaches with predictions of the near future, such as load predictions and feed-in predictions to enable curtailment actions coordinated with energy market action instead of reactive actions. However, it is a real challenge to predict the local feed-in and load for distribution grids, and it needs to be studied whether the potential benefit compensates the risks and costs for dealing with prediction errors. The control unit will be modified according to experiences from the field test: for example, it will take gradients into account in order to inhibit the issuance of fast changing control values. This measure will improve quality of control.

References

1. CEN-CENELEC-ETSI Smart Grid Coordination Group: Smart Grid Reference Architecture, SG-CG/RA TR3.0, Version 3.0, Brussels, Nov. 2012.
2. Currie, R., Macleman,D., Mclorn, G., Sims, R.: Operating the Orkney Smart Grid: Practical Experience. In Proc. 21st International Conference on Electricity Distribution, Frankfurt, Germany, 2011.
3. Dierkes, S., Wagner, A., Eickmann, J., Moser, A.: Wirk- und Blindleistungsverhalten von Verteilungsnetzen mit hoher Durchdringung dezentraler Erzeugung. In: Internationaler ETG-Kongress 2013 (ETG-FB 139), VDE Verlag, Nov. 2013.
4. Dorf, R.C., Bishop, R.H.: Modern Control Systems, Prentice Hall, ISBN-13: 9780136024583, 12th edition, 2010.
5. EEG, German Renewable Energy Sources Act, 2014. http://www.bmwi.de/BMWi/Redaktion/PDF/G/gesetz-fuer-den-ausbau-erneuerbarer-energien,property=pdf,bereich=bmwi2012,sprache=de,rwb=true.pdf (2014). Accessed 1st March 2015.
6. EN 50160:2010, Voltage characteristics of electricity supplied by public distribution networks, 2010.
7. Forschungsprojekt Nr. 44/12, Moderne Verteilernetze für Deutschland, (Verteilernetzstudie), Abschlussbericht, Studie im Auftrag des Bundesministeriums für Wirtschaft und Energie (BMWi), http://www.bmwi.de/DE/Mediathek/publikationen,did=654018.html (2014). Accessed 1st March 2015.
8. IEC – International Electrotechnical Commission. IEC 61968–11 Application integration at electric utilities – System interfaces for distribution management, 2010.
9. Marx Gómez, J., Sonnenschein, M., Vogel, U., Winter, A., Rapp, B., Giesen, N., eds: EnviroInfo 2014 – 28th International Conference on Informatics for Environmental Protection. BIS-Verlag, Oldenburg, 2014, ISBN 978-3-8142-2317-9.
10. Schütte, S., Scherfke, S., Sonnenschein, M.: Mosaik – Smart Grid Simulation API. In: International Conference on Smart Grids and Green IT Systems, SciTePress, 2012.

Chapter 7
Aligning IT Architecture Analysis and Security Standards for Smart Grids

Mathias Uslar, Christine Rosinger, Stefanie Schlegel, and Rafael Santodomingo-Berry

Abstract In this paper, an approach using the European Smart Grid Architecture Model (SGAM) in the context of the NISTIR 7628 is presented. Research has shown that both models and methodologies have particular impact, but have not yet been put into mutual context. The combination of these models makes it possible for US Smart Grid experts to reuse the SGAM model and its benefits, and vice versa, European stakeholders are encouraged to use the security analysis framework from NIST. Within this paper, we briefly introduce the methodologies including their strengths and fallbacks. We outline the necessity to make them interoperable and aligning them. Finally, the logical interface framework from NISTIR 7628 is mapped onto the SGAM and its planes, domains and zones, bridging the existing gap. In addition to those results, we outline the need for a future integration with a maturity model for security assessment and point out a roadmap and preliminary results.

Keywords SGAM • NISTIR 7628 • Critical infrastructure analysis • Risk assessment • Security

1 Introduction

One particular important aspect of a future Smart Grid, being a system-of-a-system, is the growing need for using ICT for communication between the various components involved in the processes. Particular goals to be achieved by the Smart Grid may be related to aspects like the optimization and coordination of the various elements and their operation in the transmission as well as the distribution grid [8]. Additionally, since the increasing ICT utilization and the addition of more and new actors in the energy domain; there is also a growing threat potential in the Smart Grid [12]. The importance of the aspect of (system) availability and uptime for the electric power distribution system is high. Furthermore, the dependability of the infrastructure, as well as of its basic components, is the focus of system and

M. Uslar (✉) • C. Rosinger • S. Schlegel • R. Santodomingo-Berry
OFFIS – Institute for Information Technology, Escherweg 2, 26121 Oldenburg, Germany
e-mail: mathias.uslar@offis.de; christine.rosinger@offis.de

J. Marx Gómez et al. (eds.), *Advances and New Trends in Environmental and Energy Informatics*, Progress in IS, DOI 10.1007/978-3-319-23455-7_7

interfaces at design-time. Additionally, interoperability and interchangeability have to be taken into account to ensure a meaningful analysis of both, technical and non-technical requirements [15]. To achieve this goal, one particular way is to standardize (technical) solutions like data models, interfaces, processes and communication protocols at both international and national level [20]. After the first standardization initiatives were raised by both IEC and NIST, it became apparent that standards without being real-world best practices cannot solve all open problems [14]. The NIST framework and roadmap for interoperability as well as the European initiatives – derived from and driven by the M/490 Smart Grid mandate [11] – focuses on properly using, expanding and adopting the so called IEC core standards from the IEC Smart Grid Standardization Roadmap Version 1.0 as well as various related ones. To realize a meaningful working structure to coordinate the various relationships between the standards, the Smart Grid Coordination Group (SG-CG) initiated four different groups in 2012 that should develop a report for their individual topics which were "Sustainable processes", "(First) Set of Consistent Standards", "Reference Architecture" and "Smart Grid Information Security" [2]. In the second phase of the mandate M/490, groups were merged and new ones created, leading also to both a group and report on system and component interoperability in the Smart Grid. In this paper the alignment of the NISTIR 7628 and the EU Smart Grid Architecture Model (SGAM) is shown to get a security viewpoint and realizing the security by design principle constructing an architecture for the Smart Grid. Additionally, some first ideas integrating maturity models into this approach are given.

After this introduction, Sect. 2 gives an overview over existing work in security architecture development in the Smart Grid and how existing approaches of the NISTIR 7628 and the Smart Grid Architecture Model (SGAM) can be linked. Section 3 describes the application of the linked approaches with a use case example. Afterwards, in Sect. 4 some new and further work to extend this linked approach with the usage of maturity models is shown. Finally, this paper ends with a conclusion in Sect. 5.

2 Security Architecture Development in the Smart Grid

The development from a monopolistic power market to a distributed Smart Grid – caused by the organizational unbundling and increasing use of information and communication technologies (ICT) as second infrastructure for controlling the physical Smart Grid infrastructure – increases the threat potential for information security for the energy domain [15]. For this, early considerations of information security aspects at design time–which is also called security-by-design–in the architecture development process of systems in the critical infrastructure energy is necessary [12]. Within this section, we highlight the existing work which is relevant to the ideas and preliminary work presented in this paper, also at the EnviroInfo 2014 conference where this paper is based on [13]. In addition to those results, we add the link to maturity models as one future research direction. First in Sect. 2.1, the scope of the M/490 mandate motivates the need for a common architectural viewpoint in order to foster better component and system interoperability. In addition, the work conducted by the SGIS group in scope of the mandate is briefly introduced. In Sect. 2.2 the European Smart Grid Architecture Model (SGAM) is described which gives an overall viewpoint for the architecture design. Section 2.3 shows a short overview on the NISTIR 7628 document series and a motivation why this model should be combined with the SGAM for a better security-by-design methodology. After this in Sect. 2.5, a short analysis of linking the NISTIR 7628 and the SGAM is given.

2.1 *Results from the SGIS Group*

The SGIS Group was one of the in Sect. 1 mentioned four groups involved in the first phase of the M/490 mandate. As quoted from the M/490 mandate text:

> [...] the objective of this mandate is to develop or update a set of consistent standards within a common European framework [...] that will achieve interoperability and will enable or facilitate the implementation in Europe of [...] Smart Grid services and functionalities [...]. It will answer the technical and organizational needs for sustainable "state of the art" Smart Grid Information Security (SGIS), Data protection and privacy (DPP), [...]. This will enable Smart Grid services through a Smart Grid information and communication system that is inherently secure by design within the critical infrastructure of transmission and distribution networks, as well as within the connected properties (buildings, charging station to the final nodes). [...]

The content developed by the SGIS group in their first edition report does not provide a complete and definitive answer to the mandates objective. Nevertheless, it provides a good high-level guidance on how (security and safety) standards can be used to develop a plan for Smart Grid information security. It also presents concepts useful to all Smart Grid stakeholders to integrate information security into their daily activities. The so-called toolbox mainly categorizes incidents in terms of their impact on load or generation, which might have a (negative) rippling effect on the European UCTE grid. However, it lacks risk, threat and mitigation analysis on system level like the NISTIR 7628 provides. As the method will evolve in the second phase of the EU mandate M/490, we propose a harmonization of SGIS and NISTIR 7628 based on their common denominator: the SGAM.

With a proper mapping of the NISTIR 7628 onto SGAM, new viewpoints can be taken and best practices from different continents be used. This contribution as well as the corresponding and complementary papers from the authors [19, 18] will present an overview on the combined method with an outlook on how to incorporate a maturity model.

2.2 The Smart Grid Architecture Model (SGAM)

In the context of the European Commission's Standardization Mandate M/490 [5, 21], a holistic viewpoint of an overall Smart Grid infrastructure named Smart Grid Architecture Model (SGAM) has been developed. This work is based on existing architecture approaches and subsumes the different perspectives, viewpoints and methodologies of the (current) and, hopefully, future Smart Grid concepts. Figure 7.1 depicts the SGAM structure with its layers and the sub-classes of the domains and zones are outlined.

The SGAM comprises five so-called core viewpoint layers which support energy architecture design of different domain experts. Additional, the domains and zones support a holistic view on an architecture, including the business processes, which usually are out of scope for standardization. These layers were adopted from the Gridwise Alliance Architecture Council (GWAC) stack and context-setting framework (CSF) [1]. Even if the original intention of the SGAM was the identification of standardization gaps, its simple and clear structure has turned out to be of great

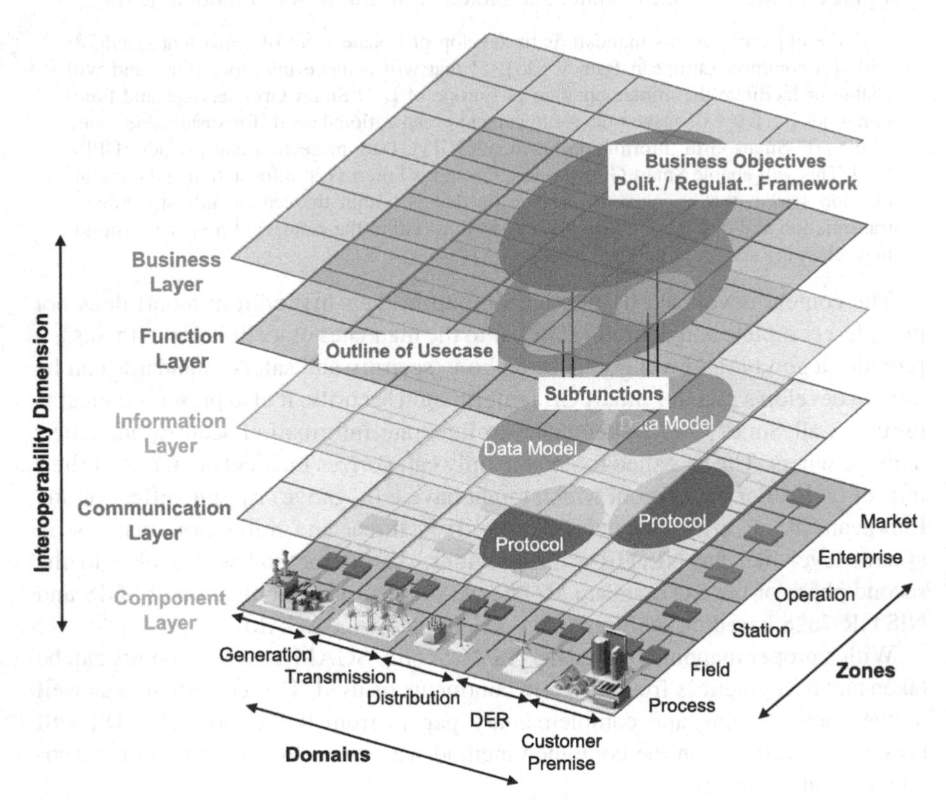

Fig. 7.1 The SGAM canvas model [1]

applicability for architecting Smart Grid systems. The utilization of the SGAM in architecting Smart Grid systems has been discussed in detail in the corresponding sources of this contribution [4].

2.3 *Security Standards: NISTIR 7628*

Information security is not just relevant for the operation of the Smart Grid as a critical infrastructure, it is also very important for user acceptance and general operations and at design time. This particularly affects technologies like smart metering, especially in the part of privacy issues. Many different standards exist in the IEC TC57 portfolio, among them there are standards especially designed for end-to-end security, see e.g. [6]; one series of particular interest is the NISTIR 7628 series [8]. The US National Institute of Standards and Technology (NIST) is a non-regulative administration office within the US Department of Commerce. The NIST developed the NISTIR 7628 guidelines in accordance with the SGIP Interoperability Framework for Smart Grids. The series covers primarily a guideline for the topic of information security for Smart Grid technologies. They start with a general motivation and introduction for the overall topic and describe the process and methods of creating the guidelines. Afterwards, one can distinguish between the individual parts or volumes. Volume 1, which is the most important in the context of this particular contribution, is titled "Smart Grid Security Strategy, Architecture, and High-Level Requirements" and covers the methodology for identifying high-level requirements in terms of threats, risks and possible mitigation. Basic information about the Smart Grid is presented and the current state-of-the-art in terms of cyber security strategies. Most important goal of the strategies is to ensure the overall reliability of grid operations as well as the confidentiality of sensitive personal information. A high-level diagram is the very core model of the report: it is used to locate and place the systems into conceptual domain spaces. Based on this taxonomy, a reference model for system interfaces and interactions is created. Within this model, 22 different logical interface categories are created and assessed to several Smart Grid domains from the NIST conceptual model. Based on those interface classes, the so called Smart Grid Cyber Security Requirements (SG-CySecReq) are defined. Volume 2 has the title "Privacy and the Smart Grid" and covers various aspects and dimension of the data privacy domain. Volume 3 is titled "Supportive analyses and references" and discusses various drawbacks and potential problems in the domains of the Smart Grid ecosystem.

2.4 *Analysis of Linking the NISTIR 7628 and SGAM*

The reference model of the NISTIR 7628 classifies its systems in different domains, as can be seen in [8]. The interfaces between the systems are called logical

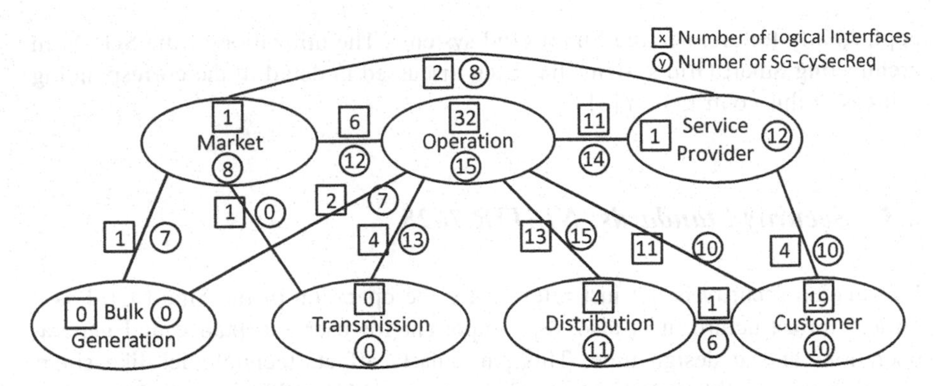

Fig. 7.2 Quantitative distribution of LIs and security requirements form NISTIR 7628

interfaces (LI). Furthermore, there are different LI categories where the systems are grouped by use cases. A quantitative analysis of the NISTIR 7628 is illustrated in Fig. 7.2. This figure shows the distribution of the different LIs in and across the particular domains. Since the LIs are associated to appropriate LI categories and furthermore LI categories are assigned to different Smart Grid Security Requirements (SG-CySecReq), there is also a quantification of the distribution of these SG-CySecReq. In Fig. 7.2 the number of LIs between systems of two domains, respectively in one domain, are depicted in blue and the number of the corresponding SG-CySecReq are displayed in red.

Concerning the SG-CySecReq the authors of the NISTIR 7628 selected 15 different SG-CySecReq which reduces the maximum of distinct SG-CySecReq. Further SG-CySecReqs can be applied equally and thus are not considered separately. In this quantitative analysis can be seen that not every domain has LIs, Bulk Generation and Transmission do not have any LIs. The most LIs are in and between the domain Operation and other domains. Thus, it can be assumed that this domain and the communication with this domain is very ICT-based and other domains are not yet penetrated by ICT. Additionally the highest numbers of SG-CySecReq is between the different LIs who are linked with the Operation domain. Hence, there is a higher demand of security in and between these domains. In the SGAM the domains and LIs are located different from the NIST conceptual model, as can be seen (without the communication connections between the systems) in Fig. 7.4 later in this paper. But the results of the quantitative analysis of the NISTIR 7628 can also be considered: Most LIs occur concerning the Operations domain in both model types. And also the classification in domains and zones shows that there is no communication in and between the domains Generation and Transmission. Additionally, there are a lot of relationships in and between the domains Distribution and Customer Premises. Within the domain DER and between this domain and others there are only few LIs. Overall, there is already a good coverage of the SGAM zone Operation and the SGAM domains Distribution and Costumer Premises. To get a good representation of the domains Generation, Transmission and DER and the zone Market the combined model has to be extended, for example with work of the EU mandate M/490. The illustration of the systems in the NISTIR 7628 has no clarity and additionally the mapping into the process organization of the utility provider is missing. The combined approach with the systems of the NISTIR 7628 mapped into the SGAM allows a good logical illustration of the coherences of the NISTIR 7628. An example for this is the LI category 18 which describes interfaces between metering equipment. The systems of this LI category are classified into the domains Customer, Operation and Service Provider of the NISTIR 7628. The mapping of this LI category 18 into the SGAM is shown in Fig. 7.3. This whole LI category lies in the Customer Premises domain of the SGAM. Within this domain, the categorization and the distribution into the different zones can get analyzed more precisely with this combined approach. Especially during design time, this can be a great advantage to get a good overview of the logical coherences in the overall architecture to determine use case-based security requirements.

2.5 *Linking Architecture and Security*

As stated beforehand, combining the two state-of-the-art models from Europe and the US should lead to a better security analysis possibility for the current SGAM methodology and, thereby, realize security-by design for the energy domain at design time. Later, this should lead to a better dependability analysis for SGAM models as well as a proper linking of logical interfaces (LI) from the NISTIR 7628 to a domain and zonal oriented viewpoint for the different architects.

Additionally, there is potential to crosscheck the NISTIR 7628 with the latest IEC Smart Grid Mapping Tool and, thus, enhancing the possibility to properly assess security standards for Smart Grids to the logical interfaces from NIST. Round-tripping between the various methodologies, tools and models will become possible. The Smart Grid Information Security (SGIS) group from the M/490 mandate currently focuses on privacy and data protection issues, therefore this only complements the work done by NIST. Part of this work was a mapping of the logical interfaces and their systems from the NISTIR 7628 onto the different domains and zones of the SGAM functional plane on basis of [1, 8]. This implies that with the logical reference model mapped onto the SGAM, even the logical interface categories with its Smart Grid Cyber Security Requirements (SG-CySecReq) can be transferred onto the SGAM model. This mapping of the LIs of the NISTIR 7628 onto the SGAM domains and zones can be seen in Fig. 7.4. The colors of the different LIs represent the respective domains as they are also used in the NISTIR 7628.

The example, shown in this paper in Sect. 3, will cover a part of this work. To properly use this model, the authors suggest a canonical model using five individual steps to integrate the methodology into the development to use it as security assessment, which can be seen in the following enumeration and will be explained in more detail in the use case example in Sect. 3 of this paper.

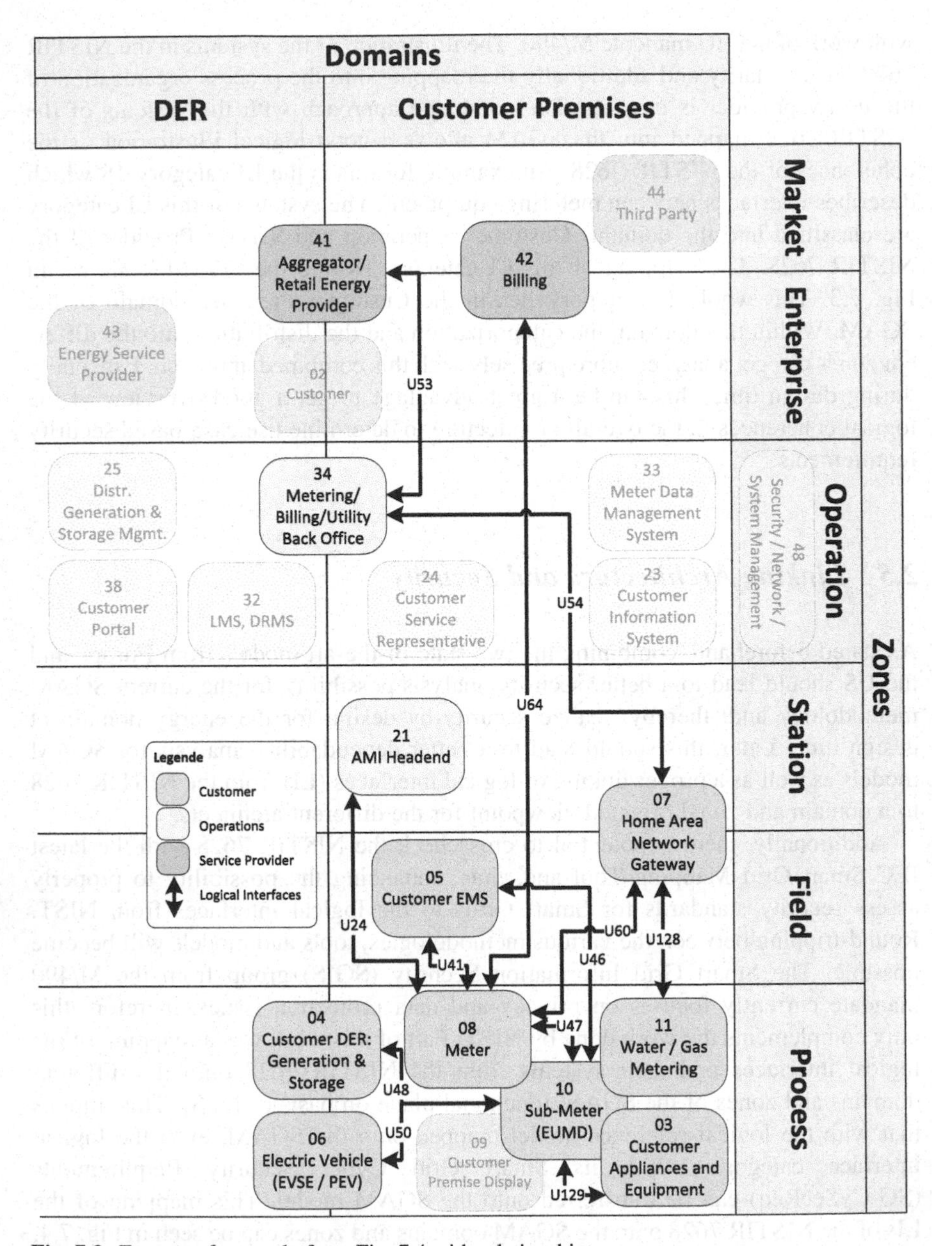

Fig. 7.3 Excerpt of example from Fig. 7.4 with relationships

1. Identifying and (formally) specifying the use case in IEC PAS 62559 templates [7].
2. Identification and mapping of systems, LIs, communication links and interface categories.
3. Integration of the systems and LIs onto the SGAM Functional Layer.

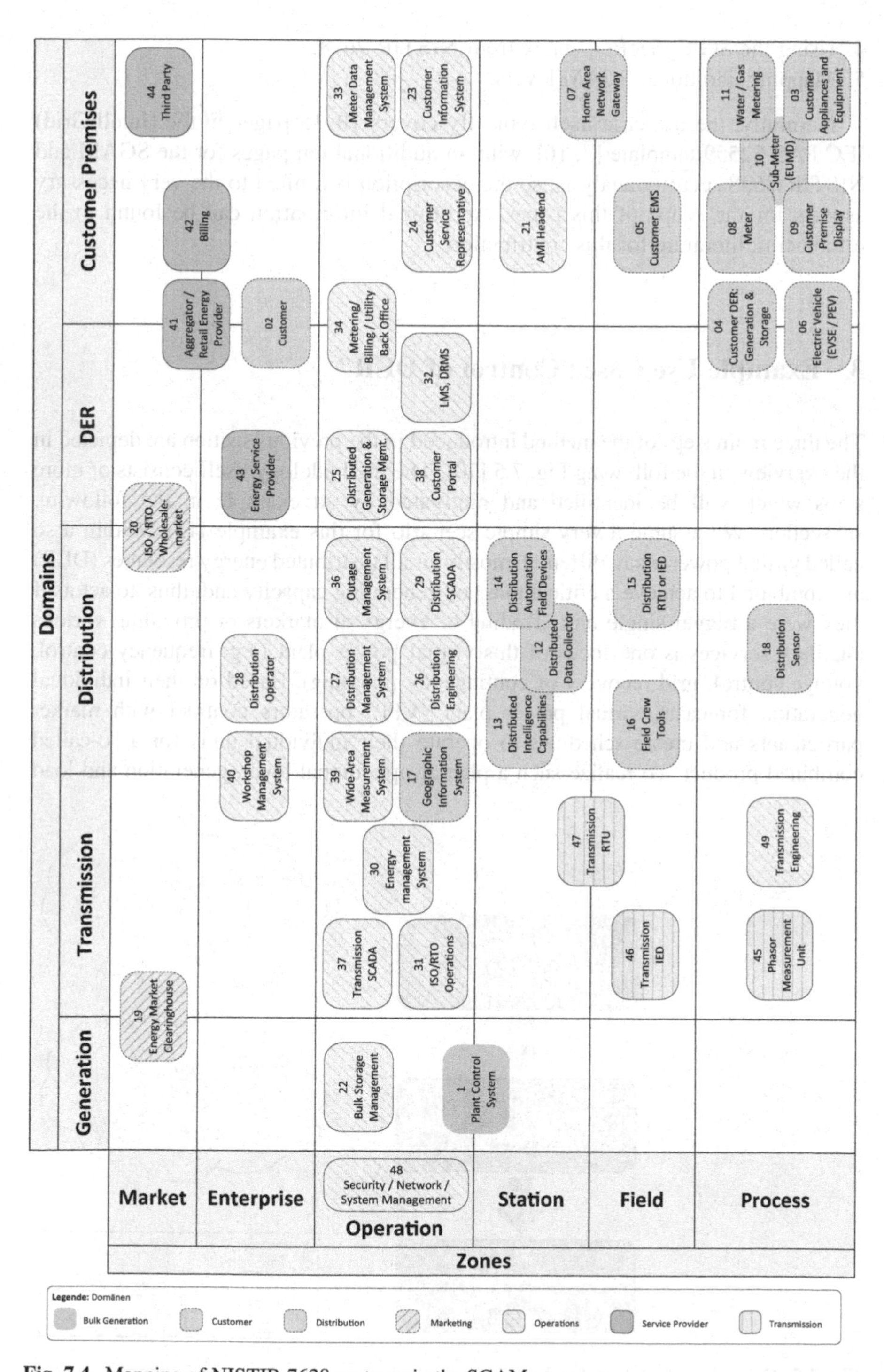

Fig. 7.4 Mapping of NISTIR 7628 systems in the SGAM

4. Using the SG-CySecReq annex from NISTIR 7628.
5. Mapping additional SGAM layers.

Normally, the use case itself typically covers 10–15 pages in the (IntelliGrid) IEC PAS 62559 template [7, 16], with an additional ten pages for the SGAM and NISTIR 7628 security analysis, so the description is limited to the very necessary aspects for the scope of this paper. Additional information can be found in the referencing literature for this contribution.

3 Example Use Case "Control of DER"

The three main steps of the method introduced in the previous section are depicted in the overview in the following Fig. 7.5 [10]. The methodology itself consists of more steps which will be identified and motivated by an example in the following subsection. We assume a very simple scenario for this example [13]. Within a so called virtual power plant, different, mostly small distributed energy resources (DER) are combined to achieve a critical mass of generating capacity and, thus, to act as if they were a bigger single unit. Trading of energy at markets or providing various ancillary services is one focus of this virtual power plant (e.g. frequency control, voltage control, grid recovery or contingency planning). Based on their individual generation forecasts, virtual power plant (VPP) operators contract with market participants and create schedules to operate their individual units for a so-called combined product. To realize such a plan at operational level, generation and load

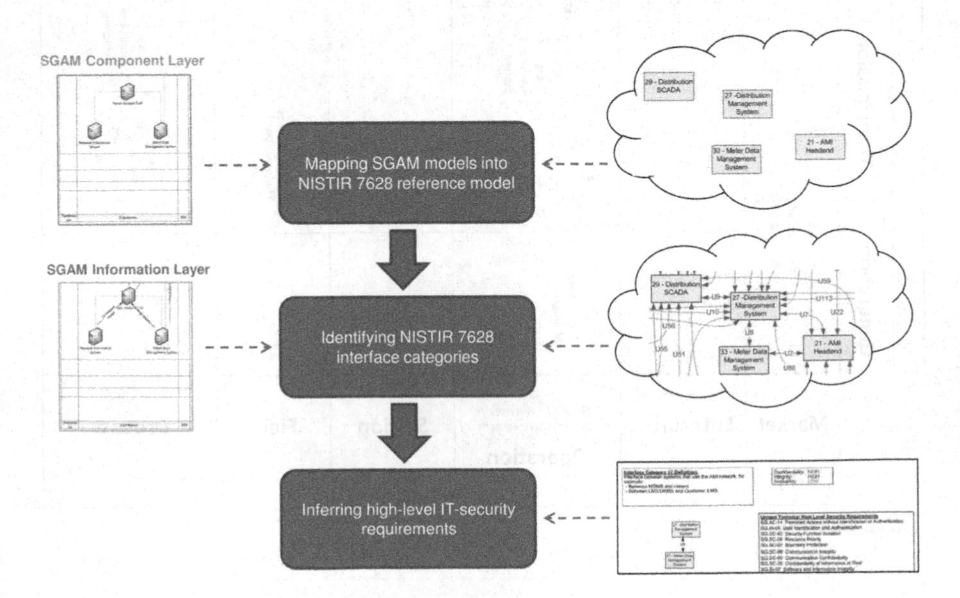

Fig. 7.5 General methodology for the security analysis

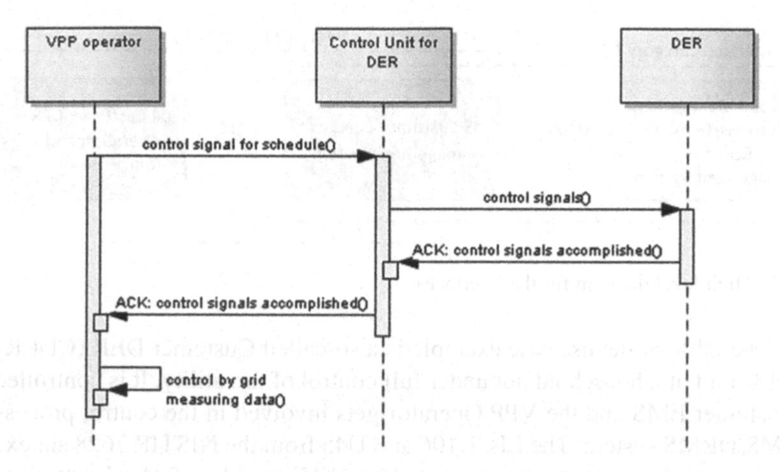

Fig. 7.6 Sequence diagram of the use case example

has to be adapted to the needs of the market bid. Typically, this is done by the direct control of the individual plants (control unit for DER) or by providing incentives to the owners, according to the forecast. In Fig. 7.6, the communication and data exchange of the systems in this use case is displayed in a so-called UML sequence diagram that is explained in the following paragraphs. Additionally, applying the aforementioned methodology, the following five steps from the next subsections have to be taken to assess security requirements from NISTIR 7628 for this use case.

3.1 Identifying and (Formally) Specifying the Use Case in IEC PAS 62559 Templates

We start using the IEC PAS 62559 template and specify the use case from the previous paragraph which is described in [7, 16]. Because of the limitation of pages in this paper the definition of the use case is reduced to the identified systems and the corresponding sequence diagram. The identified systems are the following: DER, VPP operator and Control Unit for DER. The sequence diagram of Fig. 7.6 is useful to get an overview about the communication between the systems and to identify the interfaces.

3.2 Identification and Mapping of Systems, LIs and Interface Categories

The identified systems and LIs have to be mapped on the NISTIR 7628 descriptions. Figure 7.7 shows the scenario as a so-called high-level diagram from NISTIR

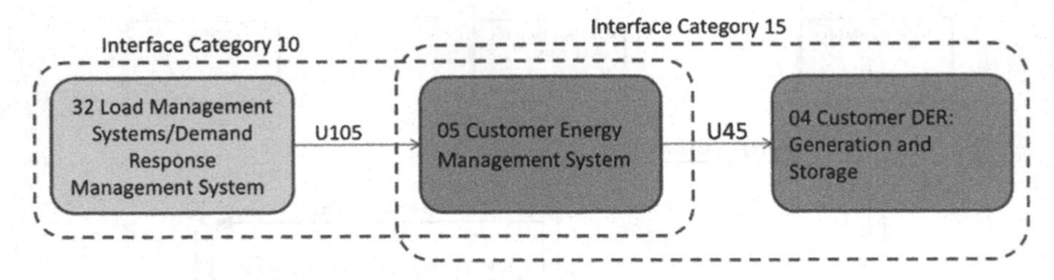

Fig. 7.7 High level diagram for the interfaces

7628. The DER of the use case example is a so-called Customer DER (CDER) – it is a DER unit at a household not under full control of the utility. It is controlled via the Customer EMS and the VPP Operator gets involved in the control process via the LMS/DRMS system. The LIs, U106 and U45 from the NISTIR 7628 annex, and their corresponding interface categories, 10 and 15, are identified using the generic blue print from the NISTIR lookup tables. The colors used in Fig. 7.7 reflect the NISTIR 7628 domains of the original LI diagrams. The system with number 32 LMS/DRMS (= yellow, domain operations) sends two different signals to the system number 5 Customer EMS (CEMS) (green = domain customer).

After an appropriate ramp-up time, the two signals for tariffs and schedules are submitted. If the time of the schedule is fulfilled, real-time measurements are used to check the fulfillment of the energy contract. If the schedule is not fulfilled due to interference from the customer on site, direct control for the DER using a control signal is initialized. Once the signals are sent to the CEMS with number 4, the CEMS decides how to react, based on predefined and engineered rule sets, and sends control signals to the CDER. After accomplishing the tasks, the CDER acknowledges to the CEMS and subsequently the CEMS acknowledges to the LMS/DRMS, as can be seen in the sequence diagram in Fig. 7.6.

3.3 Integration of the Systems and the LIs onto the SGAM Functional Layer

Within this step of the methodology, the mapping onto the SGAM layers is conducted. For this example, we take the associated systems and LIs of Fig. 7.4. Figure 7.8 provides an overview of the mapped systems as well as the corresponding generic LIs between the identified primary systems for the use case. Utilizing this kind of graphical representation makes it easier to check which domains are covered by which systems as well as to recognize the hierarchical zone they reside in. This distribution of the functional logic can be visualized and responsible persons for the individual system can be identified – this leads to easier operationalization of the responsibilities in the next step of the mapping phase.

Function Layer	Generation	Transmission	Distribution	DER	Customer Premise
Market				Interface Category 10	
Enterprise					
Operation				32 Load Management Systems/Demand Response Management System	U105
Station					
Field					05 Customer Energy Management System
Process				04 Customer DER: Generation and Storage	U45 Interface Category 15

Fig. 7.8 Mapped systems and interfaces

3.4 *Using the SG-CySecReq Annex from NISTIR 7628*

In NISTIR 7628 the interfaces are categorized and for the different categories protection goals (like CIA analysis and high-level security requirements) are documented. Based on the identified interfaces and categories in step 2 and 3, in this step the security level of the common security requirements CIA and the corresponding high-level SG-CySecReq (Smart Grid Cyber Security Requirements) can be derived. This is illustrated in Table 7.1 for the described use case and shows the resulting summary of this use case to obtain overall requirements for the communication between the LMS/DRMS and the CDER. In addition, security requirements from other standards can be used from the annex lookup tables of the NISTIR 7628 report, volume 1 and 3.

3.5 *Mapping Additional SGAM Layers*

In this last step, the identified SG-CySecReq and their systems and LIs are mapped onto the individual further SGAM planes. The business layer hosts the business view about the requirements. Therefore abstract, high-level requirements are placed here. Figure 7.9 shows what the high-level requirements from our example are, and

Table 7.1 CIA analysis and SG-CySecReq for the use case

	LI category 10	LI category 15	Result	Description of SG-CySecReq
Confidentiality	Low	Low	**Low**	–
Integrity	High	Medium	**High**	–
Availability	Medium	Medium	**Medium**	–
Smart Grid Cyber Security Requirements	AC-14	AC-14	**AC-14**	Permitted actions without identification or authentication
	IA-04	IA-04	**IA-04**	User identification and authentication
	SC-05	SC-05	**SC-05**	Denial-service protection
	SC-06	SC-06	**SC-06**	Resource priority
	SC-07	SC-07	**SC-07**	Boundary protection
	SC-08	SC-08	**SC-08**	Communication integrity
	SC-26	SC-26	**SC-26**	Confidentiality of information at rest
	SI-07	SI-07	**SI-07**	Software and information integrity
		SC-03	**SC-03**	Security function isolation
		SC-09	**SC-09**	Communication confidentiality

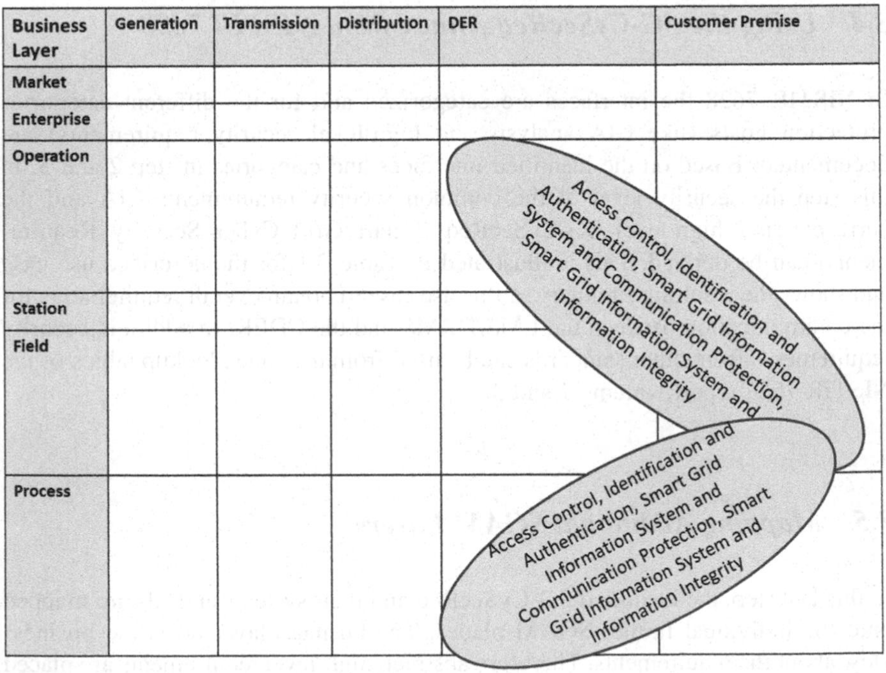

Fig. 7.9 High-level security requirements for the business layer

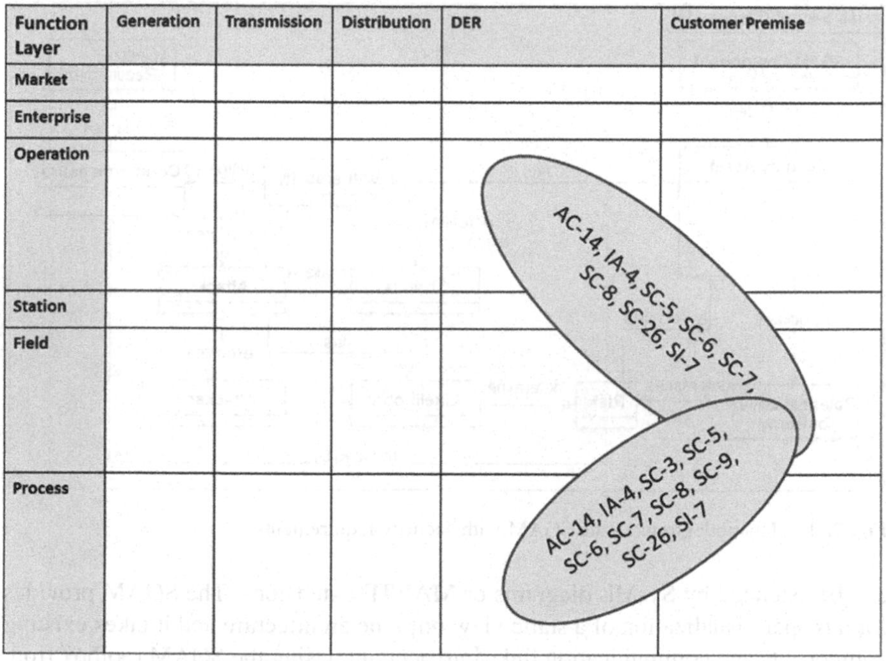

Function Layer	Generation	Transmission	Distribution	DER	Customer Premise
Market					
Enterprise					
Operation					
Station					
Field					
Process					

AC-14, IA-4, SC-5, SC-6, SC-7, SC-8, SC-26, SI-7

AC-14, IA-4, SC-3, SC-5, SC-6, SC-7, SC-8, SC-9, SC-26, SI-7

Fig. 7.10 Smart Grid Cyber Security Requirements on the function layer

where they are placed. The functions of these abstract, high-level requirements are placed on the function layer. It is a more precise description of what is required and how to put it into practice. Figure 7.10 shows the corresponding requirements from our example. These are the SG-CySecReq from the blue-print SG-CySecReq classes. What is placed on the other layers depends on the different requirements. Therefore we need to consider each requirement separately, asses how to realize it, and classify it into the component, communication or information layer.

4 Utilization of Maturity Models in Security Contexts

Because of the very need of integrating security into the development process of Smart Grid architectures from the very beginning, the preliminary work, presented in this paper, shows a combined European and American approach, whereby the advantages of both facilitate a secure architecture development in the Smart Grid domain. It addresses a tool chain, which can also be found in [18, 3, 9]. Starting with the use cases, the IntelliGrid template provides a meaningful way to document the needed information to progress with implementing a function. Functional and non-functional requirements can be captured in a way to become properly formalized. The included UML diagrams contain also dynamic behavior of the system and

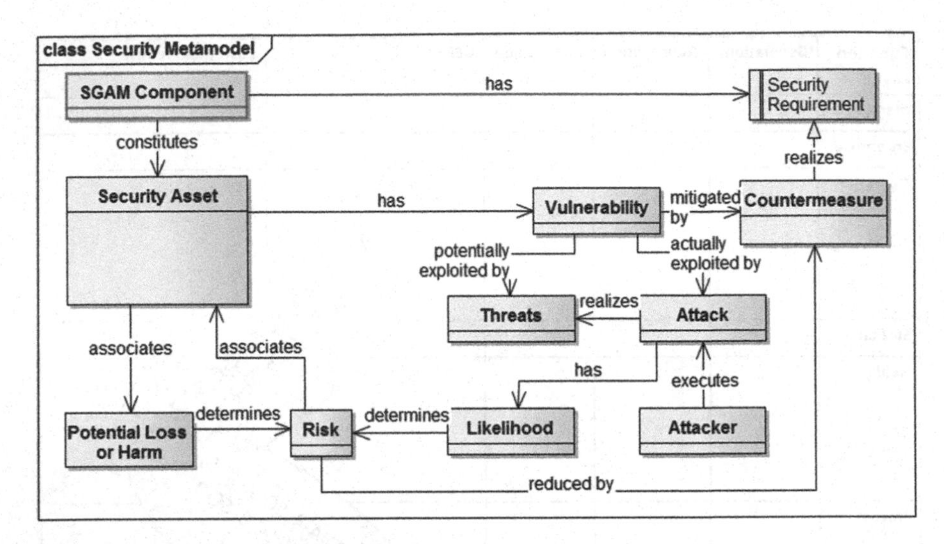

Fig. 7.11 Metamodel to integrate SGAM with security requirements

can be extended by SysML diagrams or MARTE extensions. The SGAM provides for a proper visualization of a static view onto the architecture and it takes existing components and communication links into account. Using the SGAM toolbox from the INTEGRA project, an import of use case templates becomes feasible. A mapping of SGAM components to security requirements, based on the metamodel from Fig. 7.11, can be performed. The integration of the described methodology for mapping NISTIR 7628 onto individual planes is currently under implementation within the toolbox.

In the future, additional standards should be mapped to the SGAM to obtain a comprehensive model for the development of Smart Grid architectures. Another future item to deal with is the assessment of maturity of measures taken which will be described in the next section.

4.1 Maturity Models

The Electricity Subsector Cybersecurity Capability Maturity Model (ES-C2M2) was developed to advance the current developments of Capability Maturity Models and, at the same time, improve the model's measurability and readability. Thus, it was tried to discover ways to measure maturity more precisely. The model, or rather the development of the model, was initiated by the Electric Sector Cyber security Risk Management Maturity Initiative, an initiative of the White House, and was developed under the lead of the U.S. Department of Energy (DOE) in collaboration with Homeland Security. The model is supposed to be an aggregator of existing Maturity Models (MM) and it was compiled based on the best-practices approach.

The model consists of ten domains and four Maturity Indicator Levels (MILs). The ten mentioned domains are considered as "logical essence" of cyber security practices and are listed in the following enumeration.

1. Risk management (RISK)
2. Asset, change and configuration management (ASSET)
3. Identity and access management (ACCESS)
4. Threat and Vulnerability management (THREAT)
5. Situation Awareness management (SITUATION)
6. Information sharing and Communications (SHARING)
7. Evident and Incident Response (RESPONSE)
8. Supply Chain and External Dependencies Management (DEPENDENCIES)
9. Workforce Management (WORKFORCE)
10. Cybersecurity Program Management (CYBER)

The MILs express the development status of the listed categories. MIL0 refers to the state "incomplete", MIL1 expresses that the process has been initiated, MIL2 states it is active and it is "managed" and MIL3 + x–means MIL3 and higher levels – describe all future levels of maturity. Each domain can accurately be assessed and evaluated throughout their sub-sector: (1) → Risk Management (b) Manage Cyber Security Risk → MILx and more [17].

4.2 *Ideas of Combining Security and Maturity Models*

The methodology introduced in the last section can build upon the previous described work. The SGAM can act as a base paradigm to integrate SGIS (Smart Grid Information Security of the EU mandate M/490) and NISTIR 7628 security in a centered view and take into account the quantitative and qualitative aspects from the NISTIR 7628 series. We argue that existing maturity models are still too immature to take into account more perspectives than a qualitative one. With the presented method, we can create also the quantitative perspective, needed to properly assess security and maturity levels by incorporating the individual measures taken and requirements fulfilled. The risk levels, the SGIS group takes into account mainly deal with the critical loss of both load and generation in the UCTE grid. As incidents occur, the grid stability can become endangered by those incidents. However, one could argue that incidents with relevance to security will occur before the actual event of the outage or power loss. Therefore, topics like situational awareness for individual systems and interfaces have to be considered. Within the scope of the presented Smart Grid use case in this section, situational awareness shall be considered as a concept dealing with mainly four parts, namely: logging the situational data in the RTU, monitoring the function in the distribution grid for control-loops, maintaining an overall common operations picture and managing the day to day data activities for operations. For those four aspects, maturity models have been put in place in terms of one of the dimensions from the

ES-C2M2 as described in the previous paragraph. The maturity sub-model for situational awareness takes into account those factors, which shall be used alongside a proposed tool chain in the next paragraph. Those requirements make an asset for one dimension of the overall maturity of operating. During this example, several aspects are clear. Certain dimensions of the maturity are architecture, interface or run-time based. This means, basic data for an assessment has to be gathered from various sources. The IntelliGrid Use Case template usually comprises a sequence diagram and various information about the non-functional requirements, as given for run-time operations phase. In addition, standards for security from the corresponding IEC 62559 annex can be annotated. As depicted and described in [19, 13], a combination of NISTIR 7628 and SGAM models is possible. Therefore, SGAM models can be annotated with standards from the IEC Smart Grid Mapping tool, assigned risks, and threats to interfaces as well as mitigation strategies. The only thing still missing in the state-of-the-art is the link to a proper maturity models in terms of both information sources. This activity closes this open spot by adding the maturity model from the ES-C2M2 to this spot.

4.3 Example: Situational Awareness

Based on the UML sequence diagrams e.g. data being logged and exchanged can be tracked for Set point 1: Perform Logging. We can see from the information flow (transitive) if this data is used in other systems or processes, which would lead to a MIL3 assessment. However, the system does not name the standards applied. Therefore, we use the NISTIR 7628 and SGAM modeling to assess the standards and the usual risk and mitigation strategies for this particular aspect. This enhances also our maturity model as it does not state explicitly standards to be applied at the maturity levels, e.g. ISMS based on ISO 27019 (which would be need for Set point 4: manage SITUATIONS MIL2). For the Common Operation Picture (COP), we could identify the lack of MIL3 by finding out that the systems used do not cover an aggregator for real-time data. By carefully going through our individual use cases with the corresponding SGAM and IntelliGrid models, we can easily set up a way in which those models interact to come up with a holistic tool chain.

5 Conclusion and Further Work

Within this contribution, we have outlined a new model to deal with security-by design based on artifacts used in the context of the M/490 EU mandate. Based on the IEC 62559 template and the SGAM methodology, a security analysis can be performed using the common SGIS methodology alongside the SGAM or by the methodology introduced in this paper. We have presented a multi-stage way to map the NISTIR 7628 standards series for Smart Grid security onto the SGAM layer, leading to an integrated way on how to assess risks mitigation for system security and protection goal analysis. Based on this work, a use case was motivated and presented in a brief way. We outlined the usefulness of the approach also evaluated in several FP 7 projects. In addition, the need for assessing security levels was motivated. One way to extend our existing tool chain is to use the "Electricity Subsector Cybersecurity Capability Maturity Model" ES-C2M2 from the US Department of Energy. It includes various levels for assessing cross-cutting issues towards specific use cases and systems. We presented a first link how the interaction could take place and motivated future work to be done in this context. The individual requirements from various systems can be used, e.g. using text mining or a structured mapping approach. The IEC template properly conducts a security level assessment on a convenient basis. In addition, a verification can be gathered from the NISTIR 7628 systems assessment leading to a verification whether certain risks, threats or mitigation strategies were omitted or shall be put into daily practice.

In summary, the goal of the approach was to methodologically align NIST and SGAM metamodels in order to foster a use of both methods and artefacts in context. This was achieved by mapping the NIST systems and LIs onto the SGAM, making security analysis for SGAM models possible. Benefits will be the broadening of methods for worldwide users and also the alignment of methods for standardization. However, as we pave the prevent large scale attacks on infrastructures by defining mitigation strategies at system level, the actual economic benefit is hard to measure as one cannot simply quantify the effects of attacks on the power grid in terms of damage caused as well as damage prevented. However, the applicability of security blue-prints for critical utility infrastructures will lower the amount of money spent to build up a knowledge base for each individual utility, leading to a higher security level due to shared practices in the overall domain or branch.

Acknowledgement The research leading to these results has received funding from the European Union Seventh Framework Programme (FP7/2007-2013) under grant agreement No. 308913 – DISCERN and additionally under grant agreement No. 609687 -ELECTRA.

References

1. CEN-CENELEC-ETSI: Smart Grid Coordination Group: Smart Grid Reference Architecture (2012).
2. CEN-CENELEC-ETSI Joint Working Group: Final report of the CEN/CENELEC/ETSI Joint Working Group on Standards for Smart Grids (2011).
3. Dänekas, C., Engel, D., Neureiter, C., Rohjans, S., Trefke, J., Uslar, M.: Durchgängige Werkzeugunterstützung für das EU-Mandat M/490: Vom Anwendungsfall bis zur Visualisierung. Tagungsband VDE-Kongress 2014, Frankfurt (2014).
4. Englert, H., Uslar, M.: Europäisches Architekturmodell für Smart Grids – Methodik und Anwendung der Ergebnisse der Arbeitsgruppe Referenzarchitektur des EU Normungsmandats M/490. Tagungsband VDE-Kongress 2012, Stuttgart (2012).
5. European Commission: M/490 Standardization Mandate to European Standardisation Organisations (ESOs) to support European Smart Grid deployment (2011).

6. International Electrotechnical Commission (IEC): IEC 62351 Part 1–11, Power systems management and associated information exchange – Data and Communication Security (2007–2013).
7. International Electrotechnical Commission (IEC): IEC/PAS 62559 ed. 1.0-IntelliGrid Methodology for Developing requirements for Energy Systems (2008).
8. National Institute of Standards and Technology (NIST): The Smart Grid Interoperability Panel – Cyber Security Working Group: NISTIR 7628 – Guidelines for Smart Grid Cyber Security vol. 1–3 (2010).
9. Neureiter, C., Eibl, G., Engel, D., Schlegel, S., Uslar, M.: A concept for Engineering Smart Grid Security Requirements based on SGAM models. Computer Science-Research and Development, Springer, Berlin Heidelberg (2014).
10. OFFIS: DISCERN Deliverable (D) no. 3.5: IT security concept (2014).
11. Rohjans, S., Uslar, M., Bleiker, R., Gonzalez, J.M., Specht, M., Suding, T., Weidelt, T.: Survey of Smart Grid Standardization Studies and Recommendations. First IEEE International Conference on Smart Grid Communications, Gaithersburg (2010).
12. Rosinger, C.: Informationssicherheit im Smart Grid. IT-Architekturentwicklung im Smart Grid, Springer, Berlin Heidelberg (2012).
13. Schlegel, S., Rosinger, C., Uslar, M.: Aligning IT architecture analysis and security standards for Smart Grids. Proceedings of the 28th Conference on Environmental Informatics – Informatics for Environmental Protection, Sustainable Development and Risk Management (2014), http://oops.uni-oldenburg.de/1919/1/enviroinfo_2014_proceedings.pdf
14. SMB Smart Grid Strategic Group (SG3): IEC Smart Grid Standardization Roadmap (2010).
15. Suhr, A., Rosinger, C., Honecker, H.: System Design and Architecture – Essential Functional Requirements vs. ICT Security in the Energy Domain. Internationaler ETG-Kongress, Berlin (2013).
16. Trefke, J., Gonzalez, J., Dänekas, C.: IEC/PAS 62559-Based Use Case Management for Smart Grids. Standardization in Smart Grids, Springer, Berlin Heidelberg (2013).
17. US Department of Energy, US Department of Homeland Security: Electricity Subsector Cybersecurity Capability Maturity Model (ES-C2M2) (2014).
18. Uslar, M., Rosinger, C., Schlegel, S.: Application of the NISTIR 7628 for Information Security in the Smart Grid Architecture Model (SGAM). Tagungsband VDE-Kongress, Frankfurt (2014).
19. Uslar, M., Rosinger, C., Schlegel, S.: Security by design for the Smart Grid: Combining the SGAM and NISTIR 7628. IEEE 38th International Computer Software and Applications Conference Workshops (COMPSAC), Västerås (2014).
20. Uslar, M., Schmedes, T., Lucks, A., Luhmann, T., Winkels, L., Appelrath, H.J.: Interaction of EMS related systems by using the CIM standard, Springer, Berlin (2005).
21. Uslar, M., Specht, M., Dänekas, C., Trefke, J., Rohjans, S., Gonzalez, J., Rosinger, C., Bleiker, R.: Standardization in Smart Grids: Introduction to IT-Related Methodologies, Architectures and Standards. Springer, Berlin (2013).

Chapter 8
Design, Analysis and Evaluation of Control Algorithms for Applications in Smart Grids

Christian Hinrichs and Michael Sonnenschein

Abstract In many countries, the currently observable transformation of the power supply system from a centrally controlled system towards a complex "system of systems", comprising lots of autonomously interacting components, leads to a significant amount of research regarding novel control concepts. To facilitate the structured development of such approaches regarding the criticality of the targeted system, the research and development of a distributed control concept is demonstrated by employing an integrated methodology comprising both the Smart Grids Architecture Model framework (SGAM) and the Smart Grid Algorithm Engineering process model (SGAE). Along the way, a taxonomy of evaluation criteria and evaluation methods for such approaches is presented. For the whole paper, the Dynamic Virtual Power Plants business case (DVPP) serves as motivating example.

Keywords Smart Grid • Control algorithm • Reference architecture • Process model • Evaluation criteria

1 Introduction

A significant share of global CO_2 emissions can be explained by the combustion of fossil fuels for power production. Hence, it has become politically widely accepted in Europe, to reduce national shares of fossil fuels in power production significantly. Such a politically driven evolution of the power system faces not only economical and societal challenges, but it must also address several technological challenges of ensuring a highly reliable power supply, as described in e.g. Ref. [1]. In order to address these challenges, new concepts for power grid operation are needed. The notion of Smart Grids has been introduced for this purpose. The European Technology Platform for Electricity Networks of the Future defines a Smart Grid as an "electricity network that can intelligently integrate the actions of all users connected to it – generators, consumers and those that do both – in order to efficiently deliver sustainable, economic and secure electricity supplies"

C. Hinrichs (✉) • M. Sonnenschein
Department of Computing Science, Environmental Informatics, University of Oldenburg, 26111 Oldenburg, Germany
e-mail: Christian.Hinrichs@uni-oldenburg.de; sonnenschein@informatik.uni-oldenburg.de

J. Marx Gómez et al. (eds.), *Advances and New Trends in Environmental and Energy Informatics*, Progress in IS, DOI 10.1007/978-3-319-23455-7_8

[2]. However, this implies an increased computational complexity for optimizing the coordination of these individually configured, distributed actors. A significant body of research currently concentrates on this topic, see e.g. the research agenda proposed in [3]. In turn, new possibilities are opened up for business players to offer novel control concepts also to distributed generators and consumers.

The power supply system is a critical infrastructure, therefore such approaches must be carefully studied in a secure environment before being implemented in the field. For gaining reliable results, however, this secure environment should reflect as many significant properties as possible of the targeted application area. Two relevant methodologies have been proposed to support the development of Smart Grid applications in this sense:

1. The Smart Grids Architecture Model framework (SGAM) provides a way to document static overviews of systems and actors in a Smart Grid use case [4]. As it lacks a dynamic view as well as the annotation of (non-)functional requirements for interfaces, it is complemented by the use case template IEC 62559 [5].
2. The Smart Grid Algorithm Engineering process model (SGAE) introduces guidelines for application-oriented research and development especially in control algorithms for power systems [6], see Fig. 8.1.

While the SGAM focuses on the software structural design and realization of an intended system behavior (cf. Ref. [7]), the SGAE understands itself as an

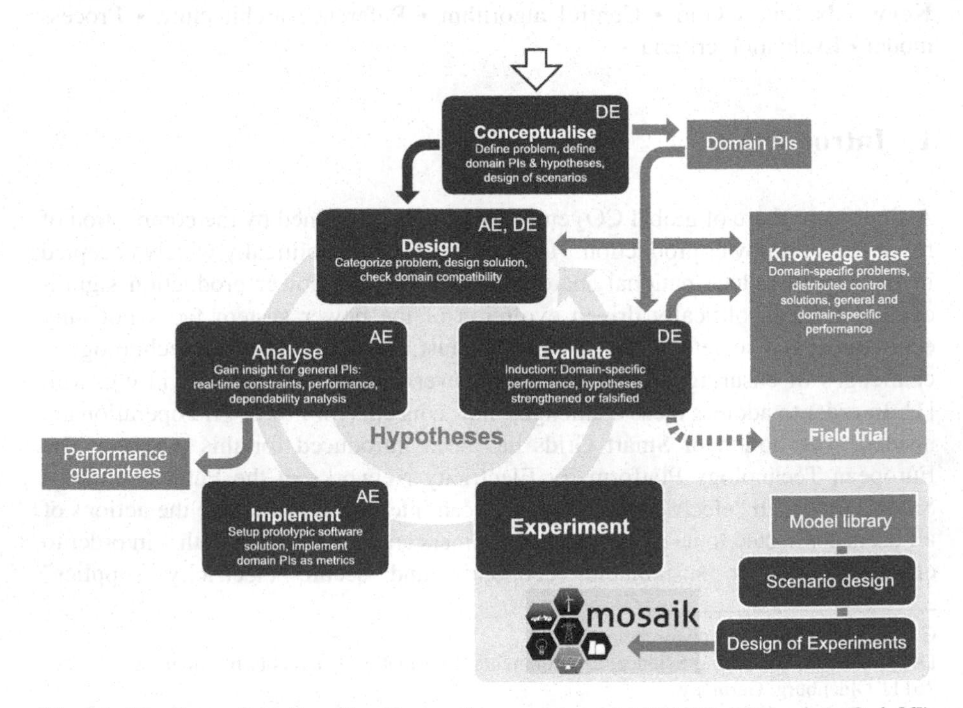

Fig. 8.1 The Smart Grid Algorithm Engineering process model (SGAE), (Taken from Ref. [6]) (Reproduced with permission)

engineering approach to develop validated algorithms for this specific domain. But as the latter is a cyclic process model that runs through the phases *Design*, *Analyze*, *Implement*, *Experiment* and *Evaluate* iteratively, the SGAM can easily be integrated in different stages of the SGAE (some of which have already been proposed in [6]).

In the contribution at hand, we demonstrate the utilization of this integrated process model for the development of a distributed control concept for the Smart Grid. Reflecting a full cycle of the SGAE, the remainder of this paper is structured as follows:

- The motivating business case "Trading Flexibility in the Smart Grid" along with a principal control concept is described in Sect. 2. This corresponds to the initial *Conceptualize* phase in the SGAE.
- For a specific use case in the motivating business case, the *Design* of a concrete algorithm as well as a suitable architecture in compliance with international standards is demonstrated by employing the SGAM framework in Sect. 3. This allows for checking the compatibility of the use case and the included control concept to the Smart Grid domain.
- In Sect. 4, the *Analysis* and *Evaluation* of the control concept is considered from a more general perspective. While these phases are usually treated separately, we take a combined approach here in order to present a novel taxonomy of evaluation criteria.
- A *Dependability Analysis* of the designed algorithm is undertaken in order to derive performance guarantees. This is done by deriving intrinsic properties of the approach formally. A brief overview on this part is given in Sect. 5.
- Section 6 discusses empirical methods for the evaluation of Smart Grid control concepts. Finally, an example of the control algorithm developed in this contribution is given.

As this contribution is an extended and revised version of [8], Sects. 4 and 6 reflect the main concepts from [8], while the remaining parts present novel content.

2 Conceptualize: Trading Flexibility as Business Case

The traditional power supply system can be seen as a centralized system. It consists of only a small number of controllable power plants. A control center acts as a central component that knows the operational constraints of the plants and performs a scheduling of the plants' operations with respect to demand and weather forecasts as well as the grid status and possibly the market situation. However, as already indicated in the introduction, such a control paradigm is not suitable for future Smart Grids anymore. It is widely accepted that the power supply system of the future will be characterized by a distributed architecture comprising autonomous components with individual sub-objectives, see e.g. Ref. [9–12]. In order to orchestrate those components towards global stability and reliability of the system, appropriate control mechanisms are necessary.

With the introduction of the *Flexibility Concept* (and therein the notions of *Flexibility Providers* and *Flexibility Operators*), the CEN-CENELEC-ETSI Smart Grid Coordination Group (SG-CG) depicts a possible architecture for such a distributed architecture [13]. A Flexibility Provider is described as an entity offering flexibility in generation, load or storage of electrical power. In contrast to the traditional power supply system, these entities can be of very small scale (e.g. individual households or appliances). Following, they do not participate directly in energy markets, but are contracted with Flexibility Operators instead, which aggregate the flexibilities of many units and make use of them in the grid or in energy markets. Hence, the Flexibility Operators act as a coordination layer between the grid/market on the one hand, and the Flexibility Providers on the other hand. From a business perspective, Flexibility Operators can be seen as Energy Service Providers (ESP), offering services that enable the customers to trade their aggregated flexibilities in the market. Of course, besides providing the required technical infrastructure, the Flexibility Operators have to employ sophisticated business logic for this task, i.e. suitable aggregation and optimization methods.

As an example of such a business logic, we refer to the concept of Dynamic Virtual Power Plants (DVPPs), which was initially introduced in [14] and has been reformulated in [15]. The concept is characterized by aggregating decentralized power producers, local storage systems and controllable loads in a self-organized way with respect to concrete products in an energy market. This way, multiple coalitions of Flexibility Providers are formed, where each coalition offers an individual power product to the market. After delivering a product, a coalition dissolves and the former participating units can then self-determinedly join the formation process of other coalitions for subsequent tradable energy products. In particular, the DVPP concept comprises the following subprocesses:

1. *DVPP setup:* Flexibility Providers are aggregated to DVPPs by coalition formation, such that the members of each DVPP agree upon trading a specific power product in the market (e.g. a certain block product in an electricity spot market). Bids for these products are then placed in the market.
2. *Predictive scheduling:* After a successful bid, a DVPP is obliged to deliver the power product. For this, the members of the DVPP have to be scheduled within their individually defined flexibility. This is done prior to the actual delivery of the product in a predictive scheduling process.
3. *Continuous scheduling:* To compensate for unforeseen changes or forecast errors, a continuous scheduling is performed during the delivery of the product. Here, the units' schedules are adapted such that product delivery is not endangered.
4. *Payoff division:* Subsequently, the revenues gained from product delivery are distributed among the DVPP members.

Referring to the Flexibility Concept of the SG-CG above, an ESP in the DVPP concept provides the technical infrastructure as well as management functions for the formation and operation of DVPPs, and thus realizes the role of a Flexibility

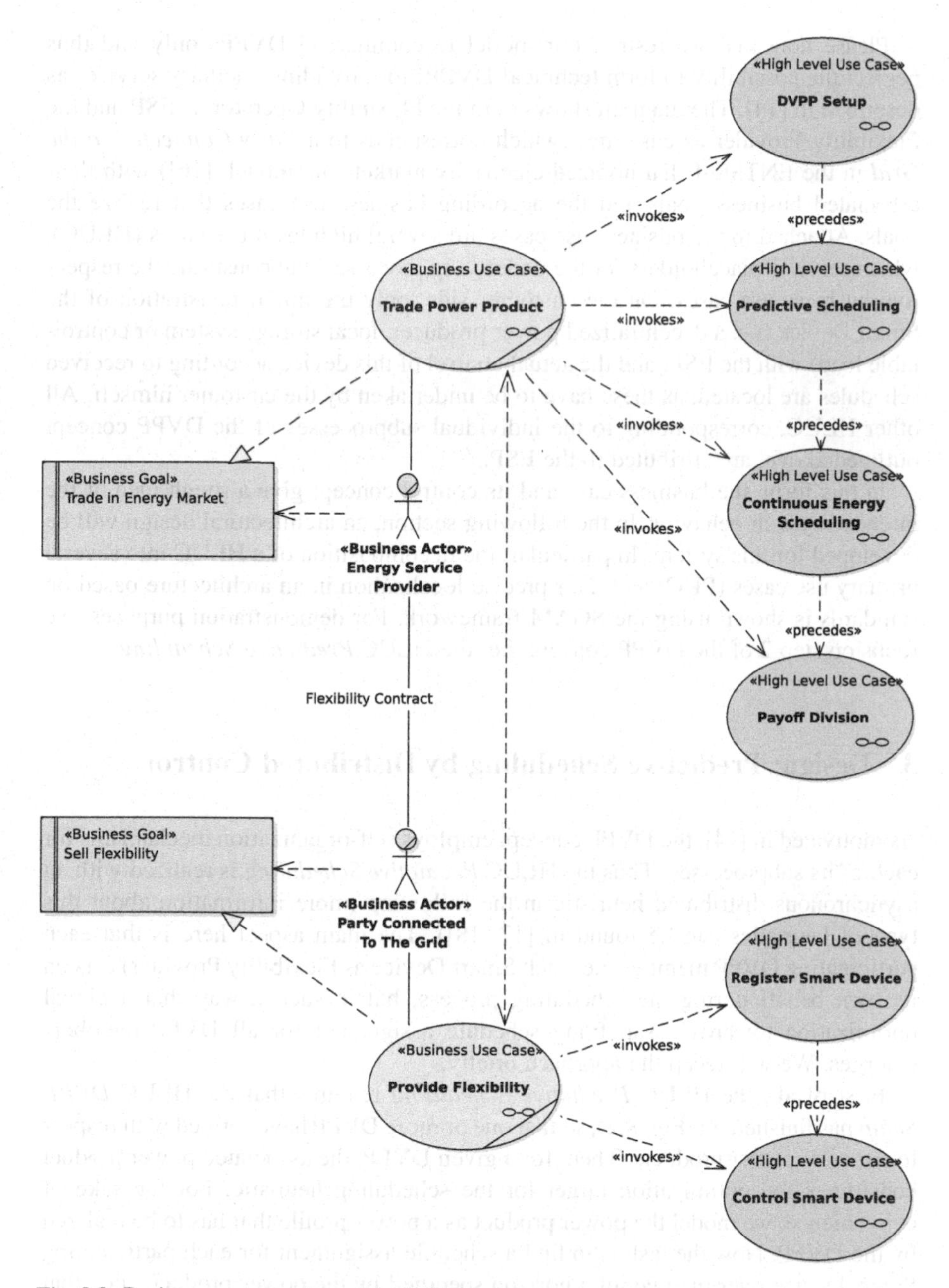

Fig. 8.2 Business case overview of the DVPP concept

Operator. A possible model of this business case is shown in Fig. 8.2 using the Unified Modeling Language (UML) notation.[1]

[1] http://www.omg.org/spec/UML

Please note that we restrict our model to commercial DVPPs only and thus neglect the possibility to form technical DVPPs for providing ancillary services as described in [14]. The diagram shows both the Flexibility Operator as ESP and the Flexibility Provider as customer (which corresponds to a *Party Connected to the Grid* in the ENTSO-E harmonized electricity market role model, [16]) with their associated business goals and the according business use cases that realize the goals. Attached to the business use cases are several high level use cases (HLUC), which serve as placeholders for the underlying processes that constitute the respective business use cases. On the customer side, only the initial registration of the Smart Device (i.e. a decentralized power producer, local storage system or controllable load) with the ESP, and the actual control of this device according to received schedules are located, as these have to be undertaken by the customer himself. All other HLUC, corresponding to the individual subprocesses of the DVPP concept outlined above, are attributed to the ESP.

In this form, the business case and its control concept give a rough idea of the intended system behavior. In the following section, an architectural design will be developed for this system. In particular, the decomposition of a HLUC into several primary use cases (PUC) and their precise localization in an architecture based on standards is shown using the SGAM framework. For demonstration purposes, we focus on step 2 of the DVPP concept, i.e. the HLUC *Predictive Scheduling*.

3 Design: Predictive Scheduling by Distributed Control

As motivated in [14], the DVPP concept employs self-organization mechanisms for each of its subprocesses. Thus the HLUC *Predictive Scheduling* is realized with an asynchronous distributed heuristic in the following (more information about this type of heuristics can be found in [17, 18]). The main aspect here is that each participating DVPP member (i.e. each Smart Device as Flexibility Provider) acts on its own behalf during the scheduling process, but in such a way that a global optimization towards a satisfying schedule assignment for all DVPP members emerges. We will recap the approach briefly.

First of all, the HLUC *Predictive Scheduling* requires that the HLUC *DVPP Setup* has finished (cf. Fig. 8.2), so that one or more DVPP have formed with respect to concrete power products. Then, for a given DVPP, the associated power product constitutes the optimization target for the scheduling heuristic. For the sake of convenience, we model the power product as a power profile that has to be realized by the DVPP. Now the task is to find a schedule assignment for each participating Smart Device over the planning horizon specified by the power product, such that the aggregation of all selected schedules matches the target power profile as close as possible. When each Smart Device then executes its assigned schedule later on, the DVPP effectively delivers the agreed power product.

This optimization task can be formulated as a distributed combinatorial optimization problem, for which a suitable heuristic has been proposed in [19]: the

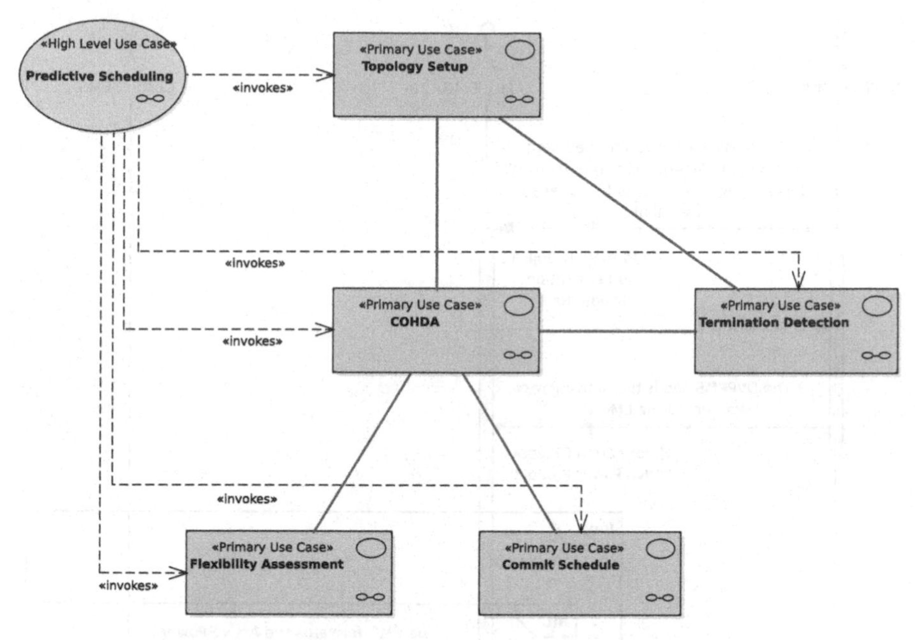

Fig. 8.3 Decomposition of the HLUC predictive scheduling into several PUCs

Combinatorial Optimization Heuristic for Distributed Agents (COHDA). Operating on a virtual communication topology, the heuristic performs the scheduling task in a fully distributed way. This is done by letting the individual Flexibility Providers coordinate based on a few simple behavioral rules for each participant, which are triggered by message exchanges between them. In Fig. 8.3, the execution of the heuristic is shown in the context of the HLUC *Predictive Scheduling*. Here, the PUC *COHDA* is surrounded by other, supportive PUCs. All PUCs are invoked by the superordinate HLUC, while the thick lines denote functional interrelations. More precisely, as COHDA relies on a virtual communication topology, the antecedent PUC *Topology Setup* has to be carried out, such that a suitable communication topology is created for the respective DVPP.

On the other hand, each participating DVPP member, i.e. each Flexibility Provider, needs a collection of feasible schedules to choose from during the scheduling process. Hence, also the PUC *Flexibility Assessment* is carried out prior to COHDA, in which a Smart Device determines its own flexibility for the planning horizon. Because the heuristic operates in a fully distributed way, the PUC *Termination Detection* might be invoked at arbitrary points in time during the operation of COHDA in order to check for (and eventually announce) the termination of the heuristic. After termination, the resulting schedules are committed to the respective Smart Devices (PUC *Commit Schedule*).

While this abstraction layer allows modeling of the interrelations of the PUCs, the concrete localization of the business logic to actual devices needs further refinement. Thus, for each PUC, the participating actors and their information exchange is identified using scenario specifications. For the PUC *COHDA*, such a

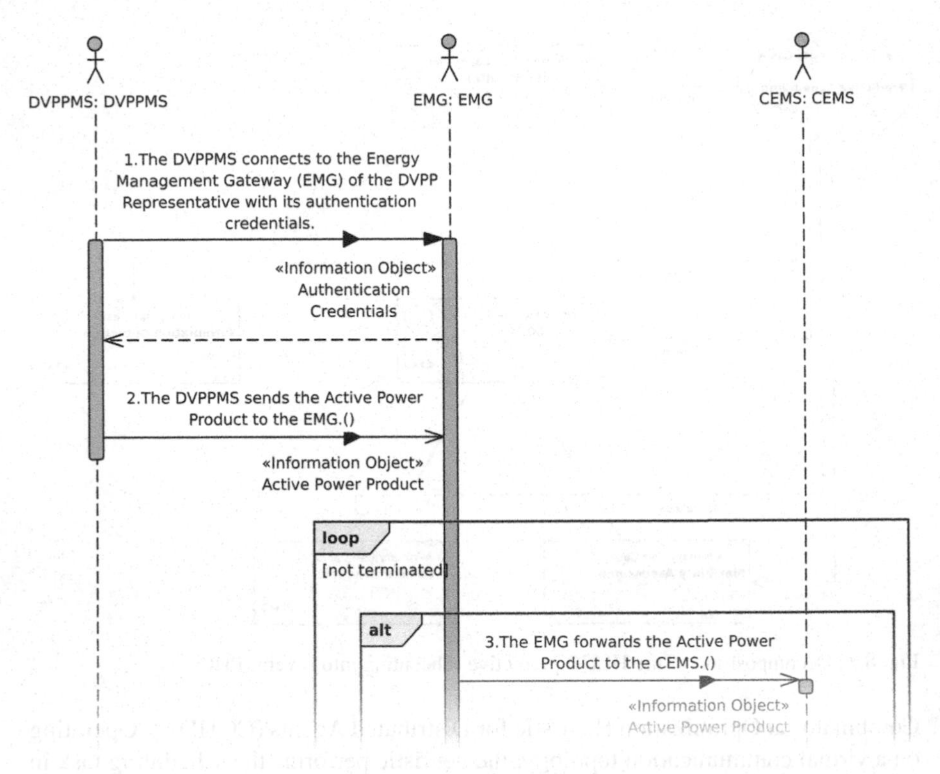

Fig. 8.4 Beginning of a sequence diagram for the COHDA scheduling approach

specification is visualized with the beginning of an according sequence diagram in Fig. 8.4.

Therein logical actors have been tailored from related generic use cases (cf. Ref. [13]) for the HLUC *Predictive Scheduling*:

- The DVPP Management System (DVPPMS) represents the ESP on a technical layer and constitutes a system that performs management operations regarding the complete lifecycle of DVPPs. This system also serves as a connection point to the energy market and thus embodies the role of a Flexibility Operator.
- The Energy Management Gateway (EMG) is an access point in the customer premises that communicates between external and internal systems.
- The Customer Energy Management System (CEMS) is responsible for gathering flexibilities in the customer premises as well as for performing optimization tasks regarding flexibility contracts.

The information flow between these actors is depicted in the figure using the standardized UML notation. An important aspect here is the attachment of *Information Objects* to the information flows, allowing modeling of the concrete data that has to be transferred between the actors.

More details on the depicted coordination steps can be found in [19, 20].

3.1 The SGAM Layers

With the modelling of the surrounding business case, its associated HLUCs, the involved PUCs as well as a detailed specification of the PUCs internal operations, resulting in the identification of involved actors and information objects, one may now begin to physically lay out the system in an architecture based on standards. The SGAM methodology defines five interoperability layers for this task [6]:

- The *Business Layer* presents a view to model interrelations regarding business, regulatory and market aspects.
- The *Function Layer* reflects the interrelations between functions and services according to use cases from a surrounding business case.
- The *Information Layer* formally describes the exchanged data between functions/components in terms of standardized information objects.
- The *Communication Layer* defines concrete protocols and mechanisms for the data exchange between physical components according to the identified information flows.
- The *Component Layer* maps logical actors of the above layers to physical components in the Smart Grid context.

Each abstraction layer spans the *Smart Grid Plane*, which allows to localize entities with respect to both electrical process and information management viewpoints. The former viewpoint is subdivided into several physical *domains* (e.g. Generation, Transmission, Distribution ...), while the latter viewpoint comprises a number of hierarchical zones (e.g. Market, Enterprise, Operation ...).

For convenience, we omit the detailed modelling of each layer for our motivating business case here, and provide a visualization of the resulting architecture in Fig. 8.5. The figure shows all interoperability layers of the SGAM framework in a stacked view. As the graphical complexity of the figure is rather high, the complete documentation of this design has been made available online.[2] The online version allows exploring the architecture interactively. The user may focus on specific layers or elements, and can access the underlying diagrams such as sequence diagrams, actor mappings etc.

In Fig. 8.5, at the business layer on top, the surrounding business case is depicted with its included HLUCs. The rest of the figure focuses on the HLUC *Predictive Scheduling* only: Obviously, this HLUC spans the from the Process zone to the Operation zone in the information management viewpoint, while being located mainly in the Distribution domain and the Customer Premises domain across the layers. Moreover, the location of the PUCs in the function layer reveals a concentration of the business logic in the Station zone under the Customer Premises domain. Accordingly, suitable components have been identified in the component layer at the bottom of the figure, which serve as hosts for the software applications that realize our business logic. Finally, the information layer and the

[2] http://www.uni-oldenburg.de/en/ui/research/topics/cohda/

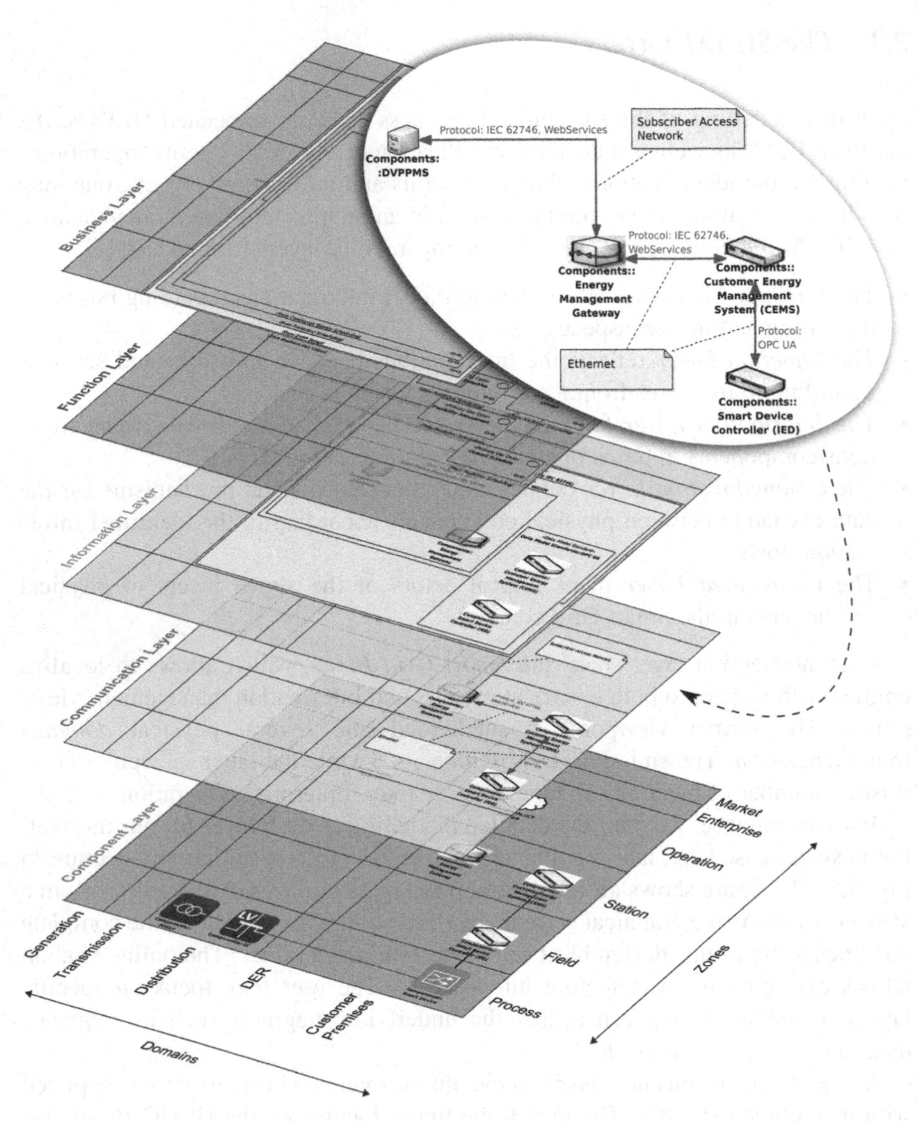

Fig. 8.5 Architecture model of the HLUC *Predictive Scheduling*, showing all interoperability layers of the SGAM framework in a stacked view. For demonstration purposes, the communication layer is shown as an extracted and magnified image without background coloring above the 3D graphic

communication layer present a mapping of international standards to the data flow between the components (see the magnified part in the graphic).

In summary, an architecture based on standards could be modelled for the HLUC *Predictive Scheduling* without major incompatibilities. Thus, despite its novel approach of a fully distributed control concept, it can be regarded as compatible to the Smart Grid application domain.

4 Analysis and Evaluation from a General Perspective

As already stated in Sect. 2, one of the main characteristics of the motivating DVPP business case is the distributed nature of its underlying algorithms. From a general perspective, a distributed algorithm or heuristic defines what, when and with whom to communicate, and what to do with received information, in order to efficiently solve a given problem or task in a distributed manner. Depending on the communication structure, an approach can further be classified as decentralized, hierarchical, distributed or fully distributed [15]. Moreover, we may distinguish synchronous from asynchronous approaches [17]. Especially in the latter, communication irregularities can have a severe impact on the overall progress, because they may change the order of actions in the system that exert influence on each other [18]. So besides performance and efficiency in terms of e.g. solution quality, run-time or communication complexity, further criteria are necessary for a proper evaluation in a particular application context. These include e.g. robustness analyses and scalability predictions with regard to different problem-specific parameters. In order to facilitate a structured evaluation approach, this section introduces a taxonomy of evaluation criteria, followed by an overview of suitable evaluation methods.

Before presenting our taxonomy of evaluation criteria, we have to define a few terms. In compliance with the SGAE process model, we understand a scenario as a specific collection of Smart Grid components, which then constitute the actors that the heuristic under evaluation operates on. These components may be configured using a set of parameters. Then an instance of such a scenario is a parameter assignment for all components within the scenario. Finally, an experiment comprises one or more computational executions of a scenario instance.

With respect to their dimensionality, we classify evaluation criteria into zeroth-, first- and higher-order criteria, cf. Fig. 8.6.

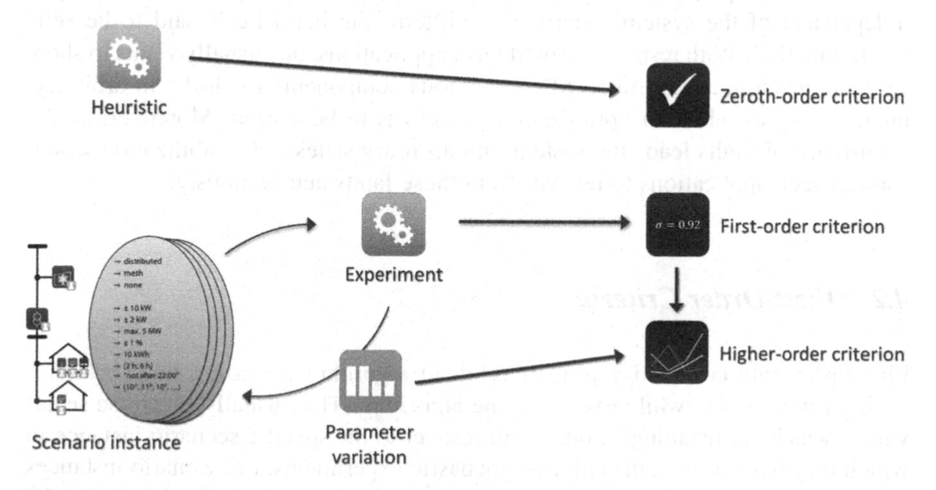

Fig. 8.6 Evaluation criteria types in the context of experimentation

In this context, a zeroth-order criterion yields a basic decision, i.e. a yes-no answer, which should generally be independent of any scenario configuration. On the other hand, a first-order criterion provides a scalar quantity, which is usually the outcome of an experiment, i.e. the interpretation of experimental data from a scenario instance. For a stochastic model, this experiment may consist of several replications with varying uncontrollable external effects, yielding a distribution instead of a single value.

Finally, higher-order criteria allow quantifying effects that occur due to interdependencies between different scenario instances and their first-order criteria, yielding higher-order quantities such as vectors or matrices as output values. For this, series of experiments are necessary, in which one or more dependent scenario parameters are strategically varied. We will now describe these types of criteria in more detail.

4.1 Zeroth-Order Criteria

One of the most basic aspects to consider when dealing with approaches that are targeted at the implementation in critical infrastructures is their *correctness* [21], which corresponds to a zeroth-order criterion in our taxonomy. In the SGAE process model, such criteria are examined in the "Analyze" phase. From a general point of view, the following distinctions can be made.

First of all, showing correctness involves asserting that if the approach yields a solution, then this solution will satisfy a given specification, e.g. it is a valid solution for the given problem (partial correctness). An additional requirement is its termination, i.e. asserting that the algorithm terminates within a finite amount of time after it has been started (total correctness). In the field of distributed heuristics, this is also known as guaranteed convergence. Moreover, if this behavior additionally is independent of the system's starting conditions, the heuristic is said to be self-stabilizing [22]. With respect to Smart Grid applications, one usually wants to show self-stabilization, as the involved autonomous components might be in arbitrary, unknown states when an optimization process is to be started. Moreover, as the occurrence of faults leads the system into arbitrary states, self-stabilization would support such applications to recover from these faults autonomously.

4.2 First-Order Criteria

First-order criteria quantify properties of an algorithm regarding their behavior during run-time, i.e. while executing the algorithm. They usually comprise scalar values which are meaningful only with respect to the specific scenario instance in which they have been collected. For stochastic experiments, i.e. scenario instances that include uncontrollable varying parameters, several executions can be performed in order to yield a distribution of values for the first-order criterion of interest rather than an insignificant single value (cf. Sect. 6.1).

The probably most evaluated criterion in this category is *performance*. The performance of an approach describes a quantification of its ability to achieve its goal [23]. Typically, this is measured in terms of solution quality, e.g. a fitness value that is calculated using an objective function. Here it is important to maintain a defined frame of reference, such that the measured value can be interpreted properly. For example, an adequate approach would be to determine the theoretically best and the theoretically worst solution for a given optimization problem as upper and lower bounds, and to normalize the fitness value to the interval that is spanned by these bounds. Apart from such general measurements, Smart Grid specific performance indicators play an important role to assess the performance of a heuristic in this field. A structured categorization of such performance indicators is yet to be defined and will be subject to future work.

Besides performance, the *efficiency* of an approach is of interest, which describes the resource requirements of an algorithm or heuristic [23]. Regarding centralized approaches, this is usually measured in terms of run-time, e.g. the amount of "steps" an algorithm takes for a given input, and memory, e.g. the amount of storage capacity an algorithm consumes while processing its input. For distributed approaches, determining the efficiency is more complicated: Regarding run-time, we have to distinguish the amount of time until the whole system terminates from the amount of "steps" the individual system components will take to reach this state. The former can be measured easily by means of real time, and will be an important information regarding the speed of the system in a specific hardware environment. The latter, however, is a more general measure as it determines the amount of work a system has to carry out. In this regard, a common practice is to count the number of calls to the objective function of the optimization problem, in each distributed component respectively. This way, both the individual work of the components as well as the overall effort can be determined in a hardware-independent manner. Finally, an additional evaluation criterion for distributed systems regarding the efficiency are communication expenses. In our motivating use case for instance, we are focusing on autonomous distributed components, which leads to a message-passing paradigm (in contrast to a shared-memory model, in which multiple components possess a common working memory, cf. Ref. [17]). Following, both the amount of exchanged messages as well as the size of these messages are significant factors for determining the efficiency of such approaches.

4.3 Higher-Order Criteria

In this category, first-order criteria are evaluated against varying input parameters, in order to quantify correlation effects, or to perform a sensitivity analysis. In contrast to repeating a single scenario instance due to stochastically, i.e. uncontrollably varying input parameters (as described for first-order criteria in the previous section), higher-order criteria are obtained by strategic variation of input parameters, such that the first-order criterion of interest is evaluated with respect to changing scenario instances, thus describing trends or gradients of first-order criteria under varying conditions (see Fig. 8.6).

A prominent higher-order criterion is the *scalability* of an approach [24]. Here, the influence of a change in magnitude of input parameters on one or more relevant first-order criteria is determined. For example, given a centralized heuristic for calculating the schedule of energy resources for a future time horizon with respect to e.g. demand predictions, one could study the effects of the length of the considered planning horizon on the run-time of the heuristic. An example regarding distributed heuristics is the influence of the amount of autonomous components that are present in the system on communication expenses.

Another important higher-order criterion is *robustness* [24], which determines the influence of incidental disturbances from the environment on one or more first-order criteria. Such disturbances could be either "dynamic" incidents at run-time like e.g. varying message delays during the execution of a distributed system, or "static" perturbations that determine the sensitivity to changing starting conditions.

It is natural that higher-order criteria are rather difficult to analyze as they include lower-order criteria in different magnitudes. On the other hand, they are especially important when targeting critical infrastructures such as the power supply system.

5 Analyze: Deriving Performance Guarantees

In the SGAE process model, the goal of the *Analyze* phase is to formally prove specific properties of the designed algorithm (i.e. the predictive scheduling in our example). This corresponds to the correctness proof of zeroth-order criteria or sometimes first-order criteria for the algorithm under development.

5.1 Methodical Aspects

In an analytical approach, evaluation criteria are quantified by mathematical calculus, i.e. inspecting the inherent design of an algorithm formally. For this, the semantic of the algorithm has to be described rigorously. An overview in this regard is given in [25]. A popular example here is the I/O automata formalization [17], which explicitly models the behavior of different components of a system through a standardized interface and thus allows for reasoning about the system's progress as a whole. Based on this, well-known proof techniques like e.g. variant functions or convergence stairs can be easily applied [22], as demonstrated below. Another approach would be to employ automatic model checkers. Due to the numerous different semantic descriptions and methods that are available in this field, we refer to [26] for an introduction.

The methods quoted above are particularly useful for zeroth-order evaluation criteria, e.g. for deriving convergence and termination properties in the *Analyze* phase of the SGAE process model. But recently, this has been adapted to first-order criteria as well. For example, in the context of self-organizing systems, [27] proposes quantitative definitions of the first-order criteria adaptivity, target orientation, homogeneity and resilience. These are based on an operational semantic in principle, which has been extended by stochastic automatons though. This allows for modeling the system's behavior not only in extreme cases (i.e. the best and worst cases as in the evaluation of zeroth-order criteria), but also in the average case, which is crucial for quantifying first-order criteria. The deduced average case behavior, however, directly depends on the chosen distribution functions for the stochastic parts of the model. As a consequence, special care must be taken in order to properly reflect the real behavior of the modeled system when employing such a method. Hence, if adequate distribution functions for a given system cannot be derived easily, an empirical study might be more appropriate in these cases. This approach is described in the following section.

5.2 *Application to the COHDA Algorithm*

The selection of properties to examine during the *Analyze* phase depends on the actual business case, but usually focuses on hard constraints like e.g. real-time requirements. For the HLUC *Predictive Scheduling*, an exemplary set of properties comprises:

- the *termination* of the scheduling algorithm at a point in time *before* the start of the delivery phase of the power product, and
- the termination in a *consistent state*, implying the calculation of a *valid solution* for the given problem, i.e. the assignment of a feasible schedule to each member of the DVPP.

In the present case, these properties can be proven for COHDA using the *Convergence Stairs* method [22]. We will sketch the proof briefly. First, the COHDA algorithm is formally described in the style of the I/O automata framework [17], which allows to reason about a system comprising interacting components. On this basis, a number of predicates (i.e. the convergence stairs) are formulated in such a way that each subsequent predicate implies its predecessor, while the last predicate altogether realizes the above defined properties. For COHDA, three predicates are needed: The first predicate regards the production of an initial valid solution after starting the heuristic, and thus covers the initial setup phase. The second predicate then considers the series of calculated solutions until a point in time, in which no more solutions are found, such that a unique final solution is calculated by some component in the system (recall that COHDA is a distributed approach in which solutions are calculated asynchronously by autonomous components in parallel). Finally, the third predicate ensures that this final solution will eventually be communicated to all components in the system, and that the system terminates in the resulting consistent state. By proving that each of these predicates completes in a finite amount of time, the second property (i.e. the termination in a consistent state) is derived. For the first property, however, the system architecture from the *Design* phase has to be considered. First, some hardware requirements are imposed on the physical components of the system, such that the first predicate will provably complete in an appropriately short amount of time (e.g. by requiring a minimal computation power of the CEMS and a communication backend which must deliver message in at least x seconds, where x depends on the remaining time until the product has to be delivered). This suffices to guarantee that the approach yields at least one valid solution for the problem in the given time.

Please note that the actual optimization towards satisfying solutions with regard to product fulfillment is not considered in the formal analysis. This is because the employed algorithm is a heuristic approach and cannot inherently guarantee any solution quality. Hence, such soft constraints are subject to the *Evaluation* phase later on. However, while examining the above termination properties, it was additionally proven that the approach exhibits the *anytime* property in the following sense: Whenever a component calculates and publishes a solution for the scheduling problem, this solution will be better than the previous solution the component has been aware of. From a global point of view, the heuristic thus produces better and better solutions (i.e. schedule assignments for the Smart Devices) over time, until no more improvement is possible. In combination with the ability to manually initiate a consistent termination of the process at any desired point in time, this property makes COHDA a highly dependable approach in the context of Smart Grid applications.

A detailed description of the full proof will be published in a subsequent paper.

6 Evaluation

Most first-order criteria and higher-order criteria have to be evaluated by empirical methods. In contrast to formal reasoning based on a rigorous semantic description of an algorithm, empirical methods are based on actually executing the algorithm, i.e. the heuristic in the scope of this paper, within a dedicated environment. From monitoring such executions, quantitative data can be recorded, whose dissection and interpretation then leads to the valuation of first- and higher-order criteria.

6.1 Methodical Aspects

Evaluation by empirical methods involves a number of subsequent steps: As a single execution of an algorithm usually does not yield enough information to deduce general conclusions about the behavior of the system in the average case, an adequate experiment design has to be defined in the first step. For first-order criteria, this includes tactical decisions, such as the number of repetitions of the executions, in order to level out random effects from uncertain environments or uncontrollable parameters. This will increase the confidence level of the deduced insights later on. For higher-order evaluation criteria, additional strategic decisions have to be made, such as defining a strategy for the intentional variation of input parameters in order to analyze the heuristic's behavior under varying conditions, cf. the *Design of Experiments* step in the SGAE process model [6]. A comprehensive overview on these topics from the perspective of simulation experiments can be found in [28]. In the context of heuristics, further care has to be taken regarding the type of scenario instance that is to be solved by a heuristic in a series of experiments [29]. While parts of this, like e.g. the magnitude of input parameters, are usually already covered in the described tactical and strategic decisions, the inherent type of an underlying problem instance might be of interest as well. On the one hand, synthetically crafted problem instances can be used. These do not reflect the targeted application field, but are constructed in such a way that specific properties are present in the problem to solve. For example, "deceptive" problem instances [30] are useful to analyze whether a given heuristic is able to overcome local optima in the search space. This way, a deep understanding of the observed effects can be gained. On the other hand, application-specific problem instances aim at reflecting the target application of an approach as close as possible, such that the system's behavior can be observed directly in its presumed environment.

In the second step, the experiment is actually carried out. This is usually done by means of simulation. Regarding our focus on heuristic approaches for Smart Grid applications in this contribution, the simulation model then comprises both the heuristic under evaluation and the environment this heuristic is executed in. Following, it is of utmost importance to build the model as realistic as needed, i.e. such that all relevant interdependencies between the (simulated) environment and the heuristic are incorporated into the model. For example, if a given distributed heuristic is said to be asynchronous based on message passing between components, possible flaws from the underlying communication technology such as message delays or buffer overflows should be anticipated. The other way around, if the outcome of a heuristic affects e.g. the power flow in an electricity grid, and the resulting effects are relevant for the evaluation, the grid must be modeled in such a way that those effects are properly accounted for. Again, [6] gives further suggestions regarding this topic. There, besides conceptual considerations, the modular Smart Grid simulation framework mosaic [31] is given as a tooling example in the SGAE process model. To permit even more realistic simulations, the framework can be coupled with hardware simulators such as the Smart Energy Simulation and Automation (SESA) lab [32].

Finally, in a third step, the preceding executions of the algorithm have to be analyzed with respect to the criteria of interest. Especially for higher-order criteria, specific metrics and suitable statistical methods can then be applied, in order to draw conclusions from the possibly vast amounts of recorded data. Examples for methods and metrics regarding various evaluation criteria can be found in [24, 28, 29].

6.2 Application to the COHDA Algorithm

With respect to the business case "Trading Flexibility", and the COHDA approach in particular, the results of an exemplary simulation scenario are presented in the following. Please note that this only serves for demonstration purposes as a proof-of-concept and by no means describes a complete empirical evaluation of the approach. For this, numerous simulation studies have already been conducted with respect to the above guidelines in the past. We refer the interested reader to [18, 20, 33] for details.

In the considered business case, the ESP acts as an intermediate between an energy market and the Flexibility Providers. The European Power Exchange (EPEX SPOT) is an example of a day-ahead spot market for active power in this sense. As a proof-of-concept of the COHDA approach, Fig. 8.7 shows the scheduling results for a simulated DVPP comprising both flexible loads and controllable small scale generators. In particular, the DVPP reflects the situation at a typical medium-voltage node in the German power grid. Thus, according to averaged registry data from four large German transmission network operators, it contains 111 geothermal heat pumps with $P_{el} \approx -2\,\text{kW}$, 4 combined heat and power plants (CHP) with $P_{el} \approx 1\,\text{kW}$ and 8 CHP with $P_{el} \approx 5\,\text{kW}$ as Smart Devices (load is depicted as negative power). For these units, we utilized simulation models for the appliances *Stiebel-Eltron WPF 10*, *Vaillant EcoPower 1.0* and *Vaillant EcoPower 4.7*, respectively. As scheduling target, the block product *Peakload* covering the hours 9–20 of a trading day was chosen from the list of standardized block products of the EPEX SPOT.[3] Reflecting the rather small net power of the DVPP, this target was set to the smallest possible magnitude of $-100\,\text{kW}$ according to the present market rules.

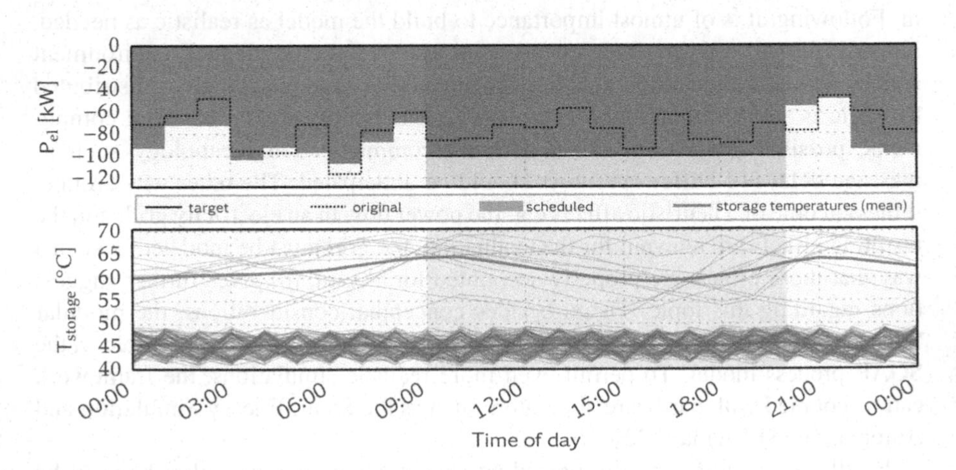

Fig. 8.7 Results of the simulation scenario "EPEX SPOT Peakload"

[3] http://www.epexspot.com/en/product-info/auction

In the figure, this target is depicted in the upper chart as a solid line. The relevant time span for this product is illustrated by a slight shading of the background.

Hence, the goal for the DVPP was to produce a constant negative amount of $-100\,\text{kW}$ of power during that time span, while being allowed to operate arbitrarily in the remaining time. The simulation covered the full trading day, i.e. 24 h. We excluded stochastic effects like e.g. uncontrollable varying thermal demand from this demonstration, thus only a single simulation run is presented. Moreover, because the HLUC *Flexibility Assessment* is not in the scope of this paper, each Smart Device was equipped with a rather simple type of flexibility representation comprising 200 randomly sampled feasible schedules for the respective device (cf. Sect. 3 and Ref. [33]). But despite this quite limited search space, the heuristic was able to find a schedule assignment that fulfills the target almost perfectly. This is visualized as aggregate power profile of the DVPP by the filled area in the upper plot. For reference, also the uncontrolled profile is shown as dotted line, which was calculated in an additional simulation run without executing the COHDA heuristic. With respect to the presented taxonomy of evaluation criteria, this corresponds to the first-order criterion *performance*: Interpreted as percentage of the target, the uncontrolled profile would in summary realize 77.35 % of the target, whereas the profile resulting from the scheduling reaches a coverage of 99.63 %. In the lower plot, the temperature trajectories of the attached hot water tanks are visible. These show that the allowed temperature ranges of the hot water tanks (between 40 °C and 50 °C for the heat pumps and 50 °C and 70 °C for the CHPs in our configuration) are not violated by the scheduling actions.

7 Conclusion

In combination with the SGAM framework, the SGAE process model provides a foundation for the structured development of Smart Grid applications. We demonstrated the utilization of the integrated process model by means of the DVPP business case, which relies on distributed algorithms and thus serves as a representative for the currently observable trend to flexible, adaptive, self-organizing control concepts in the Smart Grid domain. Within this demonstration, we introduced a taxonomy of evaluation criteria, followed by an overview of methods for valuating these criteria, in order to facilitate a structured evaluation approach in the targeted application domain. An exemplary evaluation scenario demonstrated the feasibility of the approach.

Future work in this context will be to define domain-specific performance indicators, such that Smart Grid application algorithms, especially distributed control approaches, can be developed and evaluated using standardized and accepted criteria from the problem domain.

References

1. H.-J. Appelrath, H. Kagermann, and C. Mayer, Future Energy Grid. Berlin, Heidelberg: Springer Berlin Heidelberg, 2012.
2. European Technology Platform for Electricity Networks of the Future, "Strategic Deployment Document for Europe's Electricity Networks of the Future," 2010.
3. S. D. Ramchurn, P. Vytelingum, A. Rogers, and N. R. Jennings, "Putting the 'smarts' into the smart grid," Commun. ACM, vol. 55, no. 4, p. 86, Apr. 2012.
4. CEN, CENELEC, and ETSI, "Smart Grid Reference Architecture," 2012.
5. J. Trefke, S. Rohjans, M. Uslar, S. Lehnhoff, L. Nordstrom, and A. Saleem, "Smart Grid Architecture Model use case management in a large European Smart Grid project," in IEEE PES ISGT Europe 2013, 2013, no. 978, pp. 1–5.
6. A. Nieße, M. Tröschel, and M. Sonnenschein, "Designing dependable and sustainable Smart Grids – How to apply Algorithm Engineering to distributed control in power systems," Environ. Model. Softw., vol. 56, pp. 37–51, Jun. 2014.
7. R. Santodomingo, M. Uslar, A. Goring, M. Gottschalk, L. Nordstrom, A. Saleem, and M. Chenine, "SGAM-based methodology to analyse Smart Grid solutions in DISCERN European research project," in 2014 I.E. International Energy Conference (ENERGYCON), 2014.
8. C. Hinrichs and M. Sonnenschein, "Evaluation Guidelines for Asynchronous Distributed Heuristics in Smart Grid Applications," in Proceedings of the 28th Conference on Environmental Informatics – EnviroInfo 2014 – ICT for Energy Efficiency, 2014.
9. F. F. Wu, K. Moslehi, and A. Bose, "Power System Control Centers: Past, Present, and Future," Proc. IEEE, vol. 93, no. 11, pp. 1890–1908, Nov. 2005.
10. M. D. Ilić, "From Hierarchical to Open Access Electric Power Systems," Proc. IEEE, vol. 95, no. 5, pp. 1060–1084, May 2007.
11. S. D. J. McArthur, E. M. Davidson, V. M. Catterson, A. L. Dimeas, N. D. Hatziargyriou, F. Ponci, and T. Funabashi, "Multi-Agent Systems for Power Engineering Applications—Part I: Concepts, Approaches, and Technical Challenges," IEEE Trans. Power Syst., vol. 22, no. 4, pp. 1743–1752, Nov. 2007.
12. M. Uslar, M. Specht, C. Dänekas, J. Trefke, S. Rohjans, J. M. González, C. Rosinger, and R. Bleiker, Standardization in Smart Grids. Berlin, Heidelberg: Springer Berlin Heidelberg, 2013.
13. CEN, CENELEC, and ETSI, "Sustainable Processes," 2012.
14. A. Nieße, S. Lehnhoff, M. Troschel, M. Uslar, C. Wissing, H.-J. Appelrath, and M. Sonnenschein, "Market-based self-organized provision of active power and ancillary services: An agent-based approach for Smart Distribution Grids," in 2012 Complexity in Engineering (COMPENG). Proceedings, 2012, pp. 1–5.
15. M. Sonnenschein, C. Hinrichs, A. Nieße, and U. Vogel, "Supporting Renewable Power Supply through Distributed Coordination of Energy Resources," in ICT Innovations for Sustainability. Advances in Intelligent Systems and Computing 310, L. M. Hilty and B. Aebischer, Eds. Springer International Publishing, 2015, p. 474.
16. ENTSO-E, "The harmonised electricity market role model," 2014.
17. N. A. Lynch, Distributed Algorithms. Morgan Kaufmann Publishers Inc., 1996.
18. C. Hinrichs and M. Sonnenschein, "The Effects of Variation on Solving a Combinatorial Optimization Problem in Collaborative Multi-Agent Systems," in Multiagent System Technologies, 12th German Conference, MATES 2014, Stuttgart, Germany, September 23–25, 2014.
19. C. Hinrichs, S. Lehnhoff, and M. Sonnenschein, "A Decentralized Heuristic for Multiple-Choice Combinatorial Optimization Problems," in Operations Research Proceedings 2012, 2014, pp. 297–302.
20. C. Hinrichs, S. Lehnhoff, and M. Sonnenschein, "COHDA: A Combinatorial Optimization Heuristic for Distributed Agents," in Agents and Artificial Intelligence, Communications in

Computer and Information Science, vol. 449, J. Filipe and A. Fred, Eds. Berlin, Heidelberg: Springer Berlin Heidelberg, 2014, pp. 23–29.
21. K. R. Apt, "Ten Years of Hoare's Logic: A Survey—Part I," ACM Trans. Program. Lang. Syst., vol. 3, no. 4, pp. 431–483, Oct. 1981.
22. S. Dolev, Self-Stabilization. MIT Press, 2000, p. 207.
23. E.-G. Talbi, Metaheuristics. John Wiley & Sons, Inc., 2009.
24. R. S. Barr, B. L. Golden, J. P. Kelly, M. G. C. Resende, and W. R. Stewart, "Designing and reporting on computational experiments with heuristic methods," J. Heuristics, vol. 1, no. 1, pp. 9–32, Sep. 1995.
25. N. Francez, Program Verification. Addison-Wesley, 1992, p. 312.
26. M. Müller-Olm, D. A. Schmidt, and B. Steffen, "Model-Checking: A Tutorial Introduction," in Proceedings of the 6th International Symposium on Static Analysis, 1999, pp. 330–354.
27. R. Holzer and H. De Meer, "Quantitative modeling of self-organizing properties," in Self-Organizing Systems. Lecture Notes in Computer Science, 1st ed., vol. 5918, T. Spyropoulos and K. A. Hummel, Eds. Zurich, Switzerland: Springer Berlin Heidelberg, 2009, pp. 149–161.
28. J. P. C. Kleijnen, Design and Analysis of Simulation Experiments. Springer US, 2008.
29. E. Alba, Parallel Metaheuristics. Hoboken, NJ, USA: John Wiley & Sons, Inc., 2005.
30. D. E. Goldberg, Genetic Algorithms in Search, Optimization and Machine Learning, 1st ed. Boston, MA, USA: Addison-Wesley Longman Publishing Co., Inc., 1989.
31. S. Schütte and M. Sonnenschein, "Mosaik—Scalable Smart Grid scenario specification," in Proceedings of the 2012 Winter Simulation Conference (WSC), 2012, pp. 1–12.
32. M. Büscher, A. Claassen, M. Kube, S. Lehnhoff, K. Piech, S. Rohjans, S. Scherfke, C. Steinbrink, J. Velásquez, F. Tempez, and Y. Bouzid, "Integrated Smart Grid Simulations for Generic Automation Architectures with RT-LAB and Mosaik," in 5th IEEE International Conference on Smart Grid Communications, 2014.
33. C. Hinrichs, J. Bremer, and M. Sonnenschein, "Distributed hybrid constraint handling in large scale virtual power plants," in IEEE PES ISGT Europe 2013, 2013, pp. 1–5.

Chapter 9
Multi-actor Urban Energy Planning Support: Building Refurbishment & Building-Integrated Solar PV

Najd Ouhajjou, Wolfgang Loibl, Stefan Fenz, and A. Min Tjoa

Abstract Energy strategies are needed at a city level and consequently, adequate planning tools are required to support urban energy planners in assessing their decisions (e.g. which buildings are the best to refurbish). This paper presents an ontology based approach for urban energy planning support. The approach is applied to develop a planning support system for building-refurbishment and building-integrated Photovoltaics-based energy generation. The adopted methodology works as an iterative, incremental process, where each iteration leads to the integration of a new planning decision. The process starts by the identification of the actors whose interests are affected by the decision, then developing/re-using computation models that provide answers for their questions. Different models are integrated using an ontology that represents the parts of the city that are within the scope of the questions to be answered. The system is applied within a district in the city of Vienna covering around 1200 buildings. The adopted approach provides different actors with specific information related to their view point of decision making. Further, the output is aggregated to a common level of abstraction, to be understood by all involved actors. This approach is applicable to different cities, as the ontology also integrates extension and upgrade mechanisms that provide flexibility to cope with different available data collections.

Keywords Ontologies • Energy planning support • Decision support • Energy systems modelling

N. Ouhajjou (✉) • W. Loibl
AIT Austrian Institute of Technology, Vienna, Austria
e-mail: najd.ouhajjou@ait.ac.at; wolfgang.loibl@ait.ac.at

S. Fenz • A.M. Tjoa
Vienna University of Technology, Vienna, Austria
e-mail: stefan.fenz@tuwien.ac.at; amin@ifs.tuwien.ac.at

J. Marx Gómez et al. (eds.), *Advances and New Trends in Environmental and Energy Informatics*, Progress in IS, DOI 10.1007/978-3-319-23455-7_9

1 Introduction

Energy consumption in cities results in approximately 71 % of all energy-related direct greenhouse gas (GHG) emissions [1]. Thus cities represent a rich ground for taking action to reduce the amount of GHG emissions. Therefore, decision makers, namely city administrations, building owners and planners are developing energy strategies at a city-level. Such strategies clearly state what measures to be taken, where and in what quantities and in which time horizons. Building refurbishment (in terms of thermal insulation) and building-integrated solar photovoltaics (PV) systems represent some of the measures that constitute these strategies. Their choice in this work is motivated by the fact that they are subsidized by many governments across Europe (e.g. Austria [2]), representing an incentive for their real-world implementation (once they are planned). Furthermore, these two measures are complementary in terms of their contribution in improving energy efficiency, increasing the share of renewable energy supply, and decreasing the amount of CO_2 emissions.

However, developing energy strategies is challenging because of the difficulty of assessing their impact on cities. This is due to the complexity of cities, regarding the large amount and diversity of the components and interactions they comprise. Furthermore, data at a city level is not always available at the desired levels-of-detail. Therefore, to cope with the complexity of cities, adequate planning support systems are required to simplify (explain) this complexity and automate the interactions that cannot be handled manually by energy planners.

This paper (as a detailed extension to a previous publication in EnviroInfo Conference [3]) aims to demonstrate how to develop an urban energy planning support tool that can be used in workshops (to develop energy strategies), where different actors have different concerns (questions), regarding the impact of the resulting energy strategies. Such tools are necessary regarding the participative nature of the urban energy planning process, where different stakeholders have to make sure that their interests are positively affected. The approach is illustrated through the development of a multi-actor urban energy planning support tool that considers building-integrated PV and building refurbishment (interface preview in Figs. 9.1 and 9.2).

There exists a large variety of tools that address different aspects of energy planning with a variance of scopes and addressed users. Examples of such tools include: EnerGis [4] which aims to calculate the minimum annual heat demands of buildings within a geo-referenced context. It does not provide answers to stakeholders on the impact of the developed strategies. SUNtool [5] and its later successor CitySim [6] are more on the energy simulation side and attempt to model and simulate energy flows of buildings, considering the individual properties of each building. Thus, to perform simulations, these tools require data with a high level-of-detail, which does not always exist at a city level (e.g. as it has been the case in the city of Vienna). SynCity [7] is considered as a scenario development, simulation, and optimization tool, at city level. It focuses on urban energy systems and it attempts to discover where large reductions of energy intensity can be achieved within the city. This tool is adaptable to be used in different cities; however, it does not consider providing dedicated answers to different stakeholders

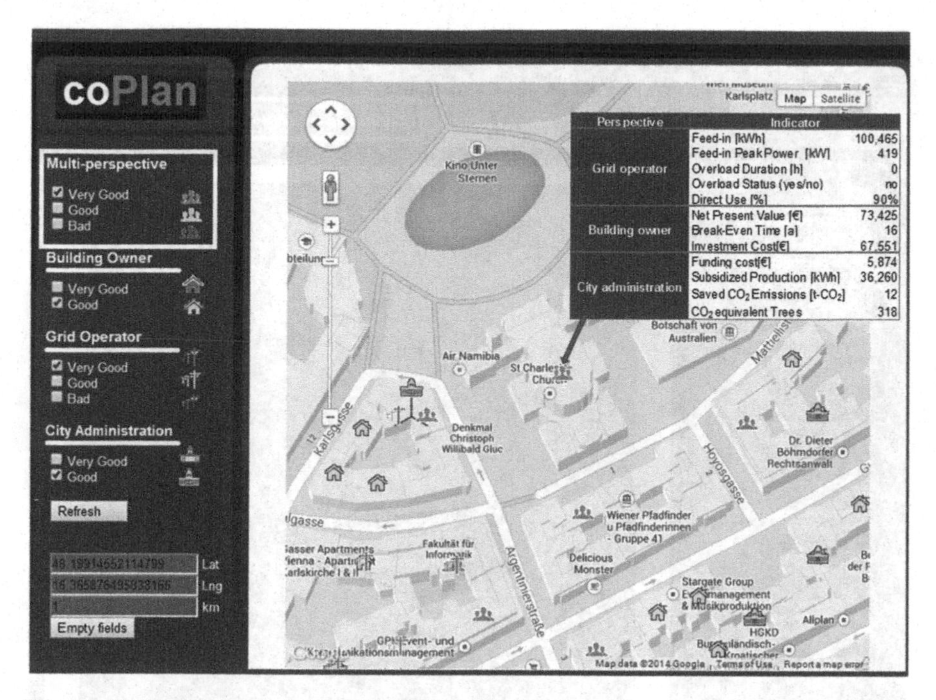

Fig. 9.1 Building-integrated solar PV planning interface

that take part in the planning process. UrbanSim [8] is an open source framework that allows constructing scenarios and simulations that can be used at a city scale. It has GIS interface and addresses not only energy in the city but also other aspects. This framework has a broader scope that addresses urban planning in general. Therefore it is rather dedicated to urban development, while the problem addressed in this article is more specific to supporting the development of energy strategies, considering their impact on involved stakeholders. CommunityViz [9] is a scenario development and decision support tool for land-use planning. As an extension of a GIS software, it has a mapping interface and offers simple wizards to create different spatially explicit scenarios of land-use as well as calculations of different user-defined indicators. This tool can be customized (more development is necessary) to integrate energy planning support in the global context of urban planning, however this is not always desired by energy planners that have a more specific scope and objectives.

The tools mentioned above fill specific gaps in urban energy planning; however, they do not address four specific problems that this paper solves. These problems, also considered as the objectives of this work (further explained in the next section) are as the following: (1) Supporting perspectives of different actors, (2) common understanding and quantifiable impact of decisions, (3) measure integration and resources negotiation, (4) system viability through robustness against data availability problems. Table 9.1 illustrates the problems (considered in this paper) that the described tools address.

Fig. 9.2 Implemented interface of the integrated measures planning support system

Table 9.1 Related tools fulfillment of problems

Tools	Addressed problems[a] 1	2	3	4
EnerGis		Partially	x	
SUNtool		Partially	x	
CitySim		Partially	x	
SynCity		Partially	x	x
UrbanSim	x	Partially	Partially	
CommunityViz	x	Partially	Partially	

[a](1) Supporting perspectives of different actors, (2) common understanding and quantifiable impact of decisions, (3) measure integration and resources negotiation, (4) robustness against data availability problems

2 Objectives

The objectives of this work (that have been achieved) are as the following:

- **Supporting the perspectives of different actors**: the decision making process must involve all the stakeholders that have potentially affected interests and provide them with specific information, from their different perspectives. For example regarding building-integrated PV, the building owner and the electric grid operator (as stakeholders) have different interests. While the building owner has financial concerns about the cost and profitability of the investment, the grid operator is more concerned about the stability of the grid. Thus, the impact of the potential implementation of the measure has to be assessed from different perspectives (of stakeholders).
- **Common understanding and quantifiable impact of decisions**: the assessment of the impact of energy strategies must be quantifiable. The output results must be aggregated to a level of abstraction that is understandable by all the different actors (i.e. stakeholders).
- **Measures integration and resources negotiation**: the assessment of the impact of energy strategies must consider the interdependencies between different components and calculations e.g. installing PV reduces the surface areas where solar thermal collectors can be installed.
- **System viability through robustness against data availability problems**: The system must be flexible to be used within different conditions of data availability and levels of detail.

These objectives represent the necessary conditions in urban energy planning support systems as described in previous related work [10]. They are mainly based on the analysis of an energy planning process and a data availability analysis in different cities.

The sustainable energy action plan (SEAP) process [11] is used as a reference process in urban energy planning, with more than 5000 cities and municipalities as users [12]. A data availability analysis was performed [10] in the cities of Vienna, Linz, Amstetten in Austria and Nanchang in China, in the context of Smart City projects [13], where data has been collected to develop energy strategies for the respective cities. The following conclusion applied: (i) the more detailed data is, the less available it becomes. (ii) The level-of-detail of available data significantly impacts the precision of the developed energy strategies. (iii) Data availability and level-of-detail of data are significantly different, varying from a city to another.

3 Methodology and Application

This section describes the methodology that has been adopted in the development of the multi-actor urban energy planning support tool. The description is complemented by a running use case (building-integrated PV) to illustrate its application. The use case considers a district (about 1200 buildings) in the city of Vienna.

The methodology is based on gradually developing and further extending an ontology, defined in this article both as an explicit specification of conceptualization (similar to a data model of a database) and as a body of knowledge (similar to a populated database) [14]. Thus, an ontology formalizes the relevant concepts and their interrelations regarding a given set of domains, then it integrates data and knowledge about these domains. The choice of ontologies in this work is motivated by their semantic richness and expressivity, which are required to model large and diverse systems such as cities. Furthermore, ontologies are flexible in terms of integrating different datasets that are not necessarily in the same format, or obey to the same data model.

The methodology is an iterative incremental process, as shown in Fig. 9.3, where each iteration starts in a scoping phase. In this specific work, this process has been run twice: once for considering PV planning, then another time for building refurbishment.

3.1 *Scoping Phase*

The actors (stakeholders) involved in the decision making process are identified. For each actor, a list of questions of interest is established followed by a breakdown of the questions to a set of expected quantifiable answers, as shown in Table 9.2.

3.2 *Data Availability Check*

Data availability check is performed to understand what datasets exist and what their levels of detail are. The data availability significantly impacts the entire process i.e. it is possible that some of the questions that are formulated in the scoping phase require answers that demand a high level-of-detail data.

In this work, data availability check has been conducted in a non-formal way. Initial information providers and sources have been identified. Expected data templates have been sent to the information providers, initiating a discussion about what data can be provided or what potential other providers can offer. What is considered as available data is what can be obtained. For example, data about electric transformers in the low voltage grids does exist but not accessible for security reasons. Other data regarding the energy demand of households exists but not accessible either, for privacy reasons. In general, data that cannot be obtained is considered (in this work) unavailable, including both the data that cannot be found or the one that cannot be accessed.

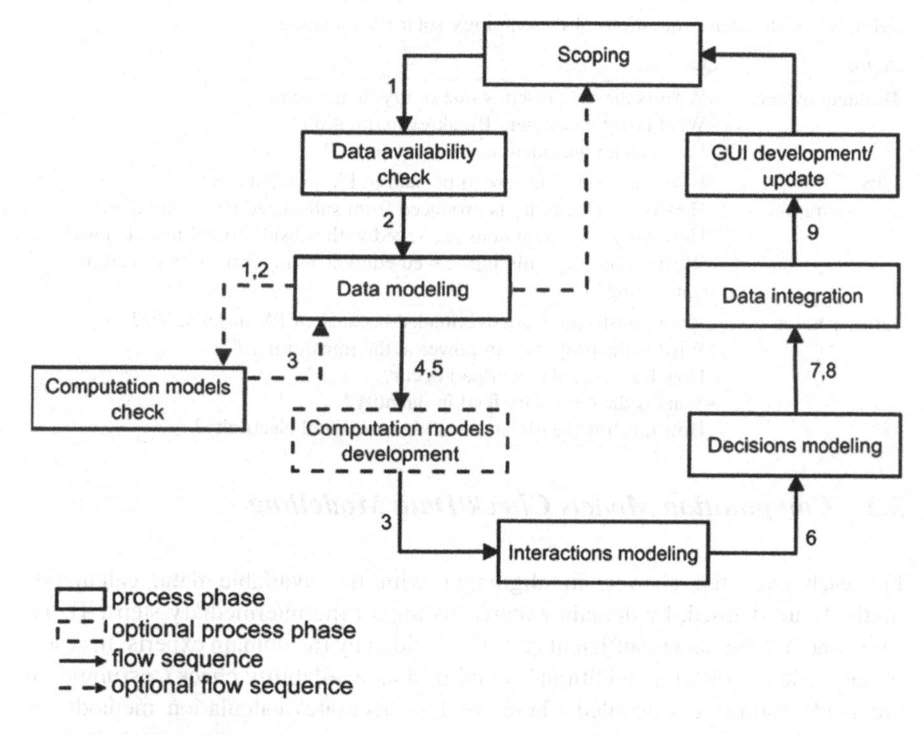

Fig. 9.3 Main phases of the development methodology

The main information provider in this case was the city of Vienna. An initial data collection has been performed which includes a solar cadaster of the city [15], building stock data, standard electricity demand profiles, demographic data, PV funding schemes, etc. In brief, the acquired data was not detailed enough or comprehensive to aim at simulations. Due to lacking detailed consumer data at building level assumptions and generalizations have been made. For instance, it has been assumed that all residential buildings have the same electricity demand profile. This assumption has been corrected by using a diversity factor.

Table 9.2 Competency questions of the ontology-solar PV planning

Actor	Question
Building owner	–What is the net present value of my investment? –What is my investment Break-even duration? –How much investment costs are required?
City administration	–How much subsidies are to be paid to PV installations? –How much electricity is produced from subsidized PV installation? –How much CO_2 emissions are saved with subsidized PV installations? –What is the CO_2-emissions-saved-equivalent in terms of trees carbon sequestering?
Grid operator	–What transformers are overloaded because of PV installations? –What is the peak feed-in power at the transformers? –How long does the overload occur? –What is the electricity feed in quantity? –How much is the direct use of the generated electricity?

3.3 Computation Models Check/Data Modelling

For each expected answer, in alignment with the available data, calculation methods are defined, by domain experts, listing all the intermediary steps. Therefore, whether the data is sufficient or not is decided by the domain experts. In case it is not sufficient (after an additional round of data availability check), assumptions are made and/or less detailed (therefore less accurate) calculation methods are developed.

Calculation methods are conditioned by the data availability. Therefore, calculating one answer could be performed according to more than one calculation method. Thus, it is possible to answer one question in different levels-of-detail. This multiple level-of-detail answering mechanism is useful when data availability is different from one city to another, or even from one place to another within the same city. This gives the flexibility to define the most accurate calculation methods that are adapted to the data availability. Figure 9.4 shows the relationship between the answers levels-of-detail (i.e. the concept *Answer LOD*) and the other concepts in the ontology.

The levels-of-detail are defined for each answer (to the questions shown in Table 9.2) in terms of time, space, and technology. For example, the question '*How much electricity is produced from subsidized PV installation*?' was answered considering a time-space-technology level-of-detail. The time level-of-detail was 15-min based i.e. considering how much electricity is generated at each 15th minute over a whole year. The space level-of-detail was building-based i.e. the answer is calculated for each single building. The technology level-of-detail was PV system-based i.e. considering parameters (e.g. efficiency) regarding the PV system as a whole (as one single component) rather than splitting it into sub-components (which would be a higher technology level-of-detail).

Based on a reference methodology of ontology development [16], the semantics extraction and classification has been performed, using the logical steps defined by

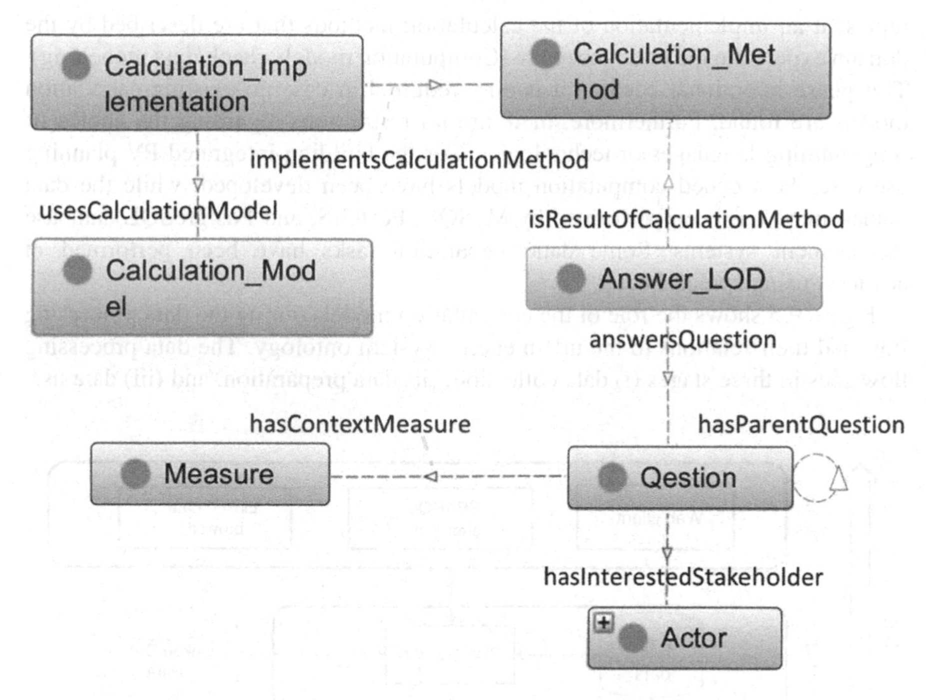

Fig. 9.4 Answers levels-of-detail integration with the ontology

domain experts. This has led, in this work, to the creation of the ontology that was further extended then used in the last phase for urban energy planning support. The keywords within each logical step are classified into their respective categories: classes, object properties, or data properties. Alternatively, in case of re-using existing computation models, their input and output data parameters are considered as the necessary semantics to be modeled.

At the end of this phase, all the semantics describing an urban energy system, as considered by the computation models, are captured. This means that the urban energy system gradually grows when more computation models are being integrated i.e. the more computation models are required, the more semantics are captured and the richer the urban energy system becomes. The need of integrating more computation models is controlled by the number of questions in which the stakeholders are interested in the scoping phase. Therefore, the richness of the semantics of the ontology representing urban energy system is related to the number and nature of the questions of the stakeholders as defined earlier (in Table 9.2).

3.4 Computation Models Development

The goal of this phase is to develop computation models to calculate the answers or any intermediary data contributing to calculate them. The computation models represent an implementation of the calculation methods that are described by the domain experts, in the previous phase (Computation models check/data modelling). This phase is optional because it is only required in case no existing calculation models are found. Furthermore, there are no restrictions regarding the choice of programming languages or technologies. For the building-integrated PV planning use case, Java coded computation models have been developed, while the data management has been performed by MySQL, PostGIS, and PostgreSQL database management systems. Some data preparation tasks have been performed in advance, using spreadsheet.

Figure 9.5 shows the role of the computation models during the data processing flow and their relations to the urban energy system ontology. The data processing flow goes in three stages (i) data collection, (ii) data preparation, and (iii) data use.

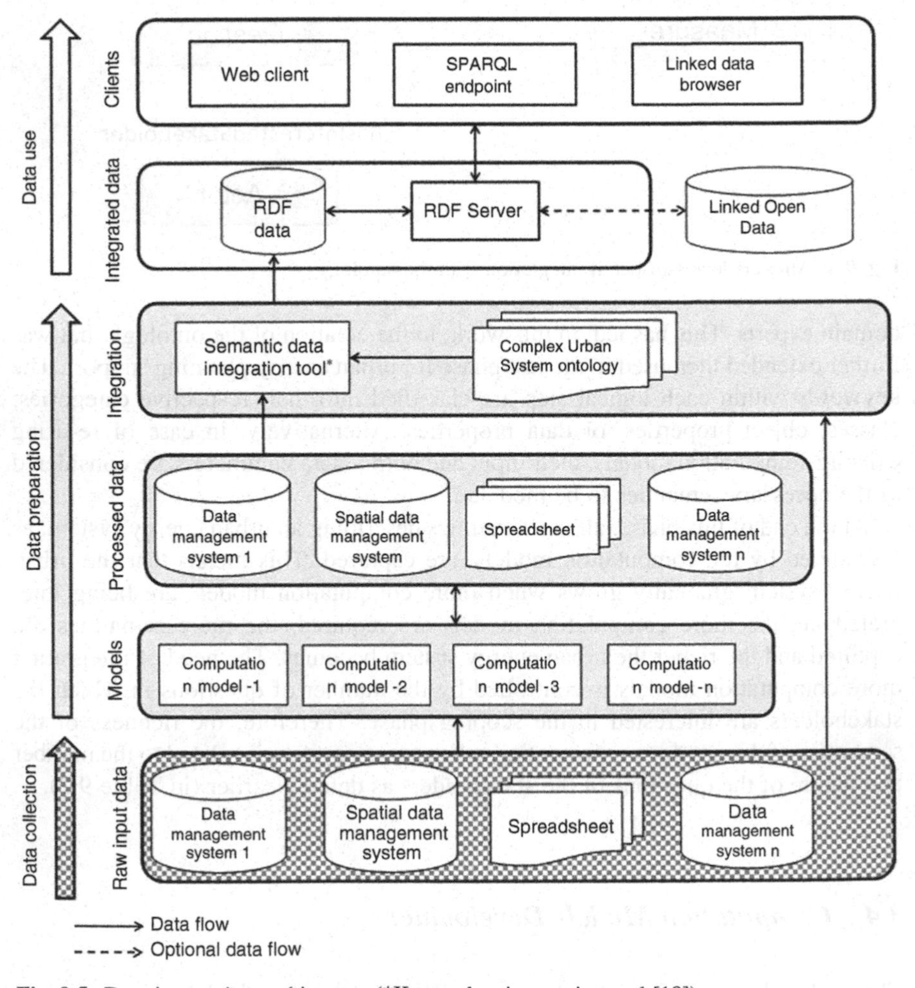

Fig. 9.5 Data integration architecture (*Karma data integration tool [18])

During the data collection stage, raw input data is stored in existing different data management systems (not necessarily integrated or "aware" of the existence of each other). The types of the data management systems range from spreadsheets, to relational databases or spatial databases. Within the data collection stage, data is harmonized regarding variable names, volume units and temporal units and physically stored in adequate data management systems; however, querying this data does not provide answers to the questions set in the scoping phase. Thus, during the data preparation stage, appropriate computation models are used to perform calculations using the raw data to reach useful information to answer the questions of the stakeholders (i.e. the competency questions of the system), listed in Table 9.2.

The computation models access the "raw data", process them, and store the synthesized/calculated data in appropriate data management systems. Each computation model can use its own data management system. The integration of data is not an issue at this stage. Data is integrated, using the ontology in later phases.

3.5 Interaction Modeling

The objective of this phase is to integrate the different computation models considering which data (describing the energy system elements) influence others, when applying measures.

The motivation to capture the interactions is to keep the developed tool open for update. The developed computation models in the previous phase could be either not re-usable in different cities (due to different data availability) or become obsolete if better data becomes available. Capturing the interactions between the components of the whole system shows which computation models are affected by which other when an update is necessary.

In this phase, more concepts are added to the ontology so that an interaction graph (regarding the data of the urban energy system) can be established. This interactions graph, which is part of the ontology, contains links between the elements of the urban energy systems i.e. which elements interact which others, when applying a measure. Thus, when applying a measure, it is transparent which computation models are used, which calculation methods are adopted, and which data of the urban energy system influences which other.

The interaction modeling phase captures the dynamics of the urban energy system. These dynamics represent what happens in the urban energy system when a computation model is run. Therefore, the level-of-detail of these dynamics is related to the one of the used computation models. The more detailed computation models are used, the more dynamics are captured within the urban energy system.

The dynamics within the ontology are modeled according to the interaction mechanisms shown in Fig. 9.6. Data properties together with their related classes form an interaction relationship. More information that can be obtained from an interaction relationship includes the computation model that triggers the interaction, and the calculation method that this computation model implements.

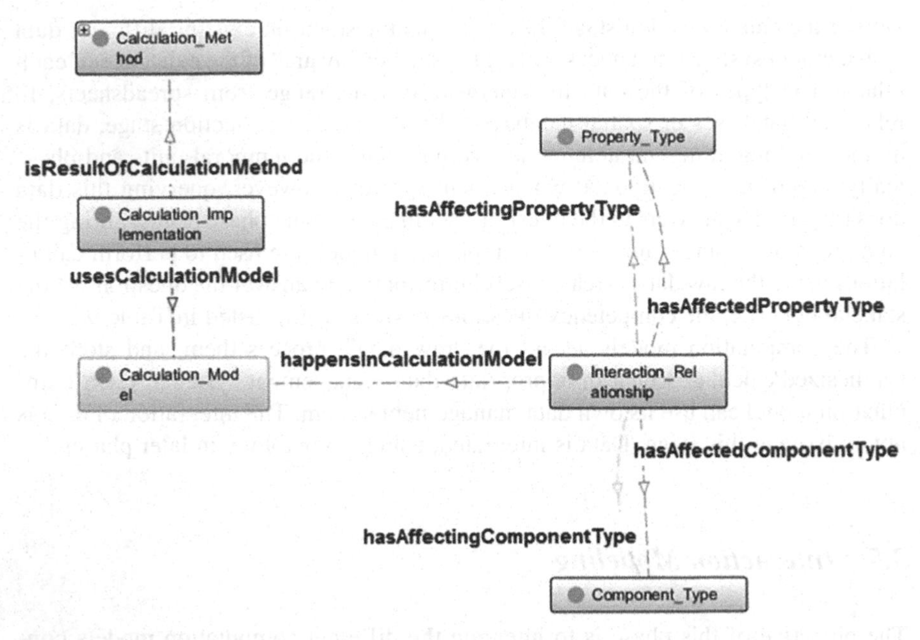

Fig. 9.6 Ontology detail-interactions of computation models

Table 9.3 Sample building-integrated PV related interactions

Interaction ID	Model ID	Component	Property
1	PV_1	Location_Use[1]	hasLocationUseElectricityDemandIndex[1]
		Location[2]	hasLocationElectricityDemand[2]
2	PV_1	Location[1]	hasLocationGFA[1]
		Location[2]	hasLocationElectricityDemand[2]
3	PV_2	Location_Use[1]	hasElectricityDemandProfileShape[1]
		Location[2]	hasElectricityDemandProfile[2]

[1]Affecting
[2]Affected component or property

By the end of this phase, the urban energy system ontology does not only include the domain concepts describing the structure of the urban energy system but also integrates concepts explaining their dynamics. In the next phase, more concepts are integrated within the urban energy system ontology so that it is aware of the opinions of the stakeholders that are involved.

An example set of interactions that occur in the urban energy system that are related to building-integrated solar PV are shown in Table 9.3. An interaction is defined between two components (affecting and affected) through a property of the affecting component and a property of the affected one. For example, the component 'Location' affects itself when the property 'hasLocationGFA' changes. This latter has an impact on the property 'hasLocationElectricityDemand'. This interaction is important to be formalized to make it transparent that the model 'PV_1' needs to be re-run in case the property 'hasLocationGFA' changes (possibly through another model), making the property 'hasLocationElectricityDemand' no longer correct.

3.6 Decision Modeling

This phase aims to capture the knowledge of the stakeholders (who are the decision makers) concerning their interpretation of the different answers, to their questions raised in the scoping phase. Thus, only the logics of the decision making are captured in this phase. These logics are used later on to classify locations in terms of their suitability for a given measure from the stakeholders' viewpoints.

The decision modeling summarizes the interpretations of the answers to the questions in Table 9.2. Therefore, this phase starts only once the answers to these questions are calculated (for each location). The calculation of these answers is performed by the developed computation models in the previous phase. For instance, the net present value when investing in a PV system (i.e. how much profit a building owner makes) is initially a monetary value. Then, depending on the range where it falls, it is summarized to "good" or "very good" so that the other stakeholders (city administration and grid operator) have a common understanding of the concerns of the building owner, without having to look at the exact figures that might be difficult to understand from their viewpoints.

A selection of indicators that the stakeholders mainly use for decision-making is described in Table 9.4. Here, the stakeholders associate an interpretation (in natural language) to value ranges, which define how satisfied they are with the answers (to the questions in Table 9.2). Table 9.4 lists the answer-ranges and their associated interpretations from the single perspectives of all the involved stakeholders in building-integrated solar PV planning. All figures of the answer-ranges are subjectively chosen and can be customized once specific stakeholders are defined.

After capturing the single stakeholder interpretations, an aggregated interpretation from all stakeholder perspectives is established. The scheme that defines the

Table 9.4 Building-integrated PV related value ranges interpretations

Stakeholder	Indicator	Value range	Interpretation
Building owner	Net present value	[10,000€, 25,000€]	Good
		[25,000€, ∞]	Very good
City administration	CO_2-equivalent trees	[200, 350]	Good
		[350, ∞]	Very good
Grid operator	Transformer overload	Yes	Not allowed
		No	Allowed
	Direct use of generation	[80 %, 90 %]	Good
		[90 %, 100 %]	Very good

Table 9.5 Multi-perspective-aggregated interpretation

Stakeholders interpretations			
City administration	Building owner	Grid operator	Aggregated interpretation
Very good	Very good	Very good	Very good
Very good	Very good	Good	
Good	Very good	Very good	
Good	Very good	Good	
Very good	Good	Very good	Good
Very good	Good	Good	
Good	Good	Very good	
Good	Good	Good	
Very good	Very good	Not allowed	Bad
Good	Good	Not allowed	
Good	Very good	Not allowed	
Very good	Good	Not allowed	

aggregated interpretations makes the building owner the highest property i.e. any locations considered as "very good" and "good" by the building owner are respectively interpreted as "very good" and "good" except in three cases: (1) a location is flagged as "not allowed" by the grid operator. Then, its interpreted aggregation is "bad". (2) A location is neither interpreted as "very good" nor "good" by city administration. Then the location has no aggregated interpretation. (3) A location is neither interpreted as "very good" nor "good" by the grid operator. Then, again, the location has no aggregated interpretation. When a location has no aggregated interpretation, the system does not mark it on the map; however, it is still possible to manually select it and view its detailed information (answers to the questions in Table 9.2). The multi-perspective aggregated interpretations are shown in Table 9.5.

The code below gives an example of the implementation of these logics within the ontology. It represents the necessary conditions so that a building is flagged as having a good potential for PV, from the building owner's perspective:

```
Declaration(Class(coPlanBuilding_Owner_Good_Potential_PV_Indicator))
Declaration(Class(coPlan:Indicator))
Declaration(DataProperty(coPlan:hasPVinvestmentNetPresentValue))
SubClassOf(
coPlan:Building_Owner_Good_Potential_PV_Indicator
coPlan:Indicator)
EquivalentClasses(
coPlan:Building_Owner_Good_Potential_PV_Indicator ObjectIntersectionOf(
ObjectIntersectionOf(
DataSomeValuesFrom(coPlan:hasPVinvestmentNetPresentValue
DatatypeRestriction(xsd:double xsd:maxInclusive "25000"^^xsd:integer))
DataSomeValuesFrom(coPlan:hasPVinvestmentNetPresentValue
DatatypeRestriction(xsd:double xsd:minExclusive "10000"^^xsd:integer)))
coPlan:Building_Owner_PV_Indicator_Bundle))
```

3.7 Data Integration

This phase constitutes a transition in the lifecycle of the ontology. It transforms from its initial state as an explicit specification of conceptualization (only containing concepts, logics, relationships, etc.) to a body of knowledge, which contains actual data, information, and knowledge.

The goal of this phase is to integrate the different results of the different developed /re-used models, with other data sources, within the context of an urban energy system (i.e. creating an instance of an urban energy system). It is important to recall that both the data to be integrated is heterogeneous. The different datasets come from different sources, and do not necessarily share the same data structure or format.

The developed ontology through the previous phases is used to integrate data from the different data sources that has been collected or produced by the computation models. Thus, all the required data in order to create an instance of an urban energy system is present. Further, the ontology, developed during the different phases, represents the "schema" of an urban energy system (including its interactions and logic). Therefore, using the ontology to integrate the collected and calculated data creates an instance of the urban energy system, which is used for decision support in urban energy planning.

A data integration tool (karma data integration tool [17]) has been used to integrate the generated and collected data. Finally, a relation between all data variable names within the data sources and the modeled semantics in the urban energy system ontology is established

The data integration result is a resource description framework (RDF) [18] file. The RDF data represents a knowledge base that embeds answers to the questions raised by the stakeholders in the scoping phase. All the semantics are preserved when using this type of data representation (RDF) i.e. descriptions of concepts, object properties, data properties, the dynamics of the urban energy system, the models that were used, etc.

By the end of the data integration phase, the data, information, and knowledge that are needed in the planning of building-integrated solar PV systems in the city are ready. Applications and browsing/querying systems can be used by urban energy planners.

The ontology as an explicit specification of conceptualization (only containing concepts, logics, relationships, etc.) describes the semantics of an urban energy system, within the scope of supporting the planning of building-integrated PV and building refurbishment. Therefore, it is possible to reuse it in different cities, as it does not contain any specific semantics to the application case (city of Vienna). However, the ontology as it becomes a body of knowledge in this phase, it is no longer usable in other cities.

As shown in the Fig. 9.5, there are two different uses of the ontology (in the data preparation then in the data use), depending on the time of its lifecycle. When the ontology is still as an explicit specification of conceptualization, it is used to perform a semantic data integration using the collected data and the data generated by the computation models. Furthermore, it is used to share the semantics of the urban energy system and achieve a common understanding of it. However, once data is integrated, the ontology becomes a body of knowledge that can be used in planning support. The knowledge base is queried by adequate software clients.

3.8 Data Use

A light web interface is developed according to the workflow that energy planners prefer to adopt. The developed interface in this work uses google maps to display the RDF data in a geo-referenced context. It is possible as well that the data is accessed through a SPARQL endpoint or a linked data browser. A sample view of the interface is show in Fig. 9.1.

The GUI design is open for discussion with energy planners/urban planners: how to present the integrated data and under which workflow, or maybe even in a decentralized participative way, where different stakeholder are involved and all having access to the interface.

Reusing the ontology (as an explicit specification of conceptualization) in different cities is possible, in case they have the same data availability as the application case it was initially designed for (Vienna). This would require re-running the computation models and repeating the data integration phase. In case the city, where it would be re-used has different data availability, some of the computation models might require an update. The interactions that have been captured in the interaction modeling phase indicates which computation models need attention, in terms of the sequence of their execution (so that a given model that alters the input of other models is not executed last). So far, this can be achieved in a non-automated way.

4 Building Refurbishment Planning Integration

Similarly applying the same methodology as described above for PV integration, building refurbishment planning has also been modelled. The different stakeholders and their respective questions are shown in Table 9.6. Considering the coarse level-of-detail of the available data in building refurbishment planning, it was only possible to perform the modeling at a census district level (group of buildings) instead of single building level, as it was the case for PV planning. Thus, computation models have been developed based on the floorspace of census districts and their distribution in terms of percentages over different building-uses, used-heating technologies, and buildings-ages classes. The main calculation method was derived based on the work of Siegl [19], who defined characteristic length (*lc*) values that correspond to building topologies in Vienna. The *lc* values define the scale of a

Table 9.6 Competency questions of the ontology-building refurbishment planning

Actor	Question
Building owner	–What is the net present value of my investment? –What is my investment break-even duration? –How much investment costs are required?
City administration	–How much subsidies are to be paid to refurbish buildings? –How much energy is saved by subsidizing building refurbishment? –How much CO_2 emissions are saved by subsidizing building refurbishment? –What is the CO_2 emissions saving-equivalent in terms of trees?

Table 9.7 Building refurbishment related value ranges interpretations

Stakeholder	Indicator	Value range	Interpretation
Building owner	Net present value [€/m2]	[60, 90]	Good
		[90, ∞]	Very good
City administration	CO_2-reduction cost [€/t-CO_2]	[815, 855]	Good
		[0, 815]	Very good

physical building and other physical properties (length: width ratio) that is associated to this value.

The indicators that building owner and the city administration (as stakeholders) use to make their decisions are described in Table 9.7. The stakeholders associate an interpretation (in natural language) to answer-ranges, which defines how satisfied they are if building refurbishment in a given census district is carried out. We note that the building owner in this use case is a conceptual stakeholder that refers to all the building owners within the census district. The selected test area shows a homogenous building and building ownership structure within the census districts: the buildings in the area to a large extent privately owned and constructed in the early 1900 years. Some blocks contain university and other education buildings which are owned by a building agency belonging to the federal state.

Table 9.7 lists the value ranges and their associated interpretations from the single perspectives of building owners and the city administration regarding building refurbishment. The figures of the answer-ranges are subjectively chosen and can be customized once specific stakeholders are defined. The figures in this specific case have been aggregated to a square-meter-level so that it is feasible to compare census districts that have different sizes.

After capturing the single perspectives interpretations, an aggregated interpretation from all the perspectives has been established. Similarly to the building-integrated PV case (detailed in the *Decision modeling* section), priority has been given to the building owner, as the investments are to be made by this stakeholder. Then, the city administration is given the second priority, as the success of the implementation of the building refurbishment does not depend directly on the city. The city administration can only subsidize this measure but cannot initiate or force it. The resulting multi-perspective aggregated interpretations are listed in Table 9.8.

Table 9.8 Building refurbishment multi-perspective-aggregated interpretation

Stakeholders interpretations		
Building owner	City administration	Aggregated interpretation
Very good	Very good	Very good
Very good	Good	
Good	Very good	Good
Good	Good	

The integration of building refurbishment planning with solar PV planning was ensured through: (a) the integration of their data, by sharing the same ontology that represents an UES that contains concepts that are part of solar PV and building refurbishment planning. (b) Ensuring the consistency of data that is shared and calculated by the different heterogeneous computation models. This is achieved in the interactions modelling phase: the output data parameters of the building refurbishment computation models were checked if they are shared as input data parameters in the PV planning computation models and vice versa. The building refurbishment involves data that is related to thermal energy while PV models are related to distributed electric energy generation. Interactions would occur only if we were considering the measure *heating system refurbishment of the buildings* e.g. replacing fossil fuel heating and using thermal heat pumps consuming electricity and using ambient air or groundwater heat. The considered measure (building refurbishment) includes only the thermal insulation of buildings. Thus, no interaction-protocols were necessary to be modelled. (c) Integration of decisions of the different actors about the same locations in terms of their suitability for solar PV installation or building refurbishment. Since the level-of-detail of the building refurbishment modeling was at a census district (group of buildings) level, the solar PV planning data was also aggregated to the census district level. Then, decisions about the integrated suitability in terms of solar PV or building refurbishment were modelled from different perspectives.

The integration of decisions of the two measures makes the resulting ontology include four different types of decisions classes. Each decision class type comprises classes implementing rules so that instances of locations (censuses) are classified according to their suitability for some measure from a certain perspective. As the following:

- **Single Perspective Single Measure Based Decision Class**: The perspective is from a single stakeholder, not considering the others and considering only one measure.
- **Multiple Perspective Single Measure Based Decision Class**: The perspective is from all stakeholders together, and it considers only one measure.
- **Single Perspective Multiple Measure Based Decision Class**: The perspective is from a single stakeholder, not considering the others but considering all the other potential measures in which the stakeholder takes part

- **Multiple Perspective Multiple Measure Based Decision Class**: The perspective is from all stakeholders together, and it considers all measures within the scope of the system (in this case building refurbishment and building-integrated solar PV) that can potentially be implemented.

Figure 9.2 below shows the implemented interface of the resulting urban energy planning support system. The interface allows browsing spatially through the city (due to data provision restrictions currently limited to one district in Vienna) and viewing locations that are suitable for building-integrated solar PV and/or building refurbishment from single – building owner or city administration- perspectives or integrated perspectives.

5 Conclusion

The developed ontology (and the ontology-based planning support system) answers questions that different stakeholders raise to understand how their different interests are affected by the potential implementation of an energy strategy (i.e. stating which locations to use for solar PV and/or building refurbishment). The questions that the ontology provides answers for are listed in Tables 9.2 and 9.6. The ontology is validated through its application within a district (containing around 1200 buildings) within the city of Vienna.

All results are geo-referenced i.e. each location is related to a set of results (answers to questions in Tables 9.2 and 9.6). Concerning building-integrated PV, the results are available at each single buildings level, however for the case of building refurbishment, given the current data availability, results are related to census districts (groups of buildings). Thus, the integrated assessment of building refurbishment with PV was only possible at the census district level.

The developed ontology fulfills the four conditions of urban energy planning support that have been set as objectives for this work. (1): Different stakeholders are provided with specific answers to their particular concerns. (2): The answers that are provided are then summarized at each location (building or group of buildings) level, as “very good”, “good”, or “bad” locations, from the perspectives of each actor, then again as “very good”, “good”, or bad locations as a common perspective. (3): As explained in the methodology section, components interactions are captured, and integrated in the ontology, allowing the possibility to check data consistency and the integration of different computation models and planning decisions. (4): The development methodology allows the flexibility in calculating each single answer in more than one level of detail, using different calculation models e.g. if more detailed data is available about a given share of the city, more detailed computation models can be used for these, while the rest is calculated using more general models that do not require detailed datasets. Mechanisms of integrating multiple levels of detail data are formalized and integrated within the ontology.

References

1. IEA, "World Energy Outlook 2008," 2008. [Online]. Available: http://www.iea.org/weo/2008.asp. [Accessed: 09-Mar-2012].
2. Photovoltaic Austria, "Federal Association of Photovoltaic Austria." [Online]. Available: http://www.pvaustria.at/. [Accessed: 12-Dec-2013].
3. N. Ouhajjou, W. Loibl, S. Fenz, and A. M. Tjoa, "Multi-Actor Urban Energy Planning Support: Building refurbishment and building-integrated Solar PV," 2014, pp. 77–84.
4. L. Girardin, F. Marechal, M. Dubuis, N. Calame-Darbellay, and D. Favrat, "EnerGis: A geographical information based system for the evaluation of integrated energy conversion systems in urban areas," *Energy*, vol. 35, no. 2, pp. 830–840, 2010.
5. D. Robinson, N. Campbell, W. Gaiser, K. Kabel, A. Le-Mouel, N. Morel, J. Page, S. Stankovic, and A. Stone, "SUNtool – A new modelling paradigm for simulating and optimising urban sustainability," vol. 81, no. 9, pp. 1196–1211, Sep. 2007.
6. D. Robinson, F. Haldi, J. K\textbackslashämpf, P. Leroux, D. Perez, A. Rasheed, and U. Wilke, "CitySim: Comprehensive micro-simulation of resource flows for sustainable urban planning," 2009.
7. J. Keirstead, N. Samsatli, and N. Shah, "SynCity: an integrated tool kit for urban energy systems modelling," *Energy Effic. Cities Assess. Tools Benchmarking Pract. World Bank*, pp. 21–42, 2010.
8. P. Waddell, "UrbanSim: Modeling urban development for land use, transportation, and environmental planning," *J. Am. Plann. Assoc.*, vol. 68, no. 3, pp. 297–314, 2002.
9. M. Kwartler and R. N. Bernard, "CommunityViz: an integrated planning support system," *Plan. Support Syst. Integrating Geogr. Inf. Syst. Models Vis. Tools*, pp. 285–308, 2001.
10. N. Ouhajjou, P. Palensky, M. Stifter, J. Page, S. Fenz, and A. M. Tjoa, "A modular methodology for the development of urban energy planning support software," presented at the IECON 2013 - 39th Annual Conference of the IEEE Industrial Electronics Society, 2013, pp. 7558–7563.
11. European Commission, *How to develop a Sustainable Energy Action Plan (SEAP)–Guidebook*. 2010.
12. Covenant of Mayors, "Covenant of Mayors," *Covenant of Mayors*, 26-Apr-2013. [Online]. Available: http://www.covenantofmayors.eu/index_en.html. [Accessed: 26-Apr-2013].
13. The Austrian Research Promotion Agency, "Smart Cities – FIT for SET \textbar FFG," 2011. [Online]. Available: http://www.ffg.at/smart-cities. [Accessed: 28-Mar-2012].
14. P. Giaretta and N. Guarino, "Ontologies and knowledge bases towards a terminological clarification," *Very Large Knowl. Bases Knowl. Build. Knowl. Shar.*, vol. 25, 1995.
15. Magistrat der Stadt Wien, "Wien Umweltgut: Solarpotenzialkataster," 09-Mar-2013. [Online]. Available: http://www.wien.gv.at/umweltgut/public/grafik.aspx?ThemePage=9. [Accessed: 09-Mar-2013].
16. M. Fernández-López, A. Gómez-Pérez, and N. Juristo, "Methontology: from ontological art towards ontological engineering," *Proc. Ontol. Eng. AAAI-97 Spring Symp. Ser.*, pp. 33–40, 1997.
17. C. A. Knoblock, P. Szekely, J. L. Ambite, A. Goel, S. Gupta, K. Lerman, M. Muslea, M. Taheriyan, and P. Mallick, "Semi-automatically mapping structured sources into the semantic web," in *The Semantic Web: Research and Applications*, Springer, 2012, pp. 375–390.
18. O. Erling and I. Mikhailov, "RDF Support in the Virtuoso DBMS," in *Networked Knowledge – Networked Media*, T. Pellegrini, S. Auer, K. Tochtermann, and S. Schaffert, Eds. Springer Berlin Heidelberg, 2009, pp. 7–24.
19. G. Siegel, "Haus 2050 – Trendanlysen zum Energiebedarf von Wohngebäauden," 2012.

Chapter 10
Beyond Eco-feedback: Using Room as a Context to Design New Eco-support Features at Home

Nico Castelli, Gunnar Stevens, Timo Jakobi, and Niko Schönau

Abstract In recent years research in Sustainable Interaction Design has put major efforts into understanding the potentials of saving energy in private households by providing energy consumption feedback. Trying to overcome pitfalls such as invisibility and immateriality, a great variety of designs with saving potentials from 5–15 %, has emerged. However, feedback mechanisms are mostly reduced to a one-dimensional view on motivating energy savings. In this paper, we argue to take a broader view on eco-support, where eco-feedback should be used in combination with eco-control and eco-automation features. All these features have in common that they aim to reduce energy consumption in practice. From such a holistic understanding of eco-support, we demonstrate how design could benefit from ubiquitous- and context-aware computing approaches to enrich feedback, increase control and automatize cumbersome and boring routines. We use the presence of a user on room level as context information. Rooms present an essential domestic ordering system that structures daily routines at home. In this paper, we show that the usage of *room-as-a-context* has fundamental implications for the design of domestic indoor localization concepts. In addition, we show how the different types of eco-support systems benefit from it. We illustrate our consideration by presenting a prototype for Android based tablets, which was used to study the design concepts in the wild.

Keywords Consumption feedback • Eco-feedback • Context-awareness • Home energy management system • Hems • Indoor-positioning

---

N. Castelli (✉) • T. Jakobi
Human Computer Interaction, University of Siegen, 57068 Siegen, Germany
e-mail: nico.castelli@uni-siegen.de; timo.jakobi@uni-siegen.de

G. Stevens
Information and Technology Management, Bonn-Rhein-Sieg University of Applied Science, 53757 Sankt Augustin, Germany
e-mail: gunnar.stevens@uni-siegen.de

N. Schönau
Information Systems and New Media, University of Siegen, 57068 Siegen, Germany
e-mail: niko.schoenau@uni-siegen.de

J. Marx Gómez et al. (eds.), *Advances and New Trends in Environmental and Energy Informatics*, Progress in IS, DOI 10.1007/978-3-319-23455-7_10

1 Introduction

Residential and commercial buildings are responsible for about 40 % of the EU's total energy consumption [1]. However, conscious sustainable use of this limited resource is hampered by a lack of visibility and materiality of consumption [2]. One of the major challenges in Sustainable Interaction Design (SID) is to enable consumers to make informed decisions about energy consumption, thereby supporting the shift to or implementation of sustainable actions [3]. Technological innovations of smart metering allow a fine-grained consumption measuring, in turn paving the way for a rapid development of energy management systems.

Early systems were largely limited to simple displays providing information about the total energy consumption of the household (smart meter systems), such as 'eco-eye' [4] or the *Wattson Energy Monitor* [5]. More sophisticated energy management systems mostly based on real-time, disaggregated energy consumption data on device level [6] with various forms of visualization and approaches like goal-setting and gamification [7], conditioning [8] and comparison [9]. They are mostly realized as web-portals [10], smartphone applications [11] or ambient, artistic design approaches like the *Power-Aware Cord* [12] or *Watt-Lite* [13].

Modern home energy management systems (HEMS) further embrace different strategies into one holistic system [14] that includes features for consumption feedback, control, and automation.

However, with the increasing volume of data, eco-support becomes more complex. To cope with this challenge current research focuses on how feedback can be made more informative and how the user can be supported to take action.

A promising approach in this topic is trying to support context awareness in enriching consumption data with personalized information. Context is a cross-cutting issue that could be used to enhance the various levels of eco-support (see Fig. 10.1). The aim of this approach is to reduce information complexity, make control easy and to provide a rich context for interpretation to make data more meaningful for the user. By reducing the complexity of information and providing a rich context, context awareness enables the user to better interpret consumption data.

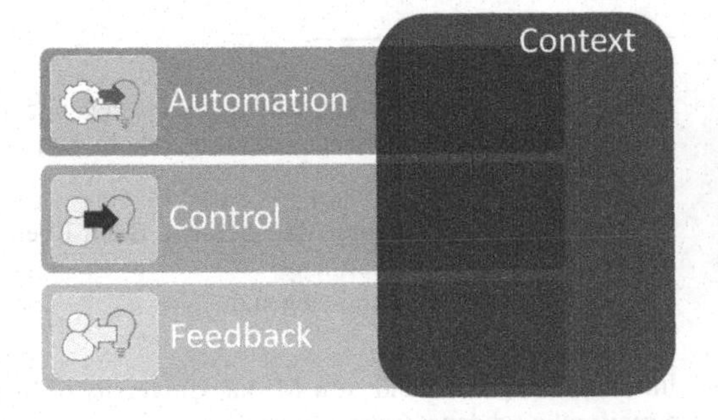

Fig. 10.1 Context as a cross-cutting topic enhancing the various levels of eco-support

Contributing to this, we present the concept of room as context information, as rooms play an important role in structuring domestic routines, thus domestic energy consumption. We discuss implications for designing more meaningful interfaces with the regard to three different approaches that support a more sustainable behavior. Based on the concept [15], we additionally show an architecture for such a context-aware home energy management system and demonstrate a fully functional prototype for android-based tablets that illustrates how room information can be used to enrich feedback mechanisms and contextualize user interfaces of mobile home energy management systems (mHEMS).

2 Eco Support Systems

In literature a manifold of approaches for designing systems to support more sustainable behavior can be found. We subsume all these different approaches under the label of *eco support systems*.

Generally, eco-support systems are defined as measures to make behavior, habits, routines, and lifestyles more sustainable. In this paper, we focus on eco-support with regard to the private energy consumption. Here, we can roughly distinguish three categories of eco-support: *Eco-Feedback, Eco-Control,* and *Eco-Automation* (see also Loviscach [16] for a similar taxonomy). All three are expected to increase energy efficiency as well as the energy sufficiency on the demand side. They are not mutual exclusive; in opposite: they complement and strengthen each other when included into a HEMS.

In the following we give a brief overview of these categories. In our discussion we show how additional context information contributes to the effectiveness and usefulness of these three categories (see Fig. 10.1).

2.1 Eco-Feedback

Eco-Feedback is defined as providing people information about their consumption of natural resources (e.g. energy, water, etc.) with the aim to motivate and persuade them to reduce their environmental impact [17]. Today, all HEMS support at least some kind of eco-feedback.

The concept is shaped by rational behavior theories like Rational Choice, Expected Utility, or the Theory of Reasoned Action (TRA) [18] as well as corresponding motivational theories in Human-Computer-Interaction (HCI) like the Persuasion Theory (PT) [19]. Such theories interprets consumer's actions as an expression of their informed and conscious judgment aiming for utility maximization [20]. Wasteful behavior, in this view, is therefore caused by an information deficit. Hence, the major aim of eco-feedback is to bridge this gap by making the consumption behavior and its consequences visible.

The impact depends on many factors like whether the feedback is given at the right time, at the right place and if it is presented in the right form [2]. Concerning the representation issues like form, aesthetics, used metrics, frequency, granularity and comparison, optional user selections should be considered [2]. Also the feedback types vary from direct feedback (learning by looking) to inadvertent feedback (learning by association) [21].

However, just to feed back the consumption data is not enough, additional context information is needed so that users connect the rather abstract data to their domestic life [22]. For instance, more advanced eco-feedback could give advices about possible savings. Below, we want to show that room-context have a great potential to improve the eco-feedback at home.

2.2 *Eco-Control*

Eco-Control is defined as improving people's behavioral control of their consumption, making it easier to act pro-environmentally. It is shaped by the critique of rational choice theories and corresponding limitations of eco-feedback. A major critique of rational theories is that they neglect the bounded rationality of humans as well as that an intention behavior gap exists. In particular, studies shown that wasteful consumption often does not stem from a lack of knowledge or awareness, but turn the intention into action [23]. For instance, devices were not switched off, because the switches are difficult to reach and/or the process of switching off costs additional mental and physical power.

In reaction theoretical models like the Theory of Planned Behavior (TPB) [24] have been elaborated and extended in several ways e.g. by taking issues of perceived and actual control more seriously [23]. Perceived control is defined as the belief that one can influence the own environment to get desired outcomes. In contrast, actual control is defined as having the means within the situation to perform a task. In other words, perceived control presents a prerequisite, while actual control presents a necessity for action taken.

The substantial difference between eco-feedback and eco-control is that the former aims to motivate pro-environmental behavior, while the latter aims to make such behavior easy. In particular, the aim of eco-control is to increase the perceived and actual control of dwellers to cut down their consumption. While motivational factors addressed by eco-feedback have been relatively well-researched, only comparatively little reliable information is available on supporting control.[1] However, a general strategy is to provide and make aware of means to energy saving ready to hand. An example of a simple, but effective eco-control is a switchable multiple socket. Such a device makes it easier to switch off all devices

[1] Also, persuasion theory [19] acknowledged that control is an important factor. However, and somehow surprisingly, persuasive eco-design mainly leaves the factor unattended.

e.g. when leaving the room or the house and thus fosters transition of this behavior into a sustainable habit.

In this paper we want to show that this kind of eco-control could be improved by mobile computing that provides direct, contextualized control options at the right time and place. Moreover, eco-feedback and eco-control could be improved, when both are united: While feedback increases the energy awareness and literacy hinting for potential savings, control options put the user in charge of managing the consumption in easy way [25].

2.3 *Eco-Automation*

Eco-Automation is defined as the computational control of devices and domestic consumers with the aim to reduce resource consumption. In the SID literature it is only rarely discussed, because at the first glance eco-automation seems to mean that humans are entirely left out of the loop. However, we argue that both the SID and Home Automation research would benefit, if we understood eco-automation as a part of an overall eco-support strategy.

For instance, one of the unsolved problems of eco-feedback and eco-control is that some kinds of eco-friendly behavior (like switching devices off to prevent stand-by consumption) called for investment of mental and physical efforts. In the case of everyday life, this often becomes boring and cumbersome, so that there is the risk that users revert to "bad" habits over time [26].

Concerning this, studies show that eco-automation can lead to savings up to 30 % [27]. In addition, they offer more comfort than unpleasant routines like switching the heating off when leaving the home, which could be delegated to the computer. Such delegation of disagreeable duties especially prevents a fallback into lazy habits, thus helping to sustain pro-environmental behavior. However, the challenge of eco-automation is not simply to just switch things off, but to do this when it makes sense [28].

Corollary, Mert and Tritthart found that consumers might fear losing control over automated devices [29]. Hence, automation should not be designed independently from feedback and the control mechanism. In particular, eco-automation should focus on those kinds of routines in which the users are in danger to revert to "bad" habits. Concerning this, the real skill in the design of eco-automation is to find a good mix of awareness mechanisms, user control and automation support to improve people's perceived control and self-efficacy [30].

In this paper, we want to show how this could be addressed by e.g. providing a rudimentary automation editor where the user could define rules that should be delegated to the computer. Here, room-context help to express automation rules that based on domestic activates (like switching the light and heating off, when leaving the home for more than 15 min).

2.4 *Using Context in Eco Design*

In general, context-aware computing is understood as the ability of an application "to discover and react to changes in the environment they are situated in"[31]. Dey et al. [32] distinguish between three kinds of context-aware systems: First, presenting context information and services to the user with the help of sensor information. For example, showing the user her current position through the placement of a marker on a map. Second, automatically executing services in reaction of a change of context. An example poses the car navigation system that calculates a new route when an exit has been missed. Third, attaching context information for later retrieval and use. Brown [33] also distinguishes three categories in a similar manner:

- Presenting information to the user
- Running a program
- Configuring the screen of the user

In Eco-design there are already a number of approaches to enrich consumption data with additional context information. For instance, Costanza et al. [34] have built an interactive feedback system, where users can add context-tags directly within the consumption feedback application. On the one side this allows a visual linkage of specific activities and energy consumption and on the other side new forms of visualizations are possible (e.g. event-centric/energy-centric forms of visualizations). Neustaedter et al. [35] use data from personal calendars to contextualize consumption data of users. Although many events and especially most of routine activities were not registered, it could be recognized that calendar entries can be used for the declaration of energy consumption (e.g. a house party explains high consumption, while eating in a restaurant would imply low consumption). Also, people's location at home helps to contextualize and individualize feedback. Jahn et al. [36] e.g. use the position of the user to present eco-information for the devices at hand. Guo et al. [37] use a RFID based check-in/check-out to get the position of dwellers to identify the individual consumption in a multi-person household.

3 Room as a Context: A Conceptual View

In previous empirical studies in a Living Lab we have explored how people make their energy consumption accountable and what kind of information they need for this [22, 25, 38]. In this research, we uncovered a need for flexibility as it became obvious that consumption and wasteful behavior have a high individual meaning for people. However, we also found common patterns regarding how such meaning was constructed e.g. by referring to domestic routines and actions as well as comparing devices, people and households. Especially room context (defined as users' presence in a specific room) turned out to be important information for the user to reconstruct activities in order to link consumption patterns to them. In the following, we conceptually outline how room context presentation can be used as useful resource in eco-support design.

3.1 Room as a Domestic Order for Everyday Activates

People live in homes and undertake activities and interact in this physical environment. Here, rooms have a special meaning when it comes to everyday-activities. Rooms often are decorated differently and serve a particular purpose. A room-structure specifies which activities are appropriate in it and what technology is available to carry them out [39]. For example, in the most cases cooking in the bedroom is unusual. Also, for architects, rooms are of central importance. The planning of electrical sockets is related to the intended use of the room and switches for lighting and heating are used to control devices on room-level. Additionally, switches for lights are usually attached next to the door such that when entering or leaving the room, one can switch on/off the required appliances.

In the 1990s, the concept of rooms gained high attention in the context of designing information and communication technology. In their investigation Harrison and Dourish [40] linked insights from architects and urban designers with their own studies to differentiate between space and place. Space, therefore, is a three-dimensional environment with objects and events that have relative positions and directions whereas places are spaces that are valued ("We are located in space, but we act in place" [40]).

3.2 Placing and Spacing: A New View on Domestic Indoor Location

The distinction between place- and space-oriented approaches leads to different requirements for locating in domestic environments. The major difference between common indoor localization solutions and room localization is that space-oriented approaches are relying on metric error measures, commonly defined by the distance between actual and estimated position. In opposite, place-oriented approaches rely on a quasi-topologic error measure defined by the ratio whether the actual room is estimated correctly or not. Figure 10.2 gives an example that good space accuracy does not necessarily imply good place accuracy. Yet, until we have specially optimized place-oriented localization techniques, existing space-oriented techniques could be used as a heuristic.

Concerning the various localization techniques, we principally can distinguish between four classes: The first group are so called beacon-based approaches that

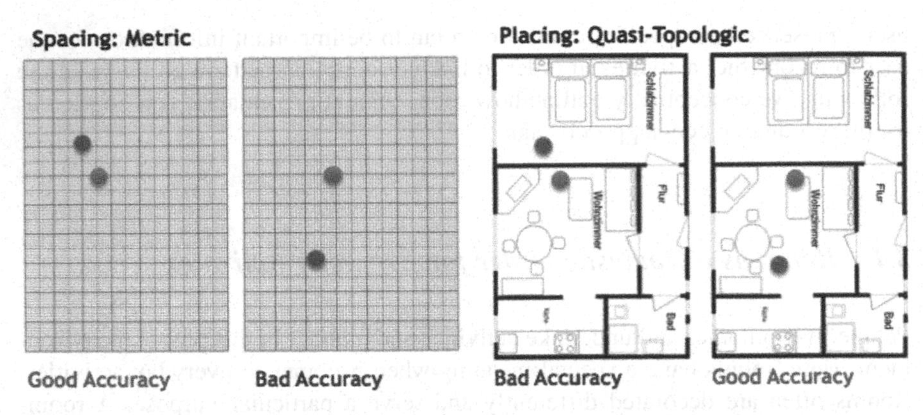

Fig. 10.2 Difference between "space-oriented" and "place-oriented" localization

use proximity detection with short-range radio communication, for example RFID or NFC. Based on a globally unique identifier, e.g. a smartphone can look up the position of the beacon (e.g. Ref. [41]). But these approaches depend on additional hardware to locate the position of the user. The second group are geometry-based approaches estimating the position e.g. by triangulation and trilateration, determining positions from measurements of angle of arrival or distance between sender and receiver. The intersection of lines or radii respectively provides the current location (cf. Ref. [42]). One disadvantage is that conventional WiFi-routers are hardly suitable, because they either need special antennas allowing angle-measurement, or, for trilateration, a much more precise measure of distance that can be provided by electromagnetic waves. The third class of indoor-positioning approaches uses accelerometers and gyroscopes of a device to log the movement: speed and direction, starting from a given position to calculate a new position. Such dead reckoning techniques suffer from a fast increasing inaccuracy as small errors add up every step [43]. The fourth group is based on fingerprinting the signal strength of e.g. WiFi routers at different places. One disadvantage is that such a system must be trained beforehand [44]. Yet, it has the great advantage that existing router infrastructures in domestic settings could be reused for the positioning.

3.3 Understanding of Energy Consumption and Energy Wastage

The interplay between technology, places and activities can be used to classify energy consumption and thereby make wastage visible. Schwartz et al. [38] have demonstrated that dwellers distinguish energy consumption between consumption of background services (typically always-on devices like the refrigerator and freezer) and activity related consumption (like using a TV for watching television, light for reading, etc.). Generally, activity based consumption is closely related to the person's presence (respectively activities which in turn are related to places [39, 40]). Therefore, the actual place of habitants in their home is a strong indicator for energy being wasted (e.g. light in a room where no one is present is a waste of energy).

We use this heuristic by identifying the presence of users in the corresponding rooms to expand existing visualizations of eco-feedback systems and to create new forms of visualization to support the user in his sustainable practices. In the following sections we conceptually describe such a system.

3.4 *Room-Context Aware Feedback*

We identified four, non-exhaustive, visualization categories where room-context information could help to make feedback more meaningful for the user:

- Analytic charts identifying spenders in the home
- Time series consumption graphs enriched by dwellers' presence information
- Person and domestic activity centered consumption visualization
- Domestic scoreboard systems

The room context information could therefore be used to identify spenders, which are defined as potential energy wasters. Analytic charts on device level allow making such spenders visible. For example, the device-level chart in Fig. 10.3, left, shows that 21 % are potential spending by splitting the overall consumption into consumption with presence and without. Such graphs help users to control their habit of switching devices off when not needed.

Further, presence information could be used to enrich time series consumption graphs in various ways. For instance, historic feedback graphs commonly show a curve of the device's consumption in a daily, weekly or monthly interval. Such graphs on a room level could be enriched by peoples' presence time in that room, e.g. by assigning a color to each dweller and coloring the graph's background accordingly for the time each person was in the room (see also in Fig. 10.3 left, "week history"). Such graphs may make it easier for dwellers to identify consumption patterns and match them with their own behavior.

The third improvement puts the focus on the user by showing the consumption of her immediate environment over time. For example, when walking through the apartment the user can see in which surrounding she has used the most energy. This person-centered visualization in combination with the previous one allows gaining new insights and surprising facts about one's own domestic energy practices. Last but not least, the room-context information could be used to define new indicators for domestic scoreboard systems like average room temperature when people are present and non-present. Further, this information could be used to personalize recommendations, tips, or statistics.

Fig. 10.3 Using room-context information to enrich eco-feedback visualization (*left*) and to adapt home control interfaces (*right*)

3.5 Room-Context Aware Home Control

In a further step, we explored, how room context information could be used to adapt home control panels. We have identified two categories, in which room-context can help to reduce the panel complexity and nudge people to switch off spenders:

- Adapt the control panel to the devices of the actual room
- Make aware about spenders outside the room

One of the current problems of control panels is the large number of switching options that can lead to a cluttered design. Architectures solve, for example, the problem of complex control panels by making use of rooms as a domestic order system: A room only includes the controls for the room. This is a smart choice as people most often are interested in controlling activity-related devices, which typically are in the person's current surrounding. Room context information helps to adopt this strategy by showing only controls of the actual room on the user interface. This radically simplifies the complexity of home control panels.

An exception to the rule above, are devices outside the room that have been forgotten to be switched off, e.g. because of laziness. To nudge people to switch off these devices, the control panel should make aware about these spending devices. Figure 10.3, right, presents our solution for this demand, where we split the control panel into two sections: The top section shows the controls in the current room. The bottom section shows the detected spenders outside the room. By focusing on the controls that are important in the current context, the panel is more structured and the number of switching options is greatly reduced.

3.6 Room-Context Aware Home Automation

As noted, some sort of consumption is linked to the presence of a person in a room, like the light in a room, the heating of a room or activity based consumption like watching TV or listening to the radio. In the previous section, we outlined how to use room-context to provide adapted control interface when people have forgotten to switch off such consumers when leaving the room. In addition to this, home automation could use this information to regulate and switch off these consumers directly without interrupting the user. The information could be used to create more sophisticated profiles of domestic behavior. Such patterns might be used by the heating regulation to decide which room should be heated and when.

However, because of the complexity of domestic life, one has to consider that the perceived control still remains at the dwellers. For example, by defining their own set of (room-context based) rules, such that users could decide which routines should be automatized and which ones should support contextualized feedback and control mechanisms.

4 Room as a Context: Realization of Design Implication

Based on our conceptual reflections, we have developed MyLocalEnergy, a fully functional prototype of a room-context aware home energy management system. Referring to eco-feedback, eco-control and eco-automation systems we demonstrate and discuss design implications, when using room as an additional contextual source.

Our prototype consists of a ZigBee-based multihop-measuring network (smart plugs) measuring consumption on socket (device) level, a raspberry pi running the energy middleware and an android device (cf. Ref. [38] for more details on this "traditional" part of our HEMS system).

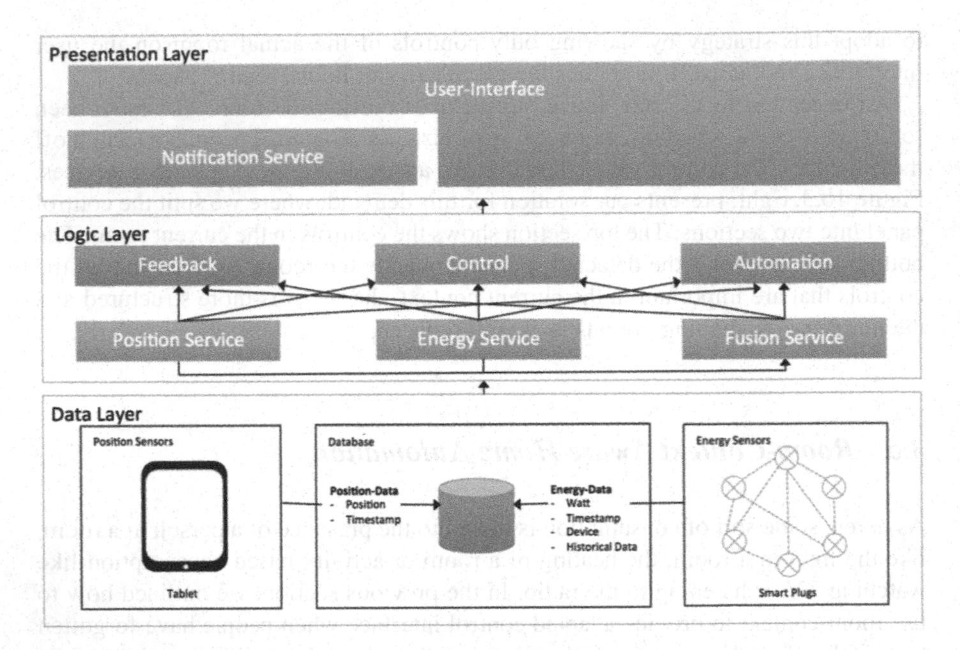

Fig. 10.4 Simplified architecture of the context-HEMS system

The system was realized as a client-server architecture where the energy- and position-data is stored in a local database on the raspberry pi. The client was implemented as a native app for all kind of Android devices, however, it was optimized for Android Tablets.

The local architecture is further based on a three-tier layer concept with a data layer, a logic layer and a presentation layer (see Fig. 10.4). The modularization allows that the representation of feedback can easily be changed and adapted to the context. In addition, the services provided by the logic layer could be used in other applications, as well. By the same token, the indoor positioning algorithm could be replaced with a better one without the need to change the presentation layer.

The data layer is responsible for storing all energy- and position-data in a database. The consumption data is linked with a timestamp to use it for direct feedback as well as for historic feedback and time series analysis.

As position data, the present room of the user is stored together with a timestamp. The positioning is mainly computed on the Android client, which tells the home server in which room the person actually is. We therefore use a fingerprinting approach based on available WiFi network signals as WiFi routers are available in most domestic environments and therefore no additional hardware is needed. Furthermore, a combination of multiple Received Signals Strengths (mRSS) provides relatively unique fingerprints. Reducing the error rate can be handled by setting up additional WiFi AP. We also minimized the mentioned training problem by providing a user interface, where users iteratively can add, edit and delete multiple measurement points and assign them to a room. The users themselves can improve system accuracy by adding additional measurement points at places that are important from their perspective. We further implemented some filters that validate the results.

The logical layer mainly implements basic services concerning the actual and historic consumption as well as the actual and historic indoor position of the dwellers. It also provides an additional service to fuse or link both types of data, e.g. listing all devices in the current room, where the user is present. The location and the consumption services are independent from each other in order to ensure a good clarity and to be able to implement changes efficiently. The services of the logical layer are provided as web-services by a Tomcat webserver that runs on our low-power home server. The web-service allows that the presentation layer can run on the same home-server (e.g. as Java Server Pages) or remotely (implemented as native Android App).

This architecture allows the implementation of higher-level services for eco-feedback, control and automation as well as more sophisticated user interfaces. In the following we discuss this part of the HEMS in more detail.

4.1 Contextualized Eco-Feedback

The most essential part of eco-feedback is to visualize the consumption so that it is meaningful and allows informed decision-making. Concerning this, we focused on how room-context helps to individualize and personalize the feedback. We further implemented various display options to address the heterogeneity among the users.

The main feedback element is a room-context aware time series consumption graph (see Fig. 10.5 left) as a variant of the design concepts outlined above. The graph either displays current live consumption or historical values together with information about the users' presence. By default, the graph shows the live consumption of the devices defined in the system. Historical values, in a freely selectable period, can be shown in the graph as well. The graph helps to uncover relations between the personal behavior and the electricity consumption. E.g. the graph in Fig. 10.5, left shows how the laptop of the user was awoken from sleep mode after he entered the room "Zimmer". As he was present in the room at the time of the consumption it can be assumed that this consumption is not "wastage".

We also use room context to adapt the screen of the user [33]. For instance, with information about the current room consumption, the share of total consumption, the average room consumption etc. The display also includes a room based "energy" flower as an ambient visualization, which changes the look depending on the consumption. It does not provide exact details, but provides emotionally perceived information [45].

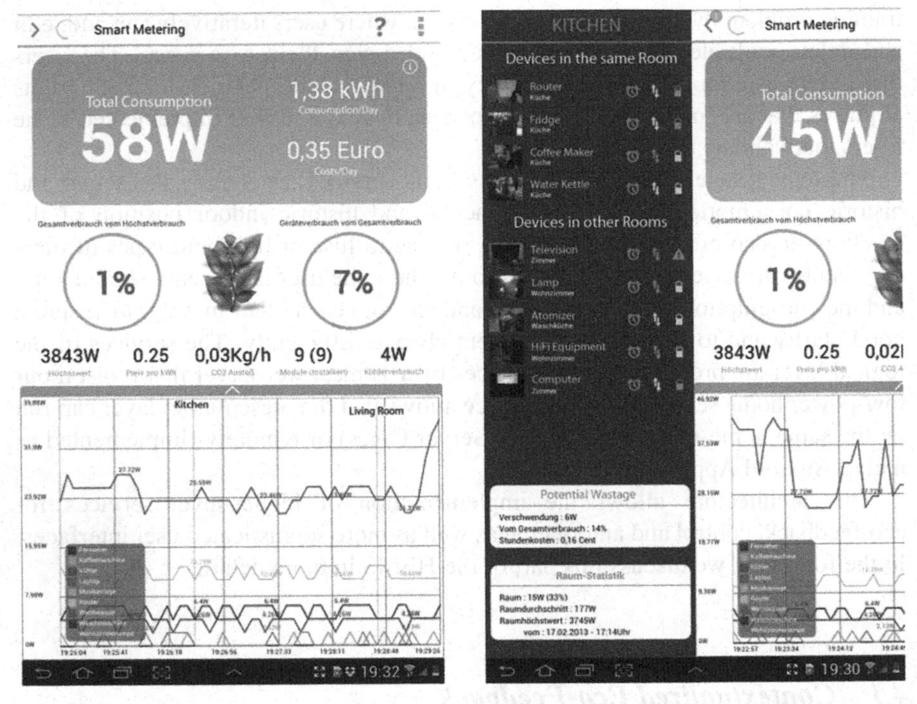

Fig. 10.5 Visualization of the Prototype on an android tablet, partially translated (*left*: home screen, *right*: home screen with open control panel)

4.2 *Contextualized Eco-Control*

We also realized some of the eco-control concepts. For instance, the system provides a context-adaptive display showing the home devices in two groups: The primary group includes all devices in the immediate environment of the user (room); the second includes all other devices (see Fig. 10.5 right). This slightly differs from the concept outlined above as some of our users wanted to switch on devices in other rooms as well. This is why we decided to display more than the appliances in the room in this view. In addition to the names of the devices the list also includes additional information like the status (using/not using), remote-switch availability and some simple automation support. Users can switch devices on/off just by clicking on the corresponding item on the list. In particular, to ensure that users still get aware about spenders, they are marked with an extra symbol in the list. In addition, an Android application notification is sent to the user if a spender is detected. To persuade pro-environmental action, the notification includes a direct control element to switch the detected spender off.

4.3 Contextualize Eco-Automation

We also implement some rudimentary automation editor to define simple automation rules. Our goal was that malfunction (e.g. if position recognition is not accurate enough) does not lead to serious problems and users do not have the feeling that the perceived control is getting lost.

The editor allows the user to define which devices should be switched off when the house is left. This rule is executed only if the position service does not detect any room and the user was outside the range of his home wireless network. The additional condition was included to reduce the rate of false negatives (the system wrongly thinks that the user leaves the home) as they have more serious consequences than false positives.

In addition, a simple editor for a timer function is included, so that devices can be switched off at a particular time (e.g. switching off a VCR after recording the television program in order to save stand-by consumption).

5 Evaluation

We split our evaluation in a technical and a conceptual part concerning overall user experiences. For the technical evaluation of the position service we used a test routine asking the user at random selected points in time, whether the actual recognized room is correct or not. We have run this routine in two different households with three WiFi networks available and collected overall 29 measuring points in 2 days. We achieved a correctness of about 85 %, which means that with an optimal establishment of the position service, a good accuracy could be achieved. The accuracy of the position determination, however, depends on the existing WiFi infrastructure and the structural conditions of the household. The WiFi networks should have sufficient signal strength and the routers should be placed in different corners and floors to get best results. The use of WiFi-extenders can distort the results, since in this case the distance to the router cannot be recognized. For the prototypical implementation, the position recognition is sufficiently accurate to examine the usefulness of the system in terms of supporting the user within a sustainable use of energy. We have not carried out a major technical evaluation, since the position determination is not the focus of this work.

We evaluated the user experience by conducting interviews and workshops with seven private living lab households [14] concerning the perceived usefulness and shortcomings using room-context to make the consumption feedback more meaningful and how such concepts should be realized. Overall, our participants appreciate the design concept and said that additional context information would help them to get a more profound understanding of their domestic consumption. Additionally, the participants agree, that their room-based position is an useful information, especially in the historical consideration of consumption data to inference on ineffective behavior. A further aspect that people regarded as practical was the better clarity provided by the distribution of the devices in the two categories in the control panel. Due to the fact that we measure up to 18 single devices for households, the usual control-panel becomes cluttered otherwise.

The people also noted that with an accurate detection of the position, some device could automatically be switched on or off, e.g. lamps. However, not all users wanted a system that would automatically switch devices on or off and "decrease" their control of an appliance. Another, point of criticism, among other detail improvements, was that participants could felt disturbed sometimes when they receive notifications after leaving a room with active devices to which they wanted to return soon again, e.g. when going to the kitchen to make a coffee.

People also expressed reservations concerning the opportunities to automatize particular routines by using context information. These reservations echoed the known control problem of home automation [29].

6 Discussion and Outlook

The first energy monitors simply fed back the more or less raw measured energy data. Today, real-time disaggregated consumption measurement is reality resulting to an explosion of data. Concerning this, major challenges in domestic settings regarding lowering the energy consumption are:

- How can we prevent an information overload given the vast amount of raw data?
- How can we make consumption feedback more meaningful for the users?

We contribute to this challenge by outlining the concept of room as a context and showing how it could be implemented. Studies on eco-feedback show that rooms have turned out to be useful for the user to support the detection of energy wastage. The aim of this paper was to demonstrate that this insight could be generalized by combining the concept of context-awareness [32] with the concept of eco-support concerning the different levels of feedback, control, and automation [16].

We think that room-context will not replace other approaches to contextualize eco-feedback in literature [34–36], but supplement them. For instance, room-context complements the device context and vice versa: When a user comes near a device, our room context-aware user interface could be adapted to a device context-aware one as outlined in [36].

In summary, this paper has outlined the potential of room-context aware HEMS. However, for the practical use several challenges have to be coped with, which we will address in our further research: Firstly, the practical value of the room-concept must be studied under realistic conditions with a larger sample and in long term. Secondly, while people most often take their smartphone and maybe their tablet with them when leaving home, they might leave devices on a desk, a sideboard, etc. when they being at home. Concerning this, future smartwatch based positioning services have great potential. Thirdly, we are aware that our solution is implicitly optimized for single households. Hence, in future we have to investigate how multi-person households adopt such design concepts and if, in which way the concepts must be extended.

References

1. European Commission: COM(2008) 772: Energy efficiency: delivering the 20% target., (2008).
2. Fitzpatrick, G., Smith, G.: Technology-Enabled Feedback on Domestic Energy Consumption: Articulating a Set of Design Concerns. IEEE Pervasive Comput. 8, 37–44 (2009).
3. Blevis, E.: Sustainable interaction design: invention & disposal, renewal & reuse. Proceedings of the SIGCHI conference on Human factors in computing systems. pp. 503–512. ACM (2007).
4. Kelsey, C., Gonzalez, V.: Understanding the use and adoption of home energy meters. Proceedings of El Congreso Latinoamericano de la Interaccion Humano-Computadora. pp. 64–71 (2009).
5. Foster, D., Lawson, S., Blythe, M., Cairns, P.: Wattsup?: Motivating Reductions in Domestic Energy Consumption Using Social Networks. Proceedings of the 6th Nordic Conference on Human-Computer Interaction: Extending Boundaries. pp. 178–187. ACM, New York, NY, USA (2010).
6. Sundramoorthy, V., Liu, Q., Cooper, G., Linge, N., Cooper, J.: DEHEMS: A user-driven domestic energy monitoring system. Internet of Things (IOT), 2010. pp. 1–8. IEEE (2010).
7. Jacucci, G., Spagnolli, A., Gamberini, L., Chalambalakis, A., Björkskog, C., Bertoncini, M., Torstensson, C., Monti, P.: Designing Effective Feedback of Electricity Consumption for Mobile User Interfaces. PsychNology J. 7, 265–289 (2009).
8. Kirman, B., Linehan, C., Lawson, S., Foster, D., Doughty, M.: There's a monster in my kitchen: using aversive feedback to motivate behaviour change. CHI'10 Extended Abstracts on Human Factors in Computing Systems. pp. 2685–2694. ACM (2010).
9. Petkov, P., Köbler, F., Foth, M., Krcmar, H.: Motivating domestic energy conservation through comparative, community-based feedback in mobile and social media. Proceedings of the 5th International Conference on Communities and Technologies. pp. 21–30. ACM (2011).
10. Erickson, T., Li, M., Kim, Y., Deshpande, A., Sahu, S., Chao, T., Sukaviriya, P., Naphade, M.: The dubuque electricity portal: evaluation of a city-scale residential electricity consumption feedback system. Proceedings of the SIGCHI Conference on Human Factors in Computing Systems. pp. 1203–1212. ACM, New York, NY, USA (2013).
11. Weiss, M., Staake, T., Mattern, F., Fleisch, E.: PowerPedia: changing energy usage with the help of a community-based smartphone application. Pers. Ubiquitous Comput. 16, 655–664 (2011).
12. Gustafsson, A., Gyllenswärd, M.: The power-aware cord: energy awareness through ambient information display. CHI'05 extended abstracts on Human factors in computing systems. pp. 1423–1426. ACM (2005).
13. Jönsson, L., Broms, L., Katzeff, C.: Watt-Lite: Energy Statistics Made Tangible. Proceedings of the 8th ACM Conference on Designing Interactive Systems. pp. 240–243. ACM, New York, NY, USA (2010).
14. Schwartz, T., Denef, S., Stevens, G., Ramirez, L., Wulf, V.: Cultivating Energy Literacy: Results from a Longitudinal Living Lab Study of a Home Energy Management System. Proceedings of the SIGCHI Conference on Human Factors in Computing Systems. pp. 1193–1202. ACM, New York, NY, USA (2013).

15. Castelli, N., Stevens, G., Jakobi, T., Schönau, N.: Switch off the light in the living room, please! –Making eco-feedback meaningful through room context information. In: Jorge Marx Gómez, N.G., Michael Sonnenschein, Ute Vogel, Andreas Winter, Barbara Rapp (ed.) BIS-Verlag. pp. 589–596. BIS-Verlag (2014).
16. Loviscach, J.: The design space of personal energy conservation assistants. PsychNology J. 9, 29–41 (2011).
17. Froehlich, J., Findlater, L., Landay, J.: The design of eco-feedback technology. Proceedings of the SIGCHI Conference on Human Factors in Computing Systems. pp. 1999–2008. ACM (2010).
18. Fishbein, M., Ajzen, I.: Belief, attitude, intention and behavior: An introduction to theory and research. (1975).
19. Fogg, B.J.: Persuasive technology: using computers to change what we think and do. Ubiquity. 2002, 5 (2002).
20. Hunecke, M.: Beiträge der Umweltpsychologie zur sozial-ökologischen Forschung: Ergebnisse und Potenziale. Expertise für die BMBF-Förderinitiative "Sozial-ökologische Forschung", Bochum (2001).
21. Darby, S.: The effectiveness of feedback on energy consumption. Rev. DEFRA Lit. Metering Billing Direct Disp. 486, 2006 (2006).
22. Schwartz, T., Stevens, G., Ramirez, L., Wulf, V.: Uncovering Practices of Making Energy Consumption Accountable: A Phenomenological Inquiry. ACM Trans Comput-Hum Interact. 20, 12:1–12:30 (2013).
23. Darnton, A.: Reference report: An overview of behaviour change models and their uses. UK Gov. Soc. Res. Behav. Change Knowl. Rev. (2008).
24. Ajzen, I.: The theory of planned behavior. Organ. Behav. Hum. Decis. Process. 50, 179–211 (1991).
25. Jakobi, T., Schwartz, T.: Putting the user in charge: End user development for eco-feedback technologies. Sustainable Internet and ICT for Sustainability (SustainIT), 2012. pp. 1–4 (2012).
26. Van Dam, S.S., Bakker, C.A., Van Hal, J.D.M.: Home energy monitors: impact over the medium-term. Build. Res. Inf. 38, 458–469 (2010).
27. Garg, V., Bansal, N.K.: Smart occupancy sensors to reduce energy consumption. Energy Build. 32, 81–87 (2000).
28. Meyer, G.: Smart home hacks. O'Reilly, Sebastopol, CA (2005).
29. Mert, W., Tritthart, W.: Get smart! Consumer acceptance and restrictions of Smart Domestic Appliances in Sustainable Energy Systems. (2009).
30. Singhvi, V., Krause, A., Guestrin, C., Garrett Jr, J.H., Matthews, H.S.: Intelligent light control using sensor networks. Proceedings of the 3rd international conference on Embedded networked sensor systems. pp. 218–229. ACM (2005).
31. Schilit, B.N., Theimer, M.M.: Disseminating active map information to mobile hosts. IEEE Netw. 8, 22–32 (1994).
32. Dey, A.K., Abowd, G.D., Salber, D.: A Conceptual Framework and a Toolkit for Supporting the Rapid Prototyping of Context-aware Applications. Hum-Comput Interact. 16, 97–166 (2001).
33. Brown, P.J.: Triggering information by context. Pers. Technol. 2, 18–27 (1998).
34. Costanza, E., Ramchurn, S.D., Jennings, N.R.: Understanding domestic energy consumption through interactive visualisation: a field study. Proceedings of the 2012 ACM Conference on Ubiquitous Computing. pp. 216–225 (2012).
35. Neustaedter, C., Bartram, L., Mah, A.: Everyday activities and energy consumption: how families understand the relationship. Proceedings of the SIGCHI Conference on Human Factors in Computing Systems. pp. 1183–1192. ACM (2013).
36. Jahn, M., Jentsch, M., Prause, C.R., Pramudianto, F., Al-Akkad, A., Reiners, R.: The energy aware smart home. Future Information Technology (FutureTech), 2010 5th International Conference on. pp. 1–8. IEEE (2010).

37. Guo, Y., Jones, M., Cowan, B., Beale, R.: Take it personally: personal accountability and energy consumption in domestic households. CHI'13 Extended Abstracts on Human Factors in Computing Systems. pp. 1467–1472. ACM (2013).
38. Schwartz, T., Denef, S., Stevens, G., Jakobi, T., Wulf, V., Ramirez, L.: What People Do with Consumption Feedback: A Long-Term Living Lab Study of a Home Energy Management System. (2013).
39. Baillie, L., Benyon, D.: Place and Technology in the Home. Comput. Support. Coop. Work CSCW. 17, 227–256 (2008).
40. Harrison, S., Dourish, P.: Re-place-ing space: the roles of place and space in collaborative systems. Proceedings of the 1996 ACM conference on Computer supported cooperative work. pp. 67–76. ACM (1996).
41. Loeffler, A., Kuznetsova, D., Wissendheit, U.: Using RFID-capable cell phones for creating an extended navigation assistance. Microwave and Optoelectronics Conference (IMOC), 2009 SBMO/IEEE MTT-S International. pp. 471–475. IEEE (2009).
42. Potortì, F., Corucci, R., Nepa, P., Barsocchi, P., Buffi, A.: Accuracy limits of in-room localisation using RSSI. Antennas and Propagation Society International Symposium, APSURSI'09. (2009).
43. Strutu, M., Caspari, D., Pickert, J., Grossmann, U., Popescu, D.: Pedestrian smartphone based localization for large indoor areas. Intelligent Data Acquisition and Advanced Computing Systems (IDAACS), 2013 I.E. 7th International Conference on. pp. 450–454. IEEE (2013).
44. Fischer, L., Hahn, N., Kesdogan, D., Hoffmann, A., Wismüller, R.: Indoor Positioning by Fusion of IEEE 802.11 Fingerprinting and Compass Bearing.
45. Yun, T.-J., Jeong, H.Y., Kinnaird, P., Choi, S., Kang, N., Abowd, G.D.: Domestic Energy Displays: An Empirical Investigation. (2010).

Part III
Smart Transportation

Chapter 11
Supporting Sustainable Mobility Using Mobile Technologies and Personalized Environmental Information: The Citi-Sense-MOB Approach in Oslo, Norway

Núria Castell, Hai-Ying Liu, Franck R. Dauge, Mike Kobernus, Arne J. Berre, Josef Noll, Erol Cagatay, and Reidun Gangdal

Abstract Urban and peri-urban growth is increasing worldwide and Europe is now one of the most urbanized continents in the world. Oslo is one of the fastest growing cities in Europe. This creates pressure on its infrastructure, including traffic and environmental urban quality. Additionally, vehicular traffic is a major contributor to CO_2 emissions, which affects climate change. It is recognized that air quality is a major factor for human health, however, although different measures have been implemented, improving air quality and lowering carbon emissions still remains an unsolved problem in Oslo.

The main objective of Citi-Sense-MOB is to demonstrate how using innovative technology to continuously measure air quality at the road level combined with innovative Information and Communication Technologies (ICT) can help to create a dynamic city infrastructure for real-time city management, access to personalized environmental information and sustainable development. The output from the project will be mobile services for citizens and authorities based on the use of near real-time data on air quality and CO_2 emissions at road level.

The societal importance of these services arises from a need to mitigate the effects of air pollution and climate change, and to combat respiratory diseases related to traffic air pollution.

N. Castell (✉) • H.-Y. Liu • F.R. Dauge • M. Kobernus
NILU, Postboks 100, 2027 Kjeller, Norway
e-mail: nuria.castell@nilu.no

A.J. Berre
SINTEF, Postboks 124, Blindern, Oslo 0314, Norway
e-mail: arne.j.berre@sintef.no

J. Noll
UNIK, Gunnar Randers vei 19, 2027 Kjeller, Norway
e-mail: josef@unik.no

E. Cagatay • R. Gangdal
Kjeller Innovation, Gunnar Randers vei 24, 2027 Kjeller, Norway
e-mail: rg@kjellerinnovasjon.no

J. Marx Gómez et al. (eds.), *Advances and New Trends in Environmental and Energy Informatics*, Progress in IS, DOI 10.1007/978-3-319-23455-7_11

In order to motivate citizens to use the information generated by the project, Citi-Sense-MOB will provide them with personalized environmental information, for instance alerting systems when pollution levels exceed a critical threshold. Furthermore, customized information will also be provided to authorities consisting of detailed air quality maps at high spatial resolution and an evaluation of possibilities to reduce CO_2 emissions by improving driving practices in public urban fleets.

Keywords Citizens Observatories • Environmental monitoring • Mobile measurements • Sustainable mobility

1 Introduction

Air pollution is one of the factors negatively affecting quality of life within cities. Many areas of Europe still have persistent problems with outdoor concentrations of particulate matter (PM), nitrogen dioxide (NO_2) and ground level ozone (O_3).

In cities, road traffic is the dominant local source of pollution, along with domestic combustion, which has been growing over the last few years [1]. At the same time, private vehicle use in Europe is growing, and a further doubling of traffic is predicted by 2025 [2]. Studies show that traffic-related air pollution may cause major adverse health effects in the population living at or near air polluted roadways [3]. Exposure to traffic related pollution has been related with heart and respiratory diseases [4]. In addition, traffic-related air pollution has been classified as carcinogenic by the WHO [5]. It is still needed more studies to characterize personal exposure to traffic-related air pollution, and to better understand the link between traffic-related air pollution and public health effects [6].

Additionally, traffic emissions also play a key role in carbon dioxide (CO_2) emissions. Energy consumption in urban areas, mostly in transport and housing, is responsible for a large percentage of CO_2 emissions. Because of their larger consumption of fossil fuels, cities emit 76 % of the world's energy-related CO_2 [7]. Consequently, cities are key players in efforts to reduce CO_2 emissions and mitigate the effects of climate change [7]. Monitoring road traffic and associated efforts to devise and evaluate strategies to reduce exhaust emissions from road traffic will benefit both air quality and climate change.

Monitoring air quality is essential for understanding how pollutants are distributed in the atmosphere and how they affect human health and the environment as a whole. However, air quality data at street level is currently scarce or non-existent.

The appearance of new low-cost sensors provides an opportunity to monitor air quality at higher spatial resolutions. The use of small and inexpensive sensors for air quality monitoring could provide a supplement to the information given by the current observation networks. Low-cost sensors are smaller, portable and easy to use. This technology is currently experiencing rapid growth, opening new opportunities for research and its application within air quality assessment and management. However, there are challenges in the use of sensor data, mainly on the issues related to precision and accuracy of the measurements, scientific understanding of these novel data and derivation of information from these data sets.

Air quality monitoring is a topic of extreme importance as pollutants have a direct effect on human health. Traditional monitoring networks are expensive and it is infeasible to install them everywhere. Several existing projects are exploring the feasibility of mobile air quality monitoring using low-cost sensors platforms to monitor the environment in high resolution [8, 9]. Here we describe the activities being developed in the EU EMMIA project Citi-Sense-MOB (Mobile Services for Environmental and Health Citizens' Observatory), launched at the end of 2013 and running for 2 years [10].

Citi-Sense-MOB focuses in mobile data collection including citizen driven data collection. Citi-Sense-MOB will employ urban bus fleets to gather continuous data at street level and involve citizens in gathering data with sensor platforms mounted on bicycles. In order to improve the information offered to the citizens it is necessary to further investigate how the less accurate information from the low-cost sensor platforms can be presented to the people and used to raise awareness and help citizens to learn about the harmful effects of exposure to pollution. New technological tools will be used and developed in Citi-Sense-MOB (e.g., sensor platforms, data collection, visualization, etc.) in order to generate information useful for citizens and authorities. In particular the citizens involved in gathering data will expect, in return, a better value-add than the current non-personalized information.

1.1 Supporting Sustainable Mobility in Oslo

Oslo is located at the top of Oslo Fjord enclosed by forests and hills. Although Oslo occupies a very large region of Norway (the area of the city is about 454 km^2), the population is considerably small, with a current population about 634,000 inhabitants (as January 2014), when compared to other European capitals. However, Oslo is one of the fastest growing cities in Europe. This creates pressure on its infrastructure, including traffic and environmental urban quality.

Today motorized traffic in Oslo creates problems with congestion, pollution and CO_2 emissions. Oslo has been persistently exceeding the annual limit value for NO_2 established in the EU air quality directive for health protection for the last 10 years [11]. The main source of NO_2 is road traffic. Sustainable and environmentally friendly mobility is an essential part of Oslo's new vision for green growth and improved quality of life [12]. Developing tools to support green initiatives is crucial. Citi-Sense-MOB will contribute to it by demonstrating innovative tools to support personalized environmental information in Oslo.

Currently, Oslo has 11 urban air quality stations monitoring main pollutants (i.e., NO, NO_2, CO, PM_{10} and $PM_{2.5}$) and offering updated information to the public every hour. Figure 11.1 displays the location of the monitoring stations. The

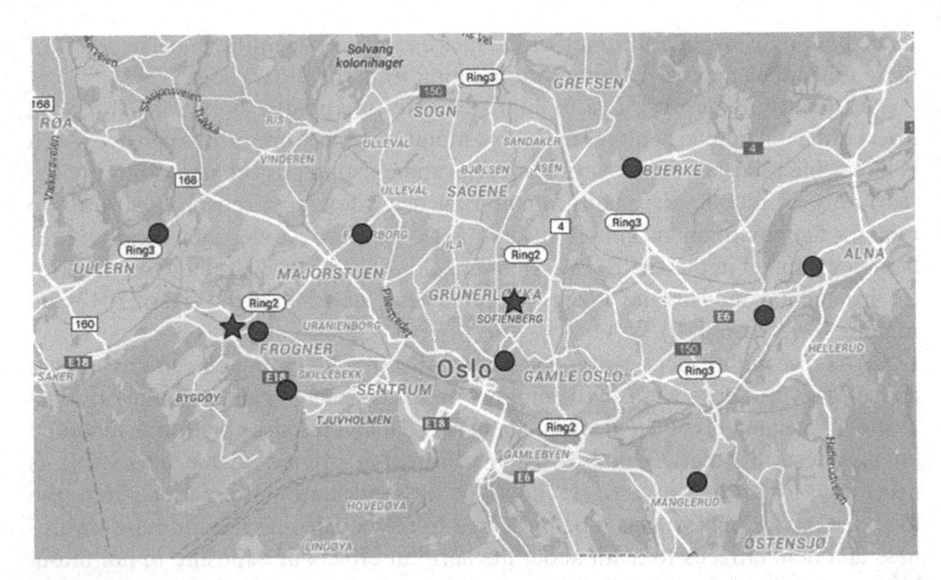

Fig. 11.1 Location of the air quality reference monitoring stations in Oslo. The locations marked with a star represent the urban background stations and the circles represent the urban traffic stations

information provided by the stations, although very accurate, it is not personalized and therefore it is not easy for the citizens to see how it is related to them.

Citi-Sense-MOB will test the feasibility of low-cost sensors mounted in mobile platforms (i.e., buses and bicycles) to complement the data gathered by the traditional monitoring network with the aim to offer a higher spatial and temporal air quality information and the possibility of getting personalized air quality information to the Oslo citizens. All the data collected will be available to the citizens in a user-friendly and visually informative layout, using both web services and mobile phone apps.

The project aims to support sustainable mobility by raising public awareness on environmental issues as the effect of air pollution on health and the environmental footprint of road transport. The services created in the project aim to contribute to mitigate the effects of air pollution and to combat diseases related to traffic air pollution, by fomenting less polluting transportation methods and helping citizens to minimize their exposure by selecting less polluted routes when walking or cycling in the city.

In order to motivate citizens to use the information generated by the project, the project will provide them with personalized and customized information specific to themselves. Such services will include alerting systems when pollution levels exceed a critical threshold that the user can configure, individual exposure along a track and advice on how to mitigate the effects arising from adverse environmental conditions. Citizens will also be able to gather data by themselves using low cost sensors and report their perception on air quality in their surroundings following a colour scale from green meaning “I don’t have any symptoms” to red indicating “I have severe symptoms and I can't perform my normal activities", turning citizens into sensors themselves.

The major benefit for the citizens will be access to personalized information of air quality levels in their immediate surroundings or at selected locations. This information will help citizens to make the decisions needed to maintain and improve their quality of life.

Customized information will also be provided to authorities consisting of detailed air quality maps at high spatial resolution and an evaluation of possibilities to reduce CO_2 emissions by improving driving practices in public urban fleets.

Citi-Sense-MOB will collaborate with another ongoing project CITI-SENSE [13] in building a Citizens' Observatory in Oslo. CITI-SENSE aims at enabling citizens to contribute to, and participate in, environmental governance by using novel technological solutions. CITI-SENSE will employ static sensors deployed in the city and personal sensors carried by people. The data from CITI-SENSE and Citi-Sense-MOB will be integrated within a common data processing system, utilising an open data platform.

Citi-Sense-MOB will also test the technological solution for urban fleets developed in the framework of the UrVamm project in Spain [14], which combines the collection of air quality information and driving patterns. This solution aims to encourage continuous learning towards more efficient driving practices with a consequent reduction in fuel consumption and CO_2 emissions.

This manuscript is an extended and more actualized version of the publication in the EnviroInfo Conference proceeding [15]. Section 2 provides an overview of the project architecture and the data management, from data collection to data dissemination; Sect. 3 discusses the implementation details and first preliminary results. Finally, Sect. 4 provides conclusions and presents details of further work building on the results from Citi-Sense-MOB.

2 Citi-Sense-MOB Architecture

Figure 11.2 illustrates the architecture of Citi-Sense-MOB. The measuring system is composed of sensors mounted on mobile platforms (i.e., buses and electrical bicycles) to monitor atmospheric gases concentrations (i.e., NO_2, NO, O_3, CO, SO_2 and CO_2) and particulate matter ($PM_{2.5}$ and PM_{10}) at road level. The buses will have an additional sensor to gather data on driving practices (e.g., instant speed, acceleration, fuel consumption, etc.). The continuously gathered data are then transmitted to a server for processing (e.g., automatic quality control, generation of maps and graphics, etc.). The data from the sensors is complemented with other available data as for instance data from the air quality reference monitoring network, air quality models, sensors from the sister project CITI-SENSE [13] and pollen data from the Norwegian Asthma and Allergy Association (NAAF) [16]. The processed data are presented in a user-friendly and visually informative layout using both web solutions and mobile phone applications. Citizens will also be

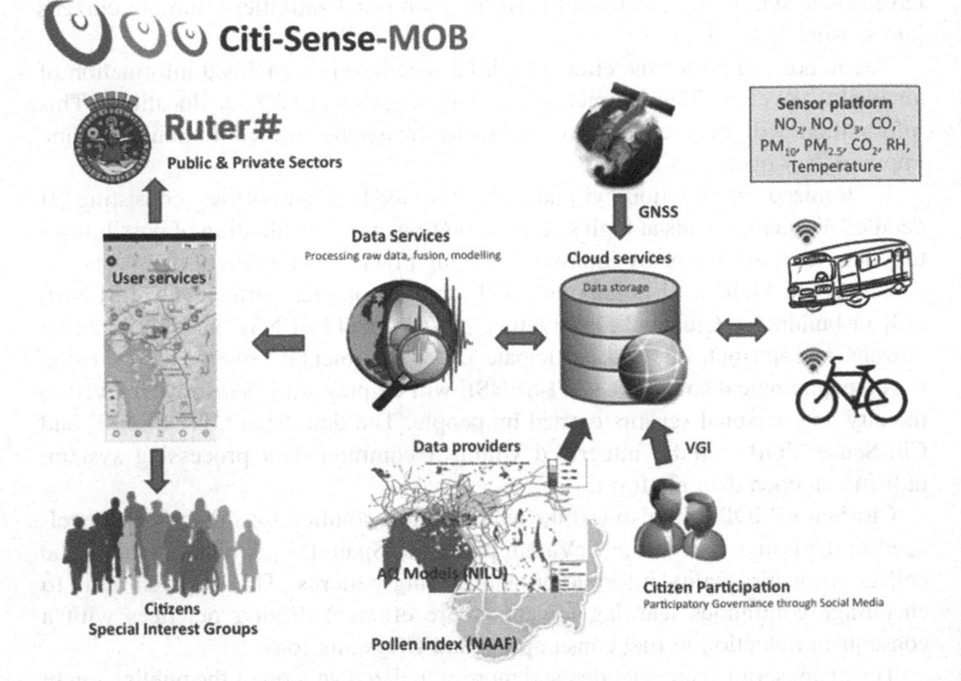

Fig. 11.2 Citi-Sense-MOB system overview

able to use their mobile phone to upload information on how they feel the air quality in their immediate surroundings (i.e., Volunteered Geographical Information, VGI). That information will allow citizens to visualize both air pollution and pollen levels and how they are affecting the people.

The following subsections explain with more detail the design of services from sensor platforms to on line data visualization.

2.1 Design of Services

Citi-Sense-MOB services have been designed to address the needs of stakeholders and to integrate the design thinking into a structured innovation process [17]. The ATONE (Actors -Touch points – Offering – Needs – Experiences) methodology with service design elements is applied in Citi-Sense-MOB. ATONE is a practitioner-based method for service-design, aimed to maximize the innovation potential at the early stages of service innovation. The method is scalable and works for short projects and major transformations [18].

The starting point was identifying the main stakeholders and the environmental applications and services they require. The project identified the following stakeholders: general public, health interest groups, local authorities, transport agencies, researchers and enterprises. In order to understand and address their needs the project has involved different target groups of users (e.g., health interest groups, transport companies, environment authorities, cyclists, etc.) into the services co-design process to provide feedback on the services.

For instance, transport is responsible for a large part of the overall greenhouse gas emissions in developed countries, and it is one of the sectors where carbon dioxide emissions continue to increase. An example of the services of interest of transport agencies and environment authorities is fostering eco-driving practices. Transport agencies are interested in reducing fuel consumption while the environment authorities are interested in reducing CO_2 emissions from urban fleets.

The sensors deployed in the bus will provide continuous geo-temporally referenced environmental and eco-driving data on fuel consumption and estimated CO_2 emissions. This generates feedback to the user, encouraging drivers to drive in a manner that creates less pollutant emissions as well as providing savings in fuel consumption. The data gathered in the project will allow transport agencies to develop their green agendas and design better planning to delivery low emissions transport projects that can be licensed by local authorities.

Other target group is the general public, the general public and in particular the sensitive groups are exposed to air pollution levels that can be harmful for their health. The major reason that impedes the general public from utilizing current air quality related services is the lack of understanding what the data is telling them. For instance, if a user is informed that the air quality is "poor", what does that mean for him/her, and how can this information be acted upon? Each individual has a unique relationship to the environment, but the information on the state of atmosphere and related hazards available today is entirely generic, and seldom personally relevant. This undermines citizens' awareness of their environment, and consequently limits their ability to recognize and change both their footprint and their exposure to the air pollution.

Citi-Sense-MOB will address this by providing the general public the possibility of getting personalized air quality information on the mobile phone. The user will then be able to check the air quality in their immediate surroundings, select less polluted routes to walk or cycle, and track their individual exposure while moving in the city.

One of the interest groups Citi-Sense-MOB will focus are the cyclists. As many other cities around the world Oslo has made a considerable investments in cycling infrastructure, including bicycle sharing programs, new cycling trails and campaigns promoting cycling as an environmentally-friendly mode of transportation [19]. Some studies show, however, that cyclists may be exposed to higher concentrations of traffic-related air pollution due to their proximity to traffic and high respiration rates [20, 21].

Citi-Sense-MOB will gather air pollution data on cycling tracks by adding air quality sensors to bicycles. The user will be involved in the process of gathering data and will be able to see the changes of air pollution level when cycling in different routes. We expect that involving citizens into the monitoring process will make users more conscious and aware of the impact of air pollution in their lives. Moreover, the data gathered will also be useful for urban planning to design cleaner bicycle lanes (e.g., use of barriers, optimal distance from the road, etc.).

2.2 Sensor Platforms

One of the pillars of Citi-Sense-MOB is the use of low-cost sensors to monitor air quality in real-time. Within the project we will employ electrochemical sensors measuring NO_2, NO, O_3, CO, SO_2 and CO_2 and optical particle counters monitoring PM_{10} and $PM_{2.5}$. The gas sensors are based on electrochemical reactions that take place within the sensor between the gas to be monitored and an electrolyte. Some of these sensors exhibit cross sensitivity to other gases and interferences with temperature, relative humidity and, in some cases, wind speed [22]. The optical particle counter (OPC) uses the principle of laser light scattering to simultaneously measure particles in various size ranges.

In Citi-Sense-MOB we will test two sensor platforms, one mounted on buses and other one mounted on electrical bicycles. The sensor platform employed in the buses has been designed and manufactured by IA-ADN [14]. The sensor platform is located inside the bus and it pumps air from outside. The air quality platform is connected to a computer system that collects data on the driving conditions and the GPS location. The data is then transmitted to a database for storing, retrieving and visualizing. The sensor platform mounted on the bicycle has been manufactured by Dunavnet [23] and adapted to be mounted on a bicycle. The sensor platform is located in the back of the bicycle and the sensors are directly exposed to the air flow. Similar to the buses, the data is then transmitted to a common database.

The Table 11.1 shows the characteristics of the air quality sensors employed in the bus and bicycle platforms. Additionally to the air quality sensors, the platforms also monitor temperature and relative humidity.

After the data are processed, it will be accessible to the citizens using a mobile app or directly via the webpage of the project. The next section describes with more detail the visualization. As mentioned before, it is one of the objectives of the project, to provide citizens with the means to understand how air pollution data relates to themselves personally. In order to do that it is important that the information is clear and easy to understand by non-experts. In the next subsection it will be explained in more detail how the visualization will be performed.

In Citi-Sense-MOB we will also involve citizens as sensors. Citizens will be able to report if they are experiencing any health symptom and the possible triggers. Citizens will be able to report using their mobile phone and answering to a short

Table 11.1 Characteristics of the air quality sensors employed in Citi-Sense-MOB

Air quality sensors	Bicycle	Bus
Ozone (O_3)	Alphasense B4	
Nitrogen dioxide (NO_2)	Alphasense B4	Electrochemical 4 electrodes
Carbon monoxide (CO)	Alphasense B4	Electrochemical 2 electrodes
Nitrogen monoxide (NO)	Alphasense B4	
Carbon dioxide (CO_2)	Alphasense IRC AT	
Particulate matter (PM10 and PM2.5)	OPC-N1	

questionnaire. The data will be geo-located and stored in the database with the possibility of being visualised in the map as other sensor data.

The data from the sensors and VGI information will be stored in a central services based on a Web Feature Service Interface Standard (WFS) server. The WFS server is designed to support multiple input and output structures.

The project will address privacy concerns. The user will control the data about themselves, and they will agree in the terms and conditions the data can be further used. Moreover, the data collected won't be related to sensitive personal information but to an agent number. When creating the collective maps a similar solution that the one followed by Drosatos et al. [24] will be developed creating aggregated maps that respect user(s)-privacy.

2.3 Online Data Visualization

It is one of the main goals of Citi-Sense-MOB to provide personalized information to the citizens. It is important that this information is user-friendly and useful. The visualization should assist in raising environmental awareness and contribute to the acceptance and development of new habits to reduce exposure to air pollution.

Displaying the detailed information of the pollutant levels in concentrations (ppb or $\mu g/m^3$) will not be user-friendly, and it will lack the necessary information on the relation between concentration and health. For that reason it was decided to employ an Air Quality Index (AQI) using a colour scale [25]. The AQI is commonly used by air quality agencies to present the pollution levels to the public. The Table 11.2 shows the Norwegian AQI. The Norwegian AQI is computed to take into account health effects, especially short-term effects as it is presented in an hourly basis [26].

The visualisation of the data will be performed using a variety of techniques. Online Visualisation will be implemented using the open source geographic information system (GIS) Server GeoServer application that will be the backend for both web-based and mobile mapping of the data provided by the project. Initially, for the online visualisation to be carried out in near-real-time, the users will be provided with maps showing the locations and measurements of all currently active sensor nodes. The sensor observations will be shown as points overlaid on a street map

Table 11.2 Norwegian air quality index. Source: Luftkvalitet.info [26]

Colour	Health relation	PM_{10} (μg/m^3)	$PM_{2.5}$ (μg/m^3)	NO_2 (μg/m^3)	SO_2 (μg/m^3)	O_3 (μg/m^3)
Red	Unhealthy	>200	>100	>200	>350	>240
Orange	Unhealthy for sensitive groups	100–200	50–100	150–200	250–350	180–240
Yellow	Moderate	50–100	25–50	100–150	150–250	100–180
Green	Good	<50	<25	<100	<150	<100

Table 11.3 Colour scale employed to represent the pollution levels, the pollen levels and people reported feelings, and its meaning

Colour scale	Meaning AQI	Meaning pollen	Meaning feeling
Green	Good	Low	I do not have any symptoms
Yellow	Moderate	Moderate	I have mild/moderate symptoms. I can do my normal activities
Orange	Unhealthy for sensitive groups	High	I have moderate/severe symptoms as eye irritation, problems breathing, etc. It is affecting my normal activities
Red	Unhealthy	Very high	I have severe symptoms. I cannot do my normal activities

(e.g., Google Maps or OpenStreetMap), and thus will provide measurement location information to the users.

The VGI information provided by citizens on how they feel will also be presented in a colour scale. In addition, additional information as for instance the pollen index will also be presented in a colour scale. In order to facilitate the interpretation of the air quality, VGI and pollen information, the same colour scale will be followed for the three indexes. The Table 11.3 shows an example of the four-colour scale that will be implemented.

The analysis of the big data set generated will help clarify relationships between air quality, pollen, meteorology and health. In the future, the identification of what particular patterns trigger symptoms in a person will help to personalize people's information and provide them with pre-alerts to help them to take the necessary actions to minimize such symptoms.

The feelings/observations reported by the citizens will also be overlaid on a street map together with the information from the sensors employing a different symbol allowing the users easily differentiate between sensor data and citizen reported data. The information reported by people is subjective and it exists also the possibility that some users abuse of the application providing false information, therefore it is important that VGI information can be easily distinguished. It is out of the scope of this project to incorporate algorithms to detect false data or users reporting false data in real-time.

Citi-Sense-MOB is developing a multi-platform app to allow citizens to get information on environmental conditions in the city of Oslo in near-real time. The GPS sensor on the mobile phone will allow the users to get automatic information related with the air quality in their immediate surroundings. Also will allow tracking air pollution exposure in daily activities by starting a track session. The mobile application will not only allow the users to get information about the air quality status but also to update voluntary geographic information as pictures, comments, perceptions, etc. That information (if the user agrees) will be publicly shared on social networks, helping to enable community empowerment.

2.4 Social Networks

Citi-Sense-MOB develops various social media platforms to foster communication between the project's partners, stakeholders and users, and to facilitate citizens' engagement, participation and network building. These social media platforms are all about engagement, participation, relationship building and dissemination. Every platform encourages its users to take part, by commenting on what they see and getting involved in conversations with others. This makes it a particularly useful vehicle both for informing users and for gaining their feedback.

The project is using on-line platforms as Facebook, Twitter, YouTube, LinkedIn, Forums and Blogs to disseminate the information and get feedback from citizens. Additionally off-line modes as meetings and workshops, newspapers and magazines, television, brochures and scientific publications are also used to reach people that is not using internet (e.g., elderly people) or other stakeholders easier to be reach by off-line modes (e.g., authorities and scientific communities).

For Citi-Sense-MOB, social media is one of the tools for succeeding with collaborative participation and citizens' empowerment. Citizens' empowerment in Citi-Sense-MOB can be regarded as a continuum, from low involvement where citizens receive relevant information to relatively high involvement, where citizens contribute by carrying sensors and reporting information.

3 Preliminary Results

3.1 Stakeholder Engagement

In Citi-Sense-MOB, there are several-type of tools for citizens' engagement and participation: (i) One is via web-based social media for public; (ii) Another through offline modes (e.g., Meetings and Workshops, Newspapers and Magazines, Television, Presentation, Brochures, Scientific publications, etc.) for those people without access to the Internet or those who prefer not accessing the internet for such participation; (iii) Third one is portable sensors that can be carried by the volunteers in their daily life; and (iv) Fourth one is mobile app developed by the project that will be used by the citizens to upload their subjective/objective observations.

Currently, in Citi-Sense-MOB, we have engaged the following stakeholders: (i) The general public who are interested in air quality related environmental issues (e.g., kindergartens, families, cyclists, etc.); (ii) Health interest groups, such as those asthma and allergy patients who suffer at certain times of the year due to ambient the air quality situation, and consequently have a critical demand for services that could enable them to take actions to mitigate their symptoms; (iii) Local governments/Public entities (i.e., Oslo municipality and other city municipalities); (iv) Transport companies (i.e., Ruter AS, Nobina AS, The Norwegian

Public Roads Administration); (v) Researchers; and (vi) Enterprises (i.e., data miners, app developers).

3.2 Social Networks

Social networks are used as a channel to communicate news on Citizens' Observatories. Currently three main social networks are used Facebook, LinkedIn and Twitter. Facebook works well for engaging citizens, offering particular advantages for reaching users for local or community focused joint citizens' observatories-related activities or dissemination. The Oslo Citizens' Observatory Facebook Group (/oslocitizensobservatory) has updated information on ongoing activities in the project Citi-Sense-MOB as well as news regarding citizens' observatories-related activities in Oslo and other cities in Norway.

The Oslo Citizens' Observatory Facebook Group allow users, among other things, to share information, establish contact between them or plan jointly activities. It will also allow conducting questionnaires to understand the needs of users better or the usability and usefulness of the services provided by the Citi-Sense-MOB project.

The LinkedIn Group (/Citizens-observatories-5164755) targets a more professional audience. This does not currently carry official branding for Citi-Sense-MOB project. However, it is highly encouraged to be used by any citizens' observatories-related activities including Citi-Sense-MOB for their professional networking.

The Twitter account (/CitiSenseMob) was created with the aim to interact with a wider audience on a different platform. Having this twitter account gives the project the ability to read/post/reply from/to interested citizens, and public organizations, companies and governmental agencies. It has the additional benefit of being able to direct efforts at focus groups and interact with them. This is specifically important in being able to improve the services that the project is going to provide.

3.3 Mobile Phone Application

In order for users to easily visualize the information on air quality a mobile phone application has been developed named Personalized air quality map (PAQ MAP). PAQ MAP will be provided as a free download application.

The five main options of the mobile application are displayed at the bottom of the screen (see Fig. 11.3) and include "air quality", "tracking", "symptoms", "information" and "settings". The option "air quality" includes the display of the AQI computed from the data gathered by the low-cost sensors and the data from the reference monitoring stations, an air quality map, the pollen index and the VGI information submitted by the citizens (see Fig. 11.3a and Table 11.3). The option "tracking" allows the users to track themselves. If a user is carrying a sensor the

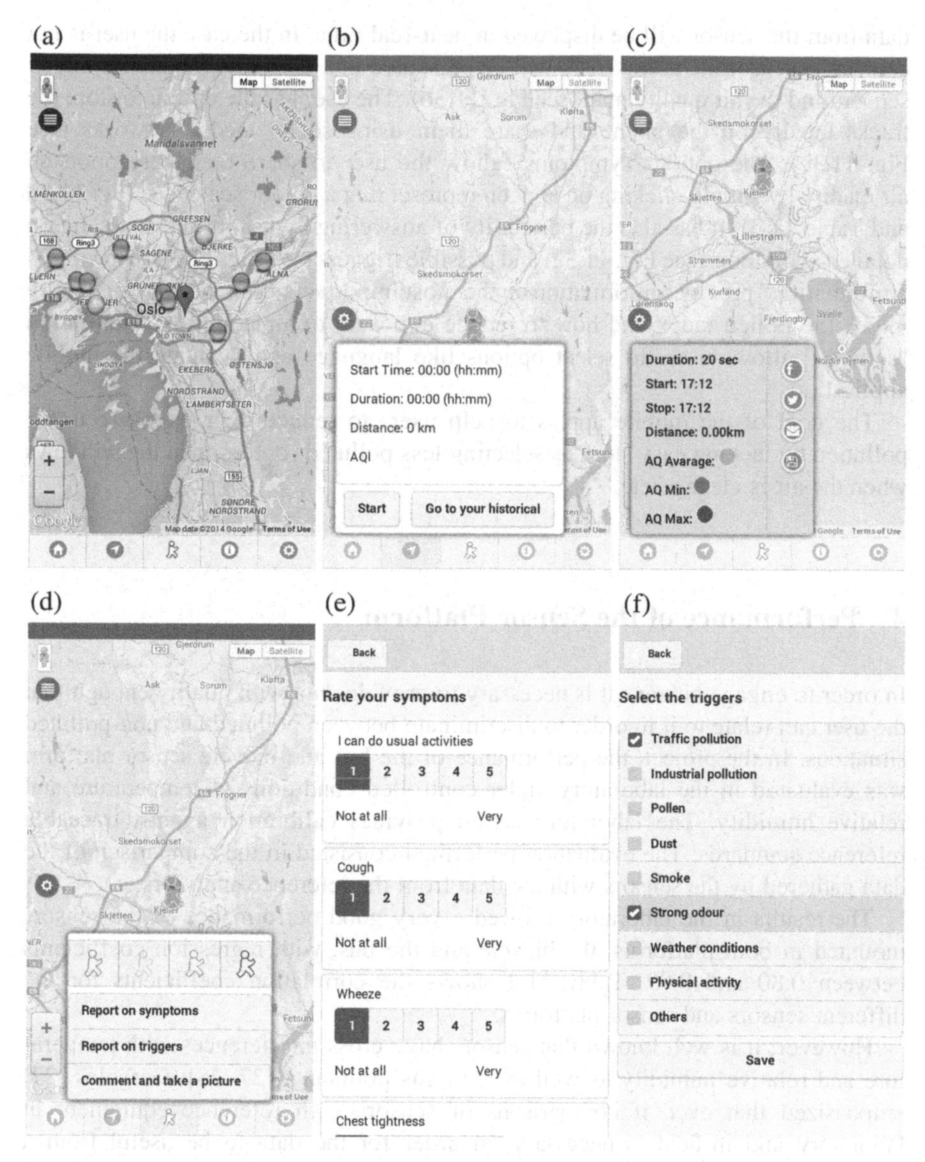

Fig. 11.3 Description of the most important functionalities in the PAQ MAP mobile phone application. (**a**) In the first menu the user can select different display options: the AQI from the sensors and the reference stations, the pollen index, the air quality map and the VGI information. (**b**) In the second menu the user can track his position and see the data in near-real time of the sensor he is carrying. It is also possible to track the position without carrying a sensor, in that case the data from the air quality map will be showed. (**d**) The tracks can be stored and shared using social networks. (**e**) In the third menu the user can report the overall feeling by simply clicking on an icon, (**f**) report on specific symptoms and symptom severity and (**d**) on possible triggers

data from the sensor will be displayed in near-real time. In the case the user is not carrying a sensor the users' track will be presented in a map with the data from the sensors and the air quality map (see Fig. 11.3b). The user has the option to store the tracks locally on the phone and share them using social media networks (see Fig. 11.3c). The option "symptoms" allow the user to share their perception on air quality by simply clicking on an icon representing a colour scale (see Fig. 11.3d and Table 11.3). It has also the possibility of answering to a short questionnaire on detailed symptoms (see Fig. 11.3e) and possible triggers (see Fig. 11.3f). The option "information" provides information on the most important pollutants, the air quality index, the pollen index and how to reduce exposure to air pollution. The option "settings" allows users to select options like language and evaluate the mobile application.

The goal of the mobile app is to help users to reduce their exposure to air pollution by tacking easy steps as selecting less polluted routes, open the windows when the air is cleaner, etc.

4 Performance of the Sensor Platform

In order to engage citizens it is necessary to provide data with quality enough that the user can relate to it in order to discriminate between polluted and non-polluted situations. In the project, the performance of the bus and bicycle sensor platform was evaluated in the laboratory under controlled conditions of temperature and relative humidity. The laboratory set-up provides calibration against traceable reference standards. The evaluation performed consisted in the comparison of the data gathered by the sensors with the data from the reference analysers.

The results in the laboratory showed a very good performance of the sensors mounted in both platforms, the bicycle and the bus, with regression coefficients between 0.80 and 0.99. Table 11.4 shows the correlation coefficients for the different sensors and sensor platforms.

However, it is well known that sensors have cross-interferences with temperature and relative humidity as well as with gas pollutants [27, 28]. Gerboles [27] emphasized that even if comparisons of sensors with reference equipment in laboratory and in-field is necessary, in order for the data to be useful from a scientific and technical point of view, it is necessary to develop methods for

Table 11.4 Regression coefficients calculated in the laboratory between the reference equipment and the platforms mounted on the bicycle and on the bus platform

Pollutants	Bus platform (r^2)	Bicycle (r^2)
CO	0.92	0.99
NO_2	0.99	0.95
NO		0.80
O_3		0.89

correcting cross-sensitivities and the effects of temperature and relative humidity, as well as, methods to correct the changes in baseline/span drift of sensors with aging.

It was seen in the laboratory that the NO_2 sensor from the bicycle platform had interferences with ozone. It is then expected that when the bicycle is monitoring NO_2 outdoors it will experience ozone interferences. Mead et al. [28] showed that it is possible to correct the NO_2-O_3 interferences if the concentrations of ozone are known. The work was conducted with a static platform and the ozone was monitored with reference equipment. In Citi-Sense-MOB the fact that the sensors are mounted on a mobile platforms supposes an increased challenge to correct the signal, as the platform is not co-located with a reference station. The signal post-processing in mobile outdoor monitoring will be investigated in the second part of the project. Post processing of the data generated by the sensors is important to improve the sensor performance.

In the platform mounted on the bus, an ozone filter was installed in the chamber in order to reduce or completely eliminate the NO_2-O_3 interferences. A more exhaustive analysis of its performance will be conducted on the second year of the project when more data is available.

4.1 Road Trials

The low-cost sensors installed on mobile platforms, namely, buses and electrical bicycles will allow to get air quality data at road level and create a high resolution map of air pollution.

Buses have been selected because they have a very well maintained vehicle infrastructure. Any malfunction on the platform can be detected and fixed when the bus is in the garage. They have well-known routes through the city, facilitating the analysis of the data, as it is possible to compare pollution patterns over time. The buses work continuously, thus generating a large amount of data for data analysis algorithms. The data collected can help increase our knowledge about pollutant concentrations at road level.

Additionally, road traffic is one of the main contributors to CO_2 emissions. Local authorities and transport agencies have green agendas that aim for a reduction in CO_2 emissions. The monitoring platform installed on buses, combining a system for air pollution data collection and monitoring of driving efficiency, will empower the drivers to adopt a more environmentally friendly driving behaviour, helping the bus company achieve fuel savings and the environment authorities to accomplish with the green agendas. Due to the number of buses running continuously in the city, reductions in their emissions should provide significant reductions in urban greenhouse gas emissions.

In addition to the vehicular platforms, sensors have also been mounted on electrical bicycles. Bicycles can cover parts of the city that are not accessible by car, for instance parks and pedestrian areas. Furthermore, bicycle lanes often run

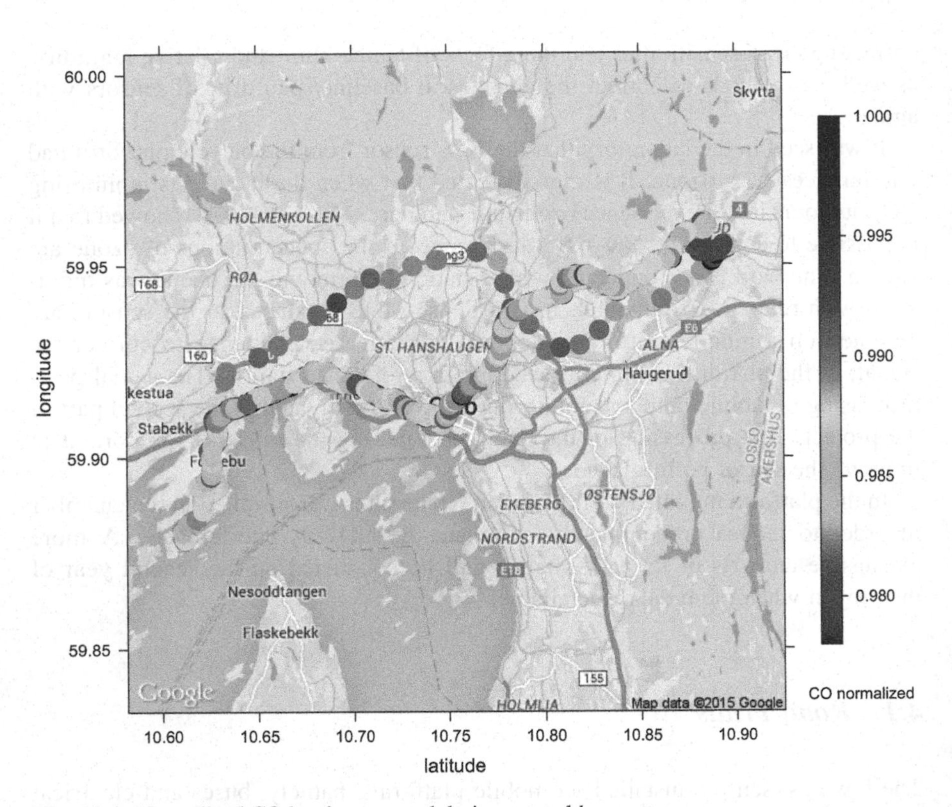

Fig. 11.4 Normalized CO levels generated during several bus routes

alongside the main roads. This will allow us to study for instance the gradient between measurements at the centre of the road, and on the side of the road, used by pedestrians.

To guarantee the quality of the data and the near-real-time communication between the sensor platform and the database, a testing phase with a suitable number of sensor units is currently being developed. During the test phase, one air quality sensor is tested on an electrical bicycle and two sensors from UrVAMM [14] have been mounted on buses and tested in their daily routes. This test phase represents a "live" test to identify unexpected issues and verify that the whole system functions as expected before the full deployment study.

Figures 11.4 and 11.5 present normalized CO levels generated during several bus routes and bicycle rides, respectively. The data collected during the first road trials has proved to be able to capture the spatial and temporal variability of urban air quality. Consequently, we expect that a denser mobile monitoring network will suppose an improvement in the spatial and temporal resolution of the urban air quality mapping.

The methodology for the analysis of air quality in Citi-Sense-MOB is based on the generation of a high volume of routes that cover as much of the city as possible. The data gathered will allow assessment of the spatial-temporal distribution of

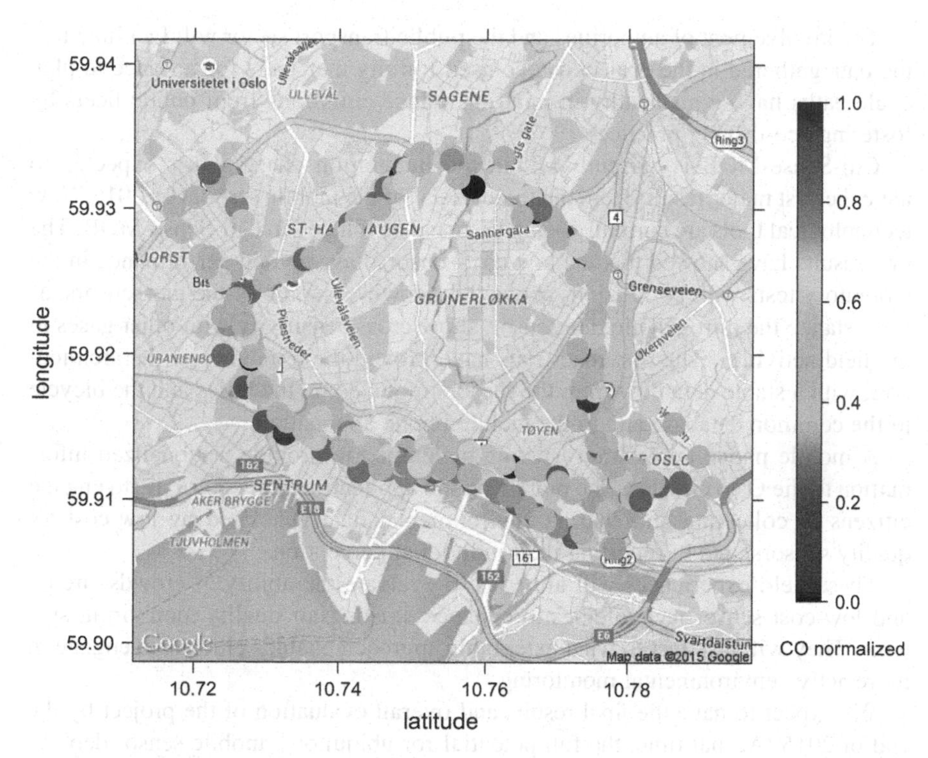

Fig. 11.5 Normalized CO levels generated during several bicycle rides

pollutant concentrations in a variety of environments where data are not available, and will help identify areas where elevated pollutant concentrations may occur. The data gathered by the air quality sensors and the position will be used to generate air quality maps at street level.

5 Conclusions

The Citi-Sense-MOB project represents an innovative development in environmental monitoring and individual exposure assessment. By deploying sensor units on mobile platforms, testing its feasibility and investigating how these data can contribute to a more comprehensive understanding of air quality monitoring, we hope to show that low-cost monitoring can contribute to create maps at road level that complement the data from the existing monitoring networks.

The participation of citizens in the project, allowing subjective observations as well as contributing with sensor measurements riding the Citi-Sense-MOB bicycle, will contribute to raise environmental awareness and help citizens mitigate exposure to air pollution.

The involvement of authorities and the public transport sector will facilitate that the data gathered in the project can be used for city managers for instance to plan cycle paths having air quality in mind or reduce emissions from public fleets by fostering eco-driving practices.

Citi-Sense-MOB is currently within the integration phase and is expected to have the first major results from the 'road trials' of the mobile sensors by 2015. New technological tools are currently being used and developed in Citi-Sense-MOB. The first results have showed that the low-cost sensors have a good performance in the laboratory tests with correlations above 0.8. However, some of the gas sensors, as for instance the nitrogen dioxide sensor, show cross-sensitivity with other gases in the field activities. The first road-trials have proved the reliability of the architecture, with a stable data flow from the sensors mounted on the buses and the bicycle to the common database and to the mobile phone application.

A mobile phone application is being developed to provide personalized information to the Oslo citizens. The mobile phone application will contribute to engage citizens in collecting and sharing environmental data generated by low-cost air quality sensors, and in reporting their individual perception.

These field experiences will allow us to evaluate the ability of crowdsourcing and low-cost sensor technologies to enhance existing air quality monitoring systems. They will also test to what extent this approach enables citizens to engage in more active environmental monitoring.

We expect to have the final results and overall evaluation of the project by the end of 2015. At that time, the full potential for ubiquitous, mobile sensor deployment and their integration into static monitoring networks will be known and we expect to demonstrate positive results for both citizen engagement, exposure assessment and city management.

Acknowledgments Citi-Sense-MOB (http://www.citi-sense-mob.eu) is a collaborative project partly funded by The European Mobile and Mobility Industries Alliance (EMMIA) strand II: Large-scale demonstrators in support of GMES and GNSS based services. We would like to thank to the support from Oslo Kommune, Ruter AS and Nobina AS.

References

1. Environmental European Agency. Air quality in Europe-2013. EEA Report No 9/2013 (2013).
2. European Commission. Reclaiming city streets for people. Chaos or quality of life?, Office for Official Publications of the European Communities (2004).
3. HEI Panel on the Health Effects of Traffic-Related Air Pollution. Traffic-related air pollution: a critical review of the literature on emissions, exposure, and health effects, HEI Special Report 17. Health Effects Institute, Boston, MA. (2010).
4. Leary P.J., Kaufman, J.D., Barr, R.G, Curl, C.L, Hough, C.L, Lima, J.A, Szpiro, A.A, Van Hee V.C, Kawut, S.M. Traffic-related air pollution and the right ventricle. The multi-ethnic study of atherosclerosis. Am. J. Respir. Crit. Care Med, vol.189, pp.1093–100 (2014).

5. WHO. Ambient (outdoor) air quality and health. Fact sheet N°313. Published by World Health Organization. http://www.who.int/mediacentre/factsheets/fs313/en/ (2014). Accessed 24 February 2015.
6. Liu, H.Y., Skjetne, E, Kobernus, M. Mobile phone tracking: in support of modelling traffic-related air pollution contribution to individual exposure and its implications for public health impact assessment. Environmental Health. vol. 12, pp.93–105 (2013).
7. European Commission. Cities of tomorrow. Challenges, visions, ways forward, European Commission, Directorate General for Regional Policy. Publications Office of the European Union, Luxemburg, 2011.
8. Milton, R., Steed, A. Mapping carbon monoxide using GPS tracked sensors. Environ. Monit. Assess. Vol. 124, pp. 1–19 (2007).
9. Dutta, P, Aoki, P.M., Kumar, N., Mainwaring, A., Myers, C., Willett, W. and Woodruff, A.. Common Sense: Participatory Urban Sensing Using a Network of Handheld Air Quality Monitors (demonstration). Proc. SenSys 2009, Berkeley, CA, Nov. 2009, 349–350 (2009).
10. Castell, N., Kobernus, M., Liu, H.Y., Schneider, P., Lahoz, W., Berre, A. J., Noll, J. Mobile technologies and services for environmental monitoring: The Citi-Sense-MOB approach. Urban Climate. Available online, (2014).
11. Miljødirektoratet. Annual average concentration of NO2 in five main cities in Norway. http://www.miljodirektoratet.no/ Accessed 24 February 2015.
12. Oslo Kommune. European green capital application form Oslo including additional information. Accessed 24 February 2015 http://www.oslo.kommune.no/getfile.php/byr%C3%A5dsavdeling%20for%20milj%C3%B8%20og%20samferdsel%20%28MOS%29/Internett%20%28MOS%29/Dokumenter/presentasjoner/Oslo_ECG_aplication.pdf
13. Engelken-Jorge M, Moreno J, Keune H, Verheyden W, Bartonova A, CITI-SENSE Consortium. Developing citizens' observatories for environmental monitoring and citizen empowerment: challenges and future scenarios. In Proceedings of the Conference for EDemocracy and Open Government (CeDEM14): 21–23 May 2014; Danube University Krems, Austria. Edited by Parycek P, Edelmann N. pp. 49–60 (2014).
14. Rionda, A., Marín, I., Martínez, D., Aparicio, F., Alija, A., García Allende, A., Miñambres, M., X. G. Pañeda., UrVAMM – A Full service for Environmental-Urban and Driving Monitoring of Professional Fleets. SmartMILE 2013, International Conference on New Concepts in Smart Cities. Gijón, Spain, (2013).
15. Castell, N., Liu, H.Y., Kobernus, M. Berre, A.J., Noll, J., Cagatay, E., Gangdal, R. Mobile technologies and personalized environmental information for supporting sustainable mobility in Oslo: The Citi-Sense-MOB approach. In : Marx Gómez, J., Sonnenschein, M., Vogel, U., Winter, A., Rapp, B., Giesen, N., eds. EnviroInfo 2014 – 28th International Conference on Informatics for Environmental Protection. BIS-Verlag, Oldenburg. ISBN 978-3-8142-2317-9 (2014).
16. NAAF. Norwegian Asthma and Allergy Association. http://www.naaf.no/ Accessed 24 February 2015.
17. Liu, H.-Y. (ed.) Report contributors: Berre, A.J., Cagatay, E., Liu, H.-Y., Noll, J., Kobernus, M., Castell, N., Fayyad, S., Khattak, W. Citi-Sense-MOB: Conceptual services design document. Kjeller, NILU OR, 19/2014 (2014).
18. ATONE: Service innovation method. http://www.service-innovation.org/wp-content/uploads/2007/12/at-one-a3-size-paper.pdf Accessed 24 February 2015.
19. Oslo Kummune: Local transport. Accessed 24 February 2015 http://www.miljo.oslo.kommune.no/getfile.php/Milj%C3%B8portalen%20%28PMJ%29/Internett%20%28PMJ%29/Dokumenter/Rapporter/3%20Kommunens%20milj%C3%B8arbeid/Indicator%202_Oslo.pdf
20. Panis, L.I, deGeus,B., Vandenbulcke,G., Willems,H., Degraeuwe,B., Bleux,N., Mishraa, V., Thomas,I., Meeusen,R., Exposure to particulate matter in traffic: a comparison of cyclists and car passengers. Atmos. Environ, vol. 44, pp. 2263–2270 (2010).

21. Hatzopoulou,M., Weichenthal,S., Barreau, G., Goldberg, W., Farrell, W., Crouse, D., Ross, N. A web-based route planning tool to reduce cyclists' exposures to traffic pollution: A case study in Montreal, Canada. Environ. Res., vol. 123, pp 58–61 (2013).
22. Aleixandre, M., Gerboles, M.,. Review of small commercial sensors for indicative monitoring of ambient gas. Chem. Eng. Trans., vol. 30, pp 169–174 (2012).
23. DunavNet: M2M, Mobile apps, Software & Game. http://www.dunavnet.eu/ Accessed 24 February 2015.
24. Drosatos, G., Efraimidis, P., Athanasiadis, I., Stevens, M., D'Hondt, E. Privacy-Preserving Computation of Participatory Noise Maps in the Cloud. Journal of Systems and Software, vol 92, pp 170–183 (2014).
25. S. van den. E, Léger, K., Nussio, F. Comparing urban air quality in Europe in real time A review of existing air quality indices and the proposal of a common alternative. Environment International, vol. 34, pp. 720–726 (2008).
26. Norwegian air quality index. www.luftkvalitet.info Accessed 24 February 2015.
27. Gerboles, M. Developments and Applications of Sensor Technologies for Ambient Air Monitoring. Workshop "Current and Future Air Quality Monitoring", Barcelona, (2012).
28. Mead, M.I., Popoola, O. Stewart, G.B., Landshoff, P., Calleja, M., Hayesb, M., Baldovi, J.J., McLeod, M.W., Hodgson, T.F., Dicks, J., Lewis, A., Cohen, J., Baron, R., Saffell, J.R., and Jones, R.L. The use of electrochemical sensors for monitoring urban air quality in low-cost, high-density networks. Atmospheric Environment, vol. 70, pp. 186–203 (2013).

Chapter 12
Spatial Information for Safer Bicycling

Martin Loidl

Abstract The need for sustainable modes of transport is obvious, especially in urban areas. Because of the large number of trips within cities and distances lesser than 5 km, the bicycle is regarded as optimal mode of transport, both for utilitarian and leisure trips. Nevertheless, safety concerns are among the most relevant factors that hamper an increasing bicycle usage. Geographical Information Systems (GIS) with their ability to model and analyze road infrastructure and users in an explicitly spatial context can significantly contribute to meet these safety concerns. They can be employed in all stages of better understanding bicycle safety as a spatio-temporal phenomenon and provide the basis for informed decisions in the context of planning, information provision and cycling promotion.

After a short introduction about why it is necessary to address safety issues in the promotion of the bicycle as sustainable mode of transport, the benefits of a spatial perspective on the road space and its users are described. The main argument is that road traffic, and with this road safety, are spatial phenomena by their very nature and thus GIS can significantly contribute to various applications that foster safety improvements for bicyclists. In order to demonstrate how spatial information can be incorporated in various contexts, several application examples and case studies, where spatial modelling and analysis are key features, are given. Based on this overview a final section provides a brief outlook of current and future research topics that aim to further make use of spatial information for safer bicycling.

Keywords Geographical information systems (GIS) • Bicycle safety • Spatial analysis

1 The Bicycle as Sustainable Mode of Transport

Numerous negative impacts of motorized traffic – from air pollution [1] to economic externalities [2] and social inequities [3] – have led to a growing demand for sustainable modes of transport; especially in densely populated, urban environments. This development has increasingly brought the bicycle, as sustainable and

M. Loidl (✉)
Department of Geoinformatics, Z_GIS, University of Salzburg, 5020 Salzburg, Austria
e-mail: martin.loidl@sbg.ac.at

J. Marx Gómez et al. (eds.), *Advances and New Trends in Environmental and Energy Informatics*, Progress in IS, DOI 10.1007/978-3-319-23455-7_12

cost-efficient mode, into the focus of researchers, planers and decision makers [4]. Masterplans for bicycle traffic and bicycle promotion initiatives from local to transnational levels are indicators for this (re-) discovery of the bicycle [5].

1.1 Bicycle Traffic and Safety Concerns

Due to extensive bicycle promotion initiatives, many cities in Europe have successfully built or extended their bicycle infrastructure. This has significantly contributed to a constantly rising number of bicyclists on the roads [6]. But still, there are some influential factors which keep people from using the bicycle for their utilitarian mobility needs, above all safety concerns [7, 8]. Although sound exposure data are rare [9], there are indications from literature and official statistics, that cycling is healthy, but dangerous related to the travelled distance [10]. These findings are in line with a recent report by the European Commission on road safety. There, the EC points to the fact, that contrary to the overall trend, the number of killed bicyclists has been increasing during the last couple of years [11]. Thus, it can be stated, that perceived and objective safety is a key issue in the context of promoting the bicycle as sustainable mode of transport [12, 13].

1.2 Improving Bicycle Safety

In order to improve safety for bicyclists, at least three issues ("safety pillars", adapted from Othman et al. [14]), which are interrelated, need to be addressed on various levels:

- First of all, the infrastructure and regulative interventions need to be designed in a way that potential risks for bicyclists are minimized [15]. This can be done, for example, by separated bike lanes, controlled intersections or actions to reduce motorized traffic and speed [16].
- Secondly, the bicyclist's physical condition, experience, compliance with road traffic rules and the technical condition of the vehicle must be taken into account. Although "individual" factors do not fully explain all incidents, it is noteworthy, that – depending on the source – 5–30 % of all fatal injuries to bicyclists are caused by single-bicycle crashes [17]. However, targeting the mentioned individual risk factors requires an integrated approach comprising actions from traffic control to awareness initiatives.
- As a third aspect, bicycle safety can be improved by context-sensitive information about the environment, such as high-resolution, near-future weather data and specific information offers about optimal (safe) routes [18]. Such routing (Fig. 12.1) recommendations aim to minimize the bicyclist's exposure to risk factors, such as primary roads with a high traffic load or roads without any bicycle infrastructure.

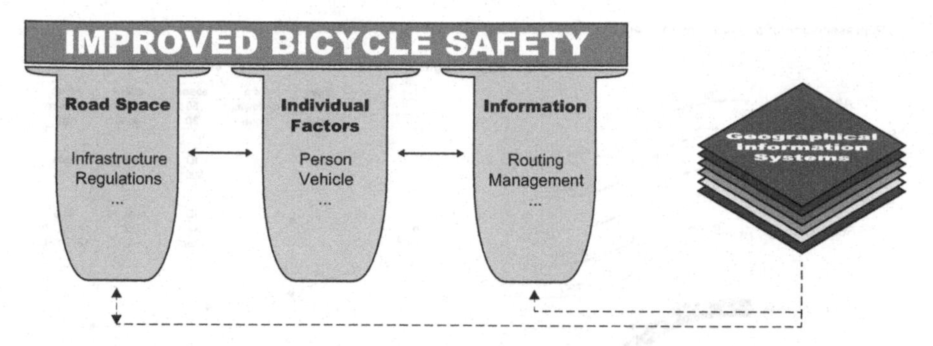

Fig. 12.1 Three approaches for the improvement of bicycle safety (adapted from Othman et al. [14])

Individual factors which potentially contribute to an improved bicycle safety cannot be directly targeted by geographical information systems. But GI systems allow for a systematic, digital representation of the road space, including the physical and legal characteristics. Additionally the spatial location can be used as a common denominator for multiple perspectives on the road space. Thus GIS can facilitate an interactive communication about the physical space in a digital, map-based environment [19, 20]. Existing authoritative, commercial or crowd-sourced data can subsequently be managed, modeled and analyzed in Geographical Information Systems and in turn serve as basis for innovative planning and information applications. Thus GIS can be – and is – employed in various planning contexts, from status-quo analysis to simulation, in the management and improvement of existing infrastructure systems, for participatory settings, such as feedback portals, and for user-specific information applications.

In the following sections it is briefly argued why and how Geographical Information Systems can contribute to a better understanding of bicycle safety and how these insights can be used for informed planning and management decisions and information applications. As safety concerns and evident safety threats hold back potential bicyclists, tackling these aspects from several sides should be part of any integrated effort to promote sustainable mobility, especially in urban environments.

2 What GIS Can Offer

Bicycle safety is a complex spatial and temporal phenomenon with multiple influential factors, such as the built infrastructure, the traffic volume, weather conditions and the individual bicyclist as an entity in physical space and time. In order to better understand this multi-facetted phenomenon with all dependencies and relations, an integrated perspective which brings together any kind of information about the road space and its users is required. Geographical Information Systems allow for such a perspective as the road space, any events and users are

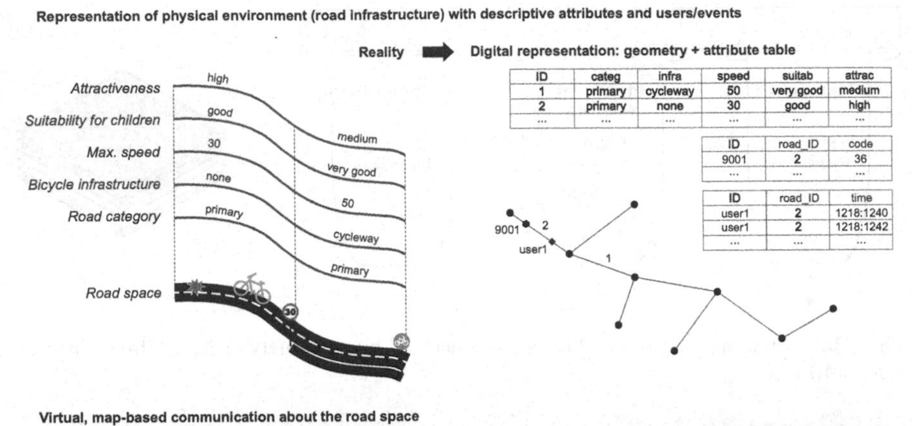

Fig. 12.2 GIS as integrated platform for multiple perspectives on the road space

spatial by their very nature or can be spatially related (geo-located). Thus they can be digitally captured, related, analyzed and mapped.

GIS is used for the digital representation of the physical environment with an (theoretically) unlimited number of descriptive attributes, for users and events in this environment and for the virtual, mainly map-based communication about this digitally represented space (see Fig. 12.2). Using the geographical location as common denominator many more information layers can be related to the digital representation of the road space as such. Through overlay techniques queries and analysis can be done over multiple layers and finally visualized in maps.

Whereas established domains dealing with mobility and transportation – such as traffic engineering, telematics or planning – have a specific, rather "technical" view, Geographical Information Systems can beyond that serve as integration platform where multiple perspectives – "technical" as well as qualitative – on the road space can be merged. GIS thus allows for a holistic approach towards a better understanding of bicycle safety and facilitates innovative applications for several domains.

Providing a sound data basis in the context of the promotion of safe cycling is of great importance for informed decisions. The consideration of explicit spatial characteristics is especially valuable in the analysis of bicycle accidents and in

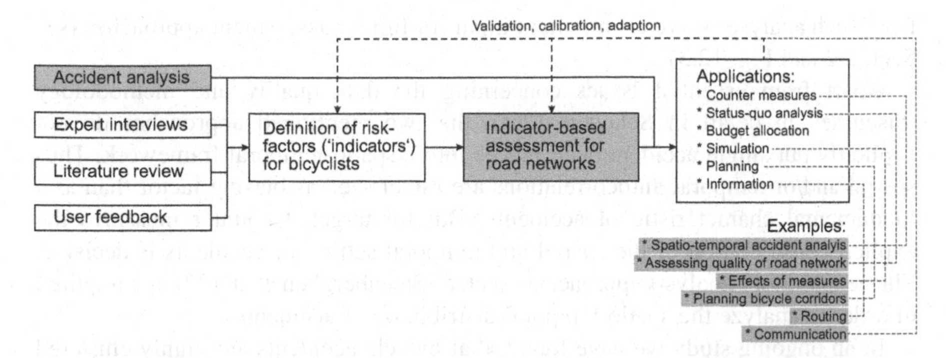

Fig. 12.3 Multi-stage workflow to provide sound data and analysis results for informed decisions in the context of bicycle safety. The employment of GIS especially allows for additional insights in accident analysis and assessment approaches (*light brown*, see Ref. [22]). The result of an integrated spatial analysis approach can then form the basis for various applications (*yellow*)

the assessment of the road network's quality in terms of bicycle safety, which in turn serve as basis for several applications (for a general introduction to GIS in the context of road safety see [21]).

In order to show the potential of an explicit consideration of space in the analysis of bicycle safety and in the promotion of safe cycling as sustainable transport mode, multiple examples are provided in the following sections. These examples can be related to or are the outcome of a multi-stage workflow which comprises several tasks that rely on GI technology and tools (see Fig. 12.3).

2.1 Spatio-Temporal Accident Analysis

Analyzing bicycle accidents with GIS helps to better understand where, when and under which physical conditions accidents did or are more likely to occur.

Currently, mainly two, explanatory approaches are applied in order to better understand the mechanisms behind bicycle accidents. Firstly, epidemiological approaches (see Ressing et al. [23] for a methodological overview) aim to calculate risk factors that lead to bicycle accidents with a certain probability. For this, mainly aggregated accident data are related to a statistical population: Delmelle and Thill [24] use census districts as aggregation level for a multivariate statistical analysis. Yiannakoulias et al. [25] calculate the risk exposure for person kilometers travelled, based on estimated distances per census districts. Secondly, in-depth analyses of detailed accident reports seek to identify individual, contributing risk factors. In an extensive study Teschke et al. [15], for example, relate accident reports, including socio-demographic variables of involved persons to different route types. In doing so, physical conditions that increase the risk for bicycle accidents (e.g. downhill grade, major streets with on-street parking) could have been identified. Findings from such analyses serve as important input for further assessment approaches (see Sect. 2.2 and Fig. 12.3).

Apart from essential issues concerning the data quality and methodology (discussed in detail in Schepers [26]), the two mentioned approaches do not explicitly put single accident occurrences into a spatio-temporal framework. Thus spatial and/or temporal autocorrelations are rather seen as biasing factor than as a fundamental characteristic of accidents. But for targeted counter measures the explicit consideration of the spatial and temporal setting of accidents is decisive. Thus exploratory analysis approaches, such as Steenberghen et al. [27], are required in order to analyze the spatio-temporal distribution of accidents.

In an ongoing study we have found, that bicycle accidents are highly clustered within the road network and that the temporal variation – in several scales – is significant. First results of a case study indicate that during the summer months bicycle accidents in the investigated city are, contrary to the "safety in numbers phenomenon" [28, 29], a function of the bicycle volume. Significantly more accidents than expected from a random distribution occur on major bicycle connections without any motorized traffic and an excellent bicycle infrastructure. In contrast to this, more accidents (than expected from random distribution) happen on radial connections with less or no bicycle infrastructure during the winter months (Fig. 12.4 shows the single locations of bicycle accidents for each season).

Counter measures, such as infrastructure adaption, traffic management or road surveillance, can be planned much more effectively when the spatial and temporal hot spots are known. In order to find the most suitable measure for individual hot spots, an in-depth analysis of the respective incidents is required. Through the preceded analysis, which functions as filter, these detailed investigations can be done much more efficiently than in a global setting.

2.2 *Assessing Quality of Road Network in Terms of Safety*

Based on findings from accident analyses, literature reviews, expert interviews and extensive user feedback, indicators that contribute to a potential safety threat for bicyclists, both perceived [30] and objective [31], can be identified and compiled in a weighted, global assessment model. As it is indicated in Fig. 12.3, the setup and calibration of the assessment model is an iterative process. Hence the result of the quality assessment depends on the indicator selection and definition of the single weights.

Compared to other assessment routines, such as in-situ expert assessment, the indicator-based assessment model has several advantages [22]: Firstly, the composition of the assessment result can be traced back to the smallest building blocks of the model. This transparency ensures a straight-forward interpretation, independent from the user. Secondly, the model is adaptable. This means that the indicator set and the weights can be adapted in a way that they reflect the actual situation as optimal as possible. Thus the model allows for example for a differentiation

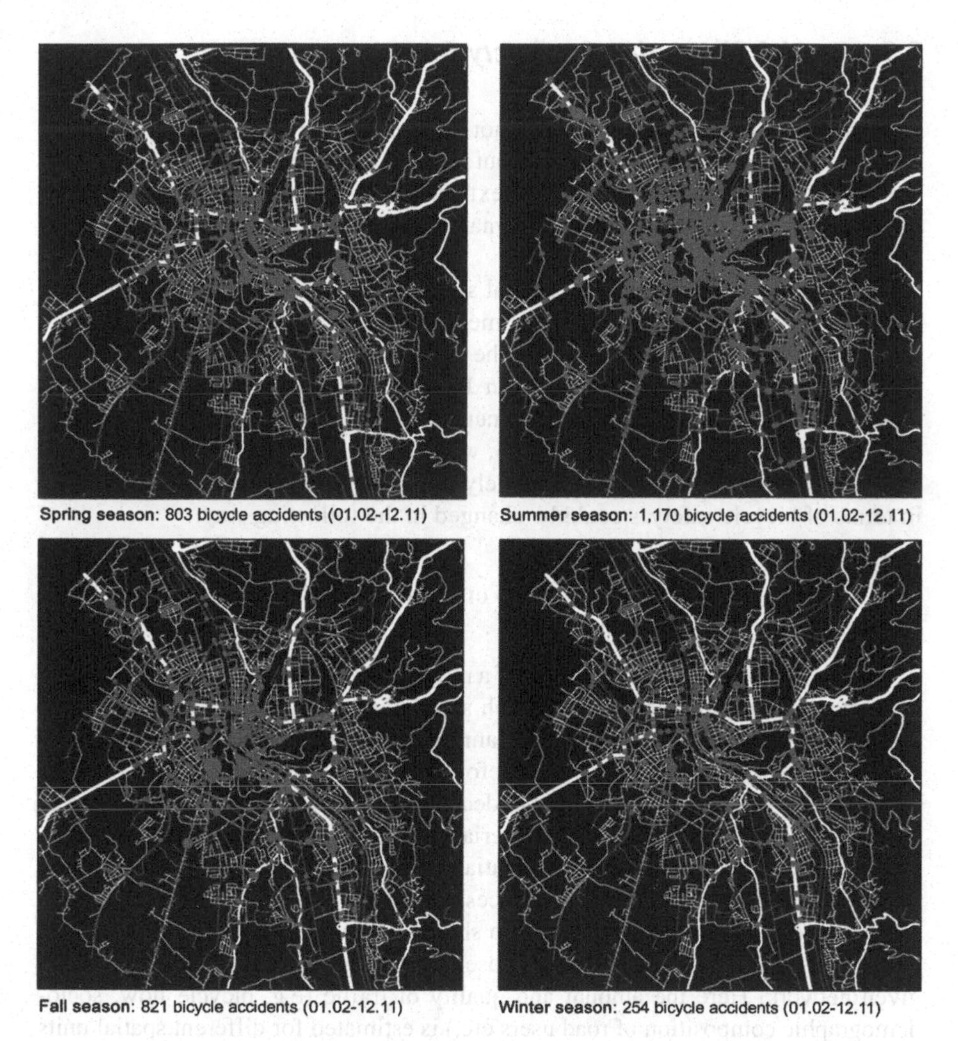

Spring season: 803 bicycle accidents (01.02-12.11)
Summer season: 1,170 bicycle accidents (01.02-12.11)
Fall season: 821 bicycle accidents (01.02-12.11)
Winter season: 254 bicycle accidents (01.02-12.11)

Fig. 12.4 Spatial distribution of bicycle accidents in the city of Salzburg (Austria) for each season. Preliminary results from an ongoing study

between rural and urban environments where the risk factors for cyclists tend to be different [32]. Thirdly, the model generates comparable outputs and facilitates comparison tasks between different regions or time slices, based on standardized inputs and analysis procedures. Related to this, the results can be reproduced and the model can be applied multiple times. Accordingly the model is perfectly suitable for simulation purposes (see next section). The assessment of a whole road network in terms of bicycle safety is the starting point of any further planning activity. It helps to gain an overview of the overall quality of the network, identifies missing connections and makes weak points in the system obvious.

2.3 Spatial Simulation for Safety Improvement

Geographical Information Systems are not only suitable to accurately represent the current situation of the road space, but also to simulate future or alternative scenarios [33]. Simulations in this context can either focus on the environment or the road users and their behavior and demands. Both aspects are being dealt with in the following.

The first application context of spatial simulation is built on models such as the aforementioned indicator-based assessment model. Here the model is used to extrapolate the effect of changes in the road space, which is reflected in the model as indicator. For example it can be simulated with which measures the quality of a given segment could be increased. An arbitrarily selected segment has, for example, an index value of 0.63, whereas the neighbouring segments have index values of 0.31 and 0.36[1] respectively. The index value – and thus the quality in terms of bicycle safety – could be changed in the following way:

- Bicycle lane, no other change: 0.44
- Physically separated bicycle way, no other change: 0.37
- Calmed road, no other change: 0.49

This example illustrates the benefit of a model-based simulation, implemented in a geographical information system. Such simulation results can serve as basis for informed decisions in the process of planning or budget allocation, as they comprehensibly test the effect of measures before investments are made. Combining the simulation of measures with additional decision factors, such as the legal scope or monetary costs, results in a multi-criteria optimization problem which is to be finally solved by decision makers. Spatial simulation and analysis provides the basis for informed decision making processes.

A second simulation approach, which should be briefly discussed in the context of bicycle safety, focuses on the road users (either grouped or individually) in a given network. Here the amount and quality of traffic (e.g. bicycle flow, socio-demographic composition of road users etc.) is estimated for different spatial units and time intervals. Knowing when and where how many and which bicyclists are on the road is important to know for several reasons. Firstly, bicycle incidents need to be related to a statistical population in order to calculate risk functions properly. Secondly, to eliminate potentially dangerous bottlenecks in the system, the capacity of existing infrastructure needs to be adapted adequately to the demand. And thirdly, the decision on building new infrastructure should be based on expected needs and demands, in order to ensure a maximum effect at reasonable costs. Based

[1] The index value is standardized between 0 (excellent) and 1 (poor) and depends on the model parameterization.

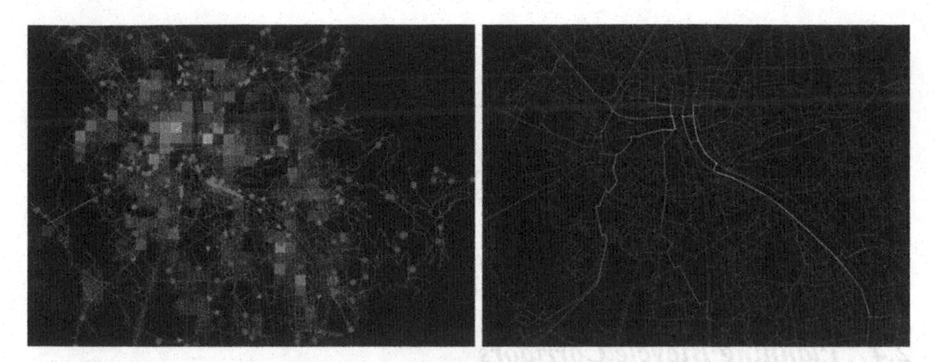

Fig. 12.5 Agent-based model simulation of bicycle traffic: randomly distributed origin-destination relations with agents (*left*) and most frequently traversed segments in the network (*right*). Preliminary results from an ongoing study

on an example from Denmark, Gudmundsson et al. [34] point to the fact, that such an evidence-based approach is rather unusual in the planning of bicycle infrastructure.

Well established methods for simulating the current or future amount of traffic for different settings and level of details already exist. They range from traffic demand models [35] to cellular automata [36], game theory [37], flow model simulation [38] and agent-based models [39]. All these methods were primarily developed for car traffic and partly for public transport. For active modes of transport only very few examples exist (e.g. Ref. [40] for pedestrians in a main station). This is mainly due to the lack of data, since there is, for example, no obligation to register bicycles and counting stations are hardly ever distributed representatively in the network.

Nevertheless, there are promising attempts to estimate bicycle flows on a macroscopic level. In an ongoing study we combine GIS with agent-based model techniques in order to simulate bicycle flows in an urban road network with a total length of roughly 800 km (see Fig. 12.5). Major advantages of this approach are the possibility to simulate different behaviors (e.g. mode choice preferences) and interactions with the road environment and other external factors (e.g. weather, other road users) while putting the agents in an explicit spatial context (e.g. characteristics of road network, location of residential and commercial areas). Beside an estimation of the number of bicyclists for the whole network, potential bottlenecks, hot-spots and central (= frequently traversed) segments can be identified. Such insights help to provide a basis for the abovementioned issues and partly tackle the problem of a significant data shortage [9]. Given the availability of real-time data of traffic flows, incidents, temporal construction sites etc., these data could be managed in a spatial data base and provided for adaptive simulation settings.

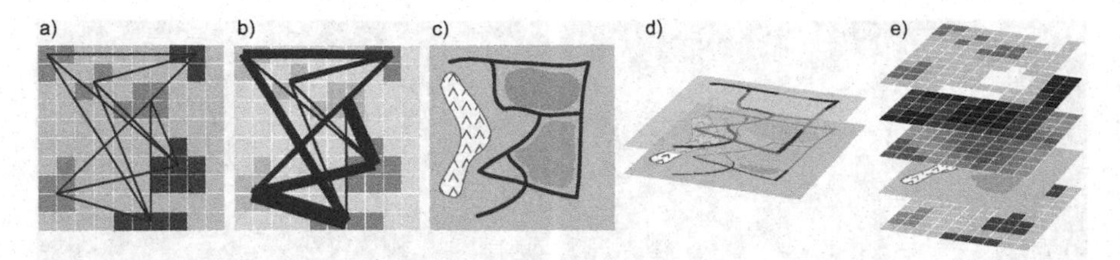

Fig. 12.6 Systematic planning approach for bicycle corridors. GIS is employed in all steps with spatial query, filter, overlay and network analysis functionalities

2.4 Planning Bicycle Corridors

Based on status-quo analyses of the network in terms of safety and a sound estimation of the spatio-temporal distribution of bicyclists, further analyses for the establishment of straight connections with high capacity ("bicycle highway") can be conducted. These connections play a key role in the establishment of the bicycle as sustainable mode of transport for daily commuters in urban agglomerations where the distance travelled and the travel time are frequently disproportionally related [41]. Apart from being a cost-efficient and sustainable alternative to motorized traffic, infrastructure which is dedicated to bicyclists contribute to a higher level of de-facto and perceived safety [42].

The planning of bicycle corridors requires several spatial analysis steps, which can be put into practice with common GIS software (see Fig. 12.6):

- Identification of major relations (functional connections) between origin and destination locations (see previous section), step (a).
- Prioritization of OD-relations based on demand, planning strategy, target group etc. step (b).
- Generation of ideal connections, either independent from and/or based on existing road network, step (c).
- Identification of existing segments which could be part of a straight connection, based on their suitability (capacity, safety, accessibility etc.), step (d).
- Optimization of multi-criteria problem (available space, monetary cost, legal scope etc.) to build missing bicycle infrastructure, step (e).

These steps can be applied in any environment and scale level. But generally, the designation of bicycle routes with high-capacity infrastructure is most efficient in urban agglomerations with overall bicycle promotion strategies. Current examples can be found in Copenhagen[2] or London.[3]

[2] http://www.supercykelstier.dk/concept (accessed 23 Dec 2014).

[3] http://www.tfl.gov.uk/modes/cycling/routes-and-maps/barclays-cycle-superhighways (accessed 23 Dec 2014).

2.5 Routing

Addressing the third safety pillar (Fig. 12.1) routing portals play a central role in the user-specific information of bicyclists. Routing services – especially when they are enriched with additional information that is relevant to bicyclists – contribute to the attractiveness of the bicycle (see e.g. Ref. [43] and [44] for the aspect of information and culture in the context of bicycle promotion) and are a perfect medium to recommend safe connections. The following brief case study demonstrates the contribution of such a routing service to an overall strategy of (safe) bicycle promotion and shows the potential of GIS as toolset and platform for several steps in a multi-stage workflow (see Fig. 12.3).

Salzburg – capital of the homonymous federal state, with approximately 150,000 inhabitants – has been following a sweeping strategy for bicycle promotion for more than two decades. A fixed sum is invested into infrastructure projects every year, resulting in a tight network of bicycle facilities with radial and tangential high-capacity connections. Through these substantial efforts, together with different accompanying actions, a modal split of approximately 20 % for the bicycle could have been reached. In order to further increase the number of bicyclists, the responsible administrative bodies have expanded their promotion strategy and started to invest into user-specific information.

The central application of this information offer is a routing service which is available over the internet (www.radlkarte.info) for both, desktop and mobile devices. The service basically provides two route options, where one is the shortest connection between origin and destination and the other the most suitable (for legal reasons it's not labeled as safest route). The latter result is based on the already mentioned indicator-based assessment model. This model calculates the potential safety threat for each segment, expressed as an index value which is further used as impedance in the route optimization algorithm. Similar to the simulation use case, the system's architecture is open to implement real time data. Additionally the service, as it is built upon a spatial data base with GI capabilities, allows for the integration of any further, spatial information layer. The application provides for example a detailed profile (derived from high-resolution laser scan elevation model), calculates travel time and energy costs, shows up-to-date departure times of public transport stops and combines the route recommendation with current weather data in the map view. So, not only the route optimization, which is a core spatial analysis task, but also the data preparation and the provision of additional information relies on GI technologies and tools. Especially the data pre-processing and the data modelling in a spatially enabled environment allows for an explicit consideration of bicycle safety and distinguishes the application from similar products [22].

The present routing platform is intended to be an additional building block of a comprehensive bicycle promotion strategy, which considers both, the infrastructure and the user-specific information offer. Such offer about safe routes aims to support everyday bicyclists in their route choice and raises awareness for the bicycle as sustainable (and safe) mode of transport.

2.6 Communication

The perception of the road space heavily depends on the transport mode of the road user [45]. In order to consider the specific perspectives and demands of bicyclists in citizen-centered, participatory applications, it is necessary to provide adequate communication tools.

Generally these tools use the geographical location as reference to ensure an effective communication. Location-based communication in the context of bicycle safety can take different conceptual, organizational and technical forms (see Fig. 12.7):

- Unidirectional, passive communication agent – server: bicyclists on the road are used as sensors [46] and contribute, for example, to the generation of real-time traffic flow visualizations. Bicycle flows can also be derived from unlabeled trajectories with segmentation algorithms [47].
- Unidirectional communication server – agent: from a server information about optimal or safe routes are communicated to the bicyclist's smartphone [18]. These recommendations are either built on models (as described in the previous section) or on crowd-sourced information.[4]
- Unidirectional feedback agent – server: the ubiquitous availability of smartphones and mobile internet connection enables bicyclists to give in-situ feedback concerning the quality of road space in terms of safety or to assess the plausibility of routing information. These feedbacks are collected on a central server and serve as valuable information for targeted infrastructure measures and other safety improvements or as inputs for the calibration of routing recommender systems (see Fig. 12.3).
- Discussion about road space: maps are ideal communication and negotiation platforms about space, either asynchronous or in real time [48]. In the context of

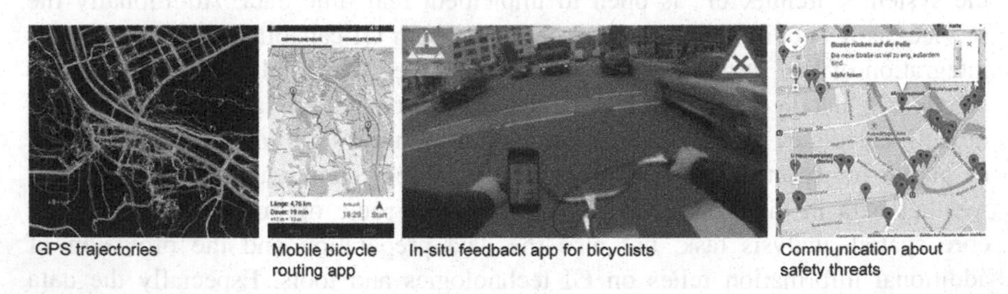

Fig. 12.7 Examples for different forms of location-based communication in the context of bicycle safety (Sources (from *left*): http://www.openstreetmap.org, http://radlkarte.info/, http://youtu.be/18eFQwsf8S4 and https://radsicherheit.berlin.de/diskussion (all accessed 29 Dec 2014))

[4] Numerous examples for these community projects can be found on the internet, e.g. http://cms.meineradspur.at/machmit.html or http://www.naviki.org (both accessed 27 Dec 2014).

bicycle safety maps could be utilized to present and discuss measures to improve the overall safety, for example where it is most urgent to build a new cycle way etc. The applicability of such an approach has been tested, just to name an example, for the installation of a new bus line by Roche et al. [49]. The reliability of citizen's judgments was investigated among others by Van Ryzin et al. [50].

Berlin launched an extensive, digital dialog about bicycle traffic in the late fall of 2013, following a community approach with options to comment, rate and discuss postings. Within only 1 month 3300 registered users contributed 5000 posts and 4000 comments. Basically all contributions had to do with safety, either directly or indirectly (see Letz and Basedow [51] for details).

Independently from the design of the communication platform, all of them use GI technology at the system's backend or for the visual interface, mostly maps. The spatial data generated this way are not only used for visualizations, but as inputs for analyses, modelling and planning purposes.

3 Conclusion and Outlook

The need for more sustainable mobility, especially in urban agglomerations, is evident. Because of the bicycle's numerous advantages in this respect, it has come into the focus of nearly all urban sustainability strategies. Mobility in general is a complex, highly dynamic system. The same holds true for any specific mode of transport, in this case the bicycle. Consequently bicycle promotion cannot consist of single, isolated measures but must follow a system approach [52], targeting various interests. Geographical Information Systems facilitate such an integrated approach, which is demonstrated in the context of bicycle safety. Referring to the threefold safety concept, sketched in Fig. 12.1, it could have been shown how GIS concepts, technologies and tools address and contribute to the planning and maintenance of safe bicycle infrastructure and to user-tailored information services. Spatial analysis and modelling are at the center of a better understanding of bicycle accidents as spatio-temporal phenomena, of integrated planning, information and communication applications. In all these applications, the geographical location serves as common denominator, allowing for a combination ("overlay") of multiple information layers.

An iterative workflow, built around the indicator-based assessment model proposed by Loidl and Zagel [22], incorporates and links several of these modelling and analysis components. The assessment approach can either be employed in a more efficient way to plan and build infrastructure ("What's the effect of building a cycleway at a particular road?") and take regulative actions ("To what degree would the road become more bicycle-friendly, if the maximum speed is lowered?")

or as basis for innovative information applications such as routing and communication platforms.

In order to further enhance the contribution of geographical information systems to the promotion of safe bicycling, the following topics should be addressed in further researches:

- Personalization. Currently bicycle route recommendations mostly address an average user, who factually does not exist. Bicyclists have a broad range of preferences: whereas daily commuters prefer direct routes, parents with young children might prefer safe connections with designated bicycle infrastructure [53]. In a configurable model environment, individual preferences and perspectives on the road space can be integrated in the pre-configured assessment model and consequently contribute to personalized information products. How individual peculiarities can be transformed to code-readable model inputs and how these inputs should be weighted is subject to research. Priedhorsky et al. [54] provide first results of an investigation of algorithms for personalized bicycle route recommendations.
- Socialization. All technical and organizational efforts that contribute to a more social environment where experiences, feedback messages and updates are shared among several instances (bicyclists, authorities, event organizers, public transportation operators etc.) could help to further increase the attractiveness of the bicycle as a sustainable and safe mode of transport. Current research topics would be, among others: inter-bicycle communication, participatory planning processes or the provision of safety-relevant real-time data (e.g. information about snow removal or temporal construction sites).
- Contextualization. Retrieving information about safe routes or giving feedback to a current situation in road space are context-sensitive activities. Frameworks for providing applications that adapt to the current context exist [55]. But to our current knowledge there is hardly any research done so far in the context of safety information for bicyclists. Again, the geographical location is highly suitable for indicating the actual environment. Through the current position context-building information, such as real-time traffic flow, weather conditions or positions of other bicyclists, can be linked, modelled, analyzed and used in the return flow to the user. Subject to research are, among others, the identification of relevant information depending on the context (e.g. providing departure times of PT stops when rain showers are expected) and the appropriate communication to the user (e.g. how to communicate the potential occurrence of black ice?).

What all these topics have in common is the explicit spatial reference. Here the power of GIS comes into play: it offers the ideal platform or framework for building relations between various instances and data and thus allows for the generation and retrieval of relevant information. In this sense, GI systems make spatial information accessible for the promotion of the bicycle as sustainable and safe mode of transport.

References

1. M. Keuken, E. Sanderson, R. van Aalst, J. Borken, and J. Scheider, "Contribution of traffic to levels of ambient air pollution in Europe," in *Health effects of transport-related air pollution*, M. Krzyzanowski, B. Kuna-Dibbert, and J. Schneider, Eds., ed Copenhagen: World Health Organization, 2005, pp. 53–84.
2. G. R. Timilsina and H. B. Dulal, "Urban Road Transportation Externalities: Costs and Choice of Policy Instruments," *The World Bank Research Observer,* vol. 26, pp. 162–191, 2011.
3. D. Laussmann, M. Haftenberger, T. Lampert, and C. Scheidt-Nave, "Social inequities regarding annoyance to noise and road traffic intensity. Results of the German Health Interview and Examination Survey for Adults," *Bundesgesundheitsblatt,* vol. 56, pp. 822–831, 2013.
4. M. Meschik, "Reshaping City Traffic Towards Sustainability Why Transport Policy should Favor the Bicycle Instead of Car Traffic," *Procedia – Social and Behavioral Sciences,* vol. 48, pp. 495–504, 2012.
5. S. Handy, B. van Wee, and M. Kroesen, "Promoting Cycling for Transport: Research Needs and Challenges," *Transport Reviews,* vol. 34, pp. 4–24, 2014.
6. J. Pucher, J. Dill, and S. Handy, "Infrastructure, programs, and policies to increase bicycling: An international review," *Preventive Medicine,* vol. 50, pp. 106–125, 2010.
7. F. Wegman, F. Zhang, and A. Dijkstra, "How to make more cycling good for road safety?," *Accident Analysis & Prevention,* vol. 44, pp. 19–29, 2012.
8. T. Lorenc, G. Brunton, S. Oliver, K. Oliver, and A. Oakley, "Attitudes to walking and cycling among children, young people and parents: a systematic review," *Journal of Epidemiology and Community Health (1979-),* vol. 62, pp. 852–857, 2008.
9. OECD, "Cycling, Health and Safety," ITF-OECD Working Group on Cycling Safety, Paris 2013.
10. C. Juhra, B. Wieskötter, K. Chu, L. Trost, U. Weiss, M. Messerschmidt, *et al.*, "Bicycle accidents – Do we only see the tip of the iceberg?: A prospective multi-centre study in a large German city combining medical and police data," *Injury,* vol. 43, pp. 2026–2034, 2012.
11. European Commission, "Road Safety Vademecum – Road safety trends, statistics and challenges in the EU 2010–2013," European Commission DG for Mobility and Transport Unit C.4 – Road Safety, Brussels 2014.
12. B. Thomas and M. DeRobertis, "The safety of urban cycle tracks: A review of the literature," *Accident Analysis & Prevention,* vol. 52, pp. 219–227, 2013.
13. M. Winters, S. Babul, J. Becker, J. R. Brubacher, M. Chipman, P. A. Cripton, *et al.*, "Safe Cycling: How Do Risk Perceptions Compare With Observed Risk?," *Canadian Journal of Public Health,* vol. 103, pp. 542–547, 2012.
14. S. Othman, R. Thomson, and G. Lannér, "Identifying critical road geometry parameters affecting crash rate and crash type," presented at the Annals of Advances in Automotive Medicine, 2009.
15. K. Teschke, M. A. Harris, C. C. O. Reynolds, M. Winters, S. Babul, M. Chipman, *et al.*, "Route Infrastructure and the Risk of Injuries to Bicyclists: A Case-Crossover Study," *American Journal of Public Health,* vol. 102, pp. 2336–2343, 2012.
16. M. A. Harris, C. C. O. Reynolds, M. Winters, P. A. Cripton, H. Shen, M. L. Chipman, *et al.*, "Comparing the effects of infrastructure on bicycling injury at intersections and non-intersections using a case–crossover design," *Injury Prevention,* 2013.
17. P. Schepers, N. Agerholm, E. Amoros, R. Benington, T. Bjørnskau, S. Dhondt, *et al.*, "An international review of the frequency of single-bicycle crashes (SBCs) and their relation to bicycle modal share," *Injury Prevention,* 2014.
18. M. Loidl, B. Zagel, S. Krampe, and J. Reithofer, "Radlkarte Salzburg – Das Radroutingportal für die Stadt Salzburg," in *AGIT*, Salzburg, 2013, pp. 456–461.
19. P. Jankowski, "Towards participatory geographic information systems for community-based environmental decision making," *Journal of Environmental Management,* vol. 90, pp. 1966–1971, 2009.

20. T. Blaschke, K. Donert, F. Gossette, S. Kienberger, M. Marani, S. Qureshi, *et al.*, "Virtual globes: serving science and society," *Information,* vol. 3, pp. 372–390, 2012.
21. R. C. Smith, D. L. Harkey, and B. Harris, "Implementation of GIS-based Highway Safety Analyses: Bridging the Gap," Turner-Fairbank Highway Research Center, McLean FHWA-RD-01-039, 2001.
22. M. Loidl and B. Zagel, "Assessing bicycle safety in multiple networks with different data models," in *GI-Forum,* Salzburg, 2014, pp. 144–154.
23. M. Ressing, M. Blettner, and S. J. Klug, "Data Analysis of Epidemiological Studies: Part 11 of a Series on Evaluation of Scientific Publications," *Deutsches Arzteblatt International,* vol. 107, pp. 187–192, 2010.
24. E. C. Delmelle and J.-C. Thill, "Urban bicyclists: spatial analysis of adult and youth traffic hazard intensity," *Transportation Research Record: Journal of the Transportation Research Board,* vol. 2074, pp. 31–39, 2008.
25. N. Yiannakoulias, S. A. Bennet, and D. M. Scott, "Mapping commuter cycling risk in urban areas," *Accident Analysis & Prevention,* vol. 45, pp. 164–172, 2012.
26. P. Schepers, "A safer road environment for cyclists," Dissertation, Transport and Planning, TU Delft, Delft, 2013.
27. T. Steenberghen, K. Aerts, and I. Thomas, "Spatial clustering of events on a network," *Journal of Transport Geography,* vol. 18, pp. 411–418, 2010.
28. P. Schepers, M. Hagenzieker, R. Methorst, B. van Wee, and F. Wegman, "A conceptual framework for road safety and mobility applied to cycling safety," *Accident Analysis & Prevention,* vol. 62, pp. 331–340, 2014.
29. P. L. Jacobsen, "Safety in numbers: more walkers and bicyclists, safer walking and bicycling," *Injury Prevention,* vol. 9, pp. 205–209, 2003.
30. J. Parkin, M. Wardman, and M. Page, "Models of perceived cycling risk and route acceptability," *Accident Analysis & Prevention,* vol. 39, pp. 364–371, 2007.
31. L. De Rome, S. Boufous, T. Georgeson, T. Senserrick, D. Richardson, and R. Ivers, "Bicycle Crashes in Different Riding Environments in the Australian Capital Territory," *Traffic Injury Prevention,* vol. 15, pp. 81–88, 2014/01/01 2013.
32. A. K. Macpherson, T. M. To, P. C. Parkin, B. Moldofsky, J. G. Wright, M. L. Chipman, *et al.*, "Urban/rural variation in children's bicycle-related injuries," *Accident Analysis & Prevention,* vol. 36, pp. 649–654, 2004.
33. S.-L. Shaw, "Geographic information systems for transportation: from a static past to a dynamic future," *Annals of GIS,* vol. 16, pp. 129–140, 2010.
34. H. Gudmundsson, E. Ericsson, M. Tight, M. Lawler, P. Envall, M. J. Figueroa, *et al.*, "The Role of Decision Support in the Implementation of "Sustainable Transport" Plans," *European Planning Studies,* vol. 20, pp. 171–191, 2012.
35. E. Cascetta, "Models for Traffic Assignment to Transportation Networks," in *Transportation Systems Engineering: Theory and Methods.* vol. 49, ed: Springer US, 2001, pp. 251–366.
36. K. Nagel and M. Schreckenberg, "A cellular automaton model for freeway traffic," *J. Phys. I France,* vol. 2, pp. 2221–2229, 1992.
37. T. Chmura and T. Pitz, "An Extended Reinforcement Algorithm for Estimation of Human Behaviour in Experimental Congestion Games," *Journal of Artificial Societies and Social Simulation,* vol. 10, p. 17, 2007.
38. D. Helbing, A. Hennecke, V. Shvetsov, and M. Treiber, "Micro- and macro-simulation of freeway traffic," *Mathematical and Computer Modelling,* vol. 35, pp. 517–547, 2002.
39. A. L. C. Bazzan and F. Klügl, "A review on agent-based technology for traffic and transportation," *The Knowledge Engineering Review,* vol. 29, pp. 375–403, 2014.
40. F. Klügl and G. Rindsfüser, "Large-Scale Agent-Based Pedestrian Simulation," in *Multiagent System Technologies.* vol. 4687, P. Petta, J. Müller, M. Klusch, and M. Georgeff, Eds., ed: Springer Berlin Heidelberg, 2007, pp. 145–156.
41. D. Banister, "The trilogy of distance, speed and time," *Journal of Transport Geography,* vol. 19, pp. 950–959, 2011.

42. E. Heinen, B. van Wee, and K. Maat, "Commuting by Bicycle: An Overview of the Literature," *Transport Reviews,* vol. 30, pp. 59–96, 2010.
43. R. Aldred and K. Jungnickel, "Why culture matters for transport policy: the case of cycling in the UK," *Journal of Transport Geography,* vol. 34, pp. 78–87, 2014.
44. P. Rietveld and V. Daniel, "Determinants of bicycle use: do municipal policies matter?," *Transportation Research Part A: Policy and Practice,* vol. 38, pp. 531–550, 2004.
45. A. Forsyth and K. Krizek, "Urban Design: Is there a Distinctive View from the Bicycle?," *Journal of Urban Design,* vol. 16, pp. 531–549, 2011.
46. M. Goodchild, "Citizens as sensors: the world of volunteered geography," *GeoJournal,* vol. 69, pp. 211–221, 2007.
47. F. Biljecki, H. Ledoux, and P. van Oosterom, "Transportation mode-based segmentation and classification of movement trajectories," *International Journal of Geographical Information Science,* vol. 27, pp. 385–407, 2012.
48. Z. Chang and S. Li, "Geo-Social Model: A Conceptual Framework for Real-time Geocollaboration," *Transactions in GIS,* vol. 17, pp. 182–205, 2013.
49. S. Roche, B. Mericskay, W. Batita, M. Bach, and M. Rondeau, "WikiGIS Basic Concepts: Web 2.0 for Geospatial Collaboration," *Future Internet,* vol. 4, pp. 265–284, 2012.
50. G. G. Van Ryzin, S. Immerwahr, and S. Altman, "Measuring Street Cleanliness: A Comparison of New York City's Scorecard and Results from a Citizen Survey," *Public Administration Review,* vol. 68, pp. 295–303, 2008.
51. B. Letz and S. Basedow, "Radfahren in Berlin – Abbiegen? Achtung! Sicher über die Kreuzung, Auswertungsbericht zur Öffentlichkeitsbeteiligung," Senatsverwaltung für Stadtentwicklung und Umwelt, Berlin 2014.
52. T. Goldman and R. Gorham, "Sustainable urban transport: Four innovative directions," *Technology in Society,* vol. 28, pp. 261–273, 2006.
53. J. Broach, J. Dill, and J. Gliebe, "Where do cyclists ride? A route choice model developed with revealed preference GPS data," *Transportation Research Part A: Policy and Practice,* vol. 46, pp. 1730–1740, 2012.
54. R. Priedhorsky, D. Pitchford, S. Sen, and L. Terveen, "Recommending routes in the context of bicycling: algorithms, evaluation, and the value of personalization," presented at the Proceedings of the ACM 2012 conference on Computer Supported Cooperative Work, Seattle, Washington, USA, 2012.
55. T. Buchholz, A. Küpper, and M. Schiffers, "Quality of context: What it is and why we need it," in *Proceedings of the workshop of the HP OpenView University Association,* 2003.

Chapter 13
Using Systems Thinking and System Dynamics Modeling to Understand Rebound Effects

Mohammad Ahmadi Achachlouei and Lorenz M. Hilty

Abstract Processes leading to an increase of demand for a resource as a consequence of increasing the efficiency of using this resource in production or consumption are known as (direct) rebound effects. Rebound effects at micro and macro levels tend to offset the reduction in resource consumption enabled by progress in efficiency. Systems thinking and modeling instruments such as causal loop diagrams and System Dynamics can be used to conceptualize the structure of this complex phenomenon and also to communicate model-based insights. In passenger transport, the rebound effect can be invoked by increased cost efficiency (direct economic rebound) and/or increase in speed (time rebound). In this paper we review and compare two existing models on passenger transport—including a model on the role of information and communication technology—with regard to the feedback loops used to conceptualize rebound effects.

Keywords Rebound effect • Energy efficiency • Systems thinking • Systems modeling • System dynamics • Causal loop diagrams • Passenger transport • ICT • Time rebound • Direct rebound

M. Ahmadi Achachlouei (✉)
Division of Environmental Strategies Research (fms), KTH Royal Institute of Technology, Stockholm SE-100 44, Sweden

Centre for Sustainable Communications (CESC), KTH Royal Institute of Technology, Stockholm SE-100 44, Sweden

Empa – Swiss Federal Laboratories for Materials Science and Technology, Technology and Society Lab, St. Gallen CH-9014, Switzerland
e-mail: Mohammad.Achachlouei@abe.kth.se

L.M. Hilty
Centre for Sustainable Communications (CESC), KTH Royal Institute of Technology, Stockholm SE-100 44, Sweden

Empa – Swiss Federal Laboratories for Materials Science and Technology, Technology and Society Lab, St. Gallen CH-9014, Switzerland

Department of Informatics, University of Zurich, Zurich CH-8050, Switzerland

J. Marx Gómez et al. (eds.), *Advances and New Trends in Environmental and Energy Informatics*, Progress in IS, DOI 10.1007/978-3-319-23455-7_13

1 Introduction

Energy efficiency is one of the main policy options to fight global climate change (e.g. see Ref. [1]). Energy efficiency helps devices and infrastructures provide the same services using less energy, and thus can be a solution for reducing greenhouse gas (GHG) emissions. However, it can also induce additional demand if the energy saved leads to a lower price of the final service. This induction is known as the rebound effects (or take-back effects) [2]. At a micro level, the rebound effect may occur through (i) creating more demand for the cheaper service (direct rebound effects), and/or (ii) increasing income available for general consumption (indirect rebound effects) [3]. At a macro level, increased efficiency in the production and use of energy will yield a series of supply and demand adjustments occurring over time [3] "with energy-intensive goods and sectors likely to gain at the expense of less energy-intensive ones" [4] (economy-wide effects).

Various methods have been employed to understand and explain the rebound effects. Economic studies have sought to estimate the magnitude of these rebound effects [4, 5]. Quasi-experimental approaches have been used to measure the demand for the service before and after an efficiency improvement [6]. Moreover, sociological and psychological studies have addressed the ways in which efficient solutions are being used in people's everyday life (e.g. see Refs. [7, 8]).

To better understand the complexity of rebound effects of investments in efficiency improvements, system-theoretical approaches and dynamic models have to be used. Systems thinking and modeling instruments, such as causal loop diagrams and System Dynamics simulation, by highlighting the causal structure of the system and feedback effects, provide effective approaches to conceptualizing such complex phenomena and communicating model-based insights [9–11].

A number of studies have employed systems thinking and modeling instruments to address rebound effects. Hilty et al. [12] used such tools (at a quantitative level) to include time rebound and direct economic rebound in their study of effects of information and communication technology (ICT) on environmental sustainability. Stepp et al. [13] employed causal loop diagrams to highlight the potential unintended consequences and rebound effects when studying the role of feedback effects in GHG mitigation policies in the transport sector. Peeters [14] used causal loop diagrams to investigate the positive and negative role (and rebound effects) of technological progress in the context of GHG emissions from tourism transport. In a recent study, Dace et al. [15] built causal loop diagrams and a System Dynamics simulation model to analyze effects of eco-design policy on packaging waste management systems, showing how tax can help counteract a rebound effect originated from eco-design.

In passenger transport, the direct rebound can be induced through increases in fuel efficiency or other improvements reducing the variable cost per person-kilometer (direct economic rebound) and/or increase in speed of transport modes (time rebound). In this paper, using the domain of passenger transport as an example, we review and compare two existing models on passenger transport (developed in the context of environmental impacts of transport), including a model on the effects of ICT on transport. We investigate two research questions:

(i) What are the main feedback loops used in modeling rebound effects (with a focus on direct economic rebound and time rebound)?
(ii) What is the contribution of systems thinking and the System Dynamics approach to the analysis of rebound effects?

The current paper is organized as follows. First, brief definitions of the concepts of rebound effects, elasticity of demand, System Dynamics, and causal loop diagrams are presented. Section 3 introduces the two models discussed in this paper. Sections 4 and 5 provide a detailed discussion of the feedback loops in Model 1 and Model 2 used to model rebound effects. Section 5 discusses the main findings and the contribution of systems thinking and modeling to the rebound literature. Section 7 concludes the paper.

2 Main Concepts

2.1 Rebound Effects

Rebound effect is an "umbrella term for a variety of economic responses to improved energy efficiency and 'energy-saving' behavioral change" [16]. Rebound effects can be categorized as follows [17, 18] (See Ref. [19] for a brief history of rebound analysis):

- *Direct economic rebound effects*: When cheaper energy (or energy efficiency improvement in using energy-intensive goods) induces price reductions that trigger an increase in the demand for the cheaper good. (The good can be a tangible good or an intangible service.)
- *Time rebound effects* (a kind of direct rebound), which is based on time efficiency in consumption: If people can consume a product or service in less time, they tend to demand more of it.
- *Indirect rebound effects* (income rebound): If the consumer saves money on one good (because it is used more efficiently and its price goes down) her disposable income is higher than the income she can spend—because she didn't use the money for the purpose, she can use it for something else that also requires energy for its provision.
- *Economy-wide rebound effects*, which appear when declining energy prices induce a reduction in the prices of intermediate and final goods throughout the economy, and cause structural changes in production patterns and consumption habits.

In this paper we only address the direct economic rebound and time rebound in the field of passenger transport. If the efficiency increase is enabled by ICT, both direct and indirect rebound effects are subsumed under the so-called third-order effects of ICT [20].

Rigorous definitions of rebound effects have been provided by Khazzum [21], Berkhout et al. [2] and Sorrell and Dimitropoulos [4]. The magnitude of the direct economic rebound effect depends on demand elasticity. The efficiency elasticity of useful work can be taken as a measure of the direct rebound effect [4]:

$$\eta_\varepsilon(S) = \frac{\partial S}{\partial \varepsilon}\frac{\varepsilon}{S} \tag{13.1}$$

where S represents the useful work (which can be measured by a variety of thermodynamic or physical indicators [4]; e.g. passenger-kilometers in the case of transport), and ε represents the energy efficiency of an energy-transforming system (which can be defined as $\varepsilon = S / E$, where E represents the energy input required for a unit output of useful work).

Economic elasticity of demand with regard to price, or price elasticity of demand (PED) is defined as the percentage change in demand divided by the percentage change in price (See Ref. [22] for an overview of price elasticities of transport demand). Use efficiency can be expressed via price as an input for calculating price elasticity of demand. If for example a vehicle is used more efficiently by transporting more persons at a time, the cost per passenger kilometer is lower, which can lead to an increase in demand. Given the elasticity of demand (η) for energy services (such as transport) with regard to energy cost of the energy service (P_S) or with regard to price of energy (P_E), we can extend the definition of direct rebound effect (Eq. (13.1)) as follows (see Ref. [4] for details):

$$\eta_\varepsilon(S) = -\eta_{P_S}(S) \tag{13.2}$$

and

$$\eta_\varepsilon(S) = -\eta_{P_E}(S) \tag{13.3}$$

Moreover, we can consider the effect of more demand originated from improved time efficiency (θ), "the influence of time costs on the rebound effect and the existence of a parallel rebound effect with respect to time" [4]. For example, one may choose faster transport modes that are more energy-intensive and more polluting; or one may travel longer or more frequently because of better time utilization using mobile ICT. The time rebound effect can be defined using an extended version of Eq. (13.1) (see Ref. [4] for details):

$$\eta_\varepsilon(S) = -\eta_{P_S}(S) + [\eta_{P_T}(S)\, \eta_\theta(P_T)\, \eta_\varepsilon(\theta)] \tag{13.4}$$

where the additional term in square brackets (compared to Eq. (13.2)) is the product of the elasticity of demand for useful work (e.g. transport) with respect to time costs

($\eta_{PT}(S)$), the elasticity of time costs with respect to time efficiency ($\eta_{\theta}(P_T)$) and the elasticity of time efficiency with respect to energy efficiency ($\eta_{\varepsilon}(\theta)$) [4].

Data availability of any elasticity values mentioned in Eqs. (13.1, 13.2, 13.3, and 13.4)—see more definitions in [4]—can guide us to choose the most appropriate equation for the magnitude of direct rebound effect in a given analysis task.

2.2 Systems Thinking and Modeling

Systems thinking and modeling, as a non-linear way of thinking about the sources of and the solutions to modern problems [23], "offers a holistic way of appreciating all dimensions of a complex problem" [24]. This approach has been applied to "almost any problem area because of its generality" [25] such as strategy and organizations, production and operations, ecology and agriculture, medicine and health, and sustainable development. In their review of theoretical aspects and applications of systems thinking and modeling (or "the systems approach"), Mingers and White [25] conclude that while this approach "may not be well established institutionally, in terms of academic departments, it is incredibly healthy in terms of the quantity and variety of its applications" [25].

In this paper, we focus on two systems approaches: Causal loop diagrams (CLDs) and System Dynamics simulation modeling (Sometimes the CLD is referred to as "qualitative System Dynamics" [26]). Although it is common to first build a CLD and then a System Dynamics simulation model, it is also possible to use only CLDs without formal computer simulation modeling "to assist issue structuring and problem-solving" [26] and when "the aim of the project is simply greater understanding of the situation, or where reliable quantitative information is not available" [25]. CLDs can provide insight into policy issues by inferring, rather than calculating [26].

The main concept in CLDs and System Dynamics is the representation of system behavior over time via feedback loop structures and cause-and-effect analysis and, used for theory building, policy analysis and strategic decision support [9–11]. A feedback approach "supports closing open sequences of causes and expected effects, thus overcoming barriers in traditional linear thinking, namely the tendency to neglect unintended consequences, which often stand at the root of policy implementation failure" [27].

A feedback loop is a closed path of causal influences and information, forming a circular-causal loop of information and action. If the tendency in the loop is to reinforce the initial action, the loop is called a positive or reinforcing feedback loop; if the tendency is to oppose the initial action, the loop is called a negative, counteracting, or balancing feedback loop [11].

In a CLD, relationships between variables are depicted using arrows with a positive (+) or negative (−) sign placed besides the arrowhead to indicate link polarity (e.g. see Figs. 13.1 and 13.2 in the next sections). A positive link polarity

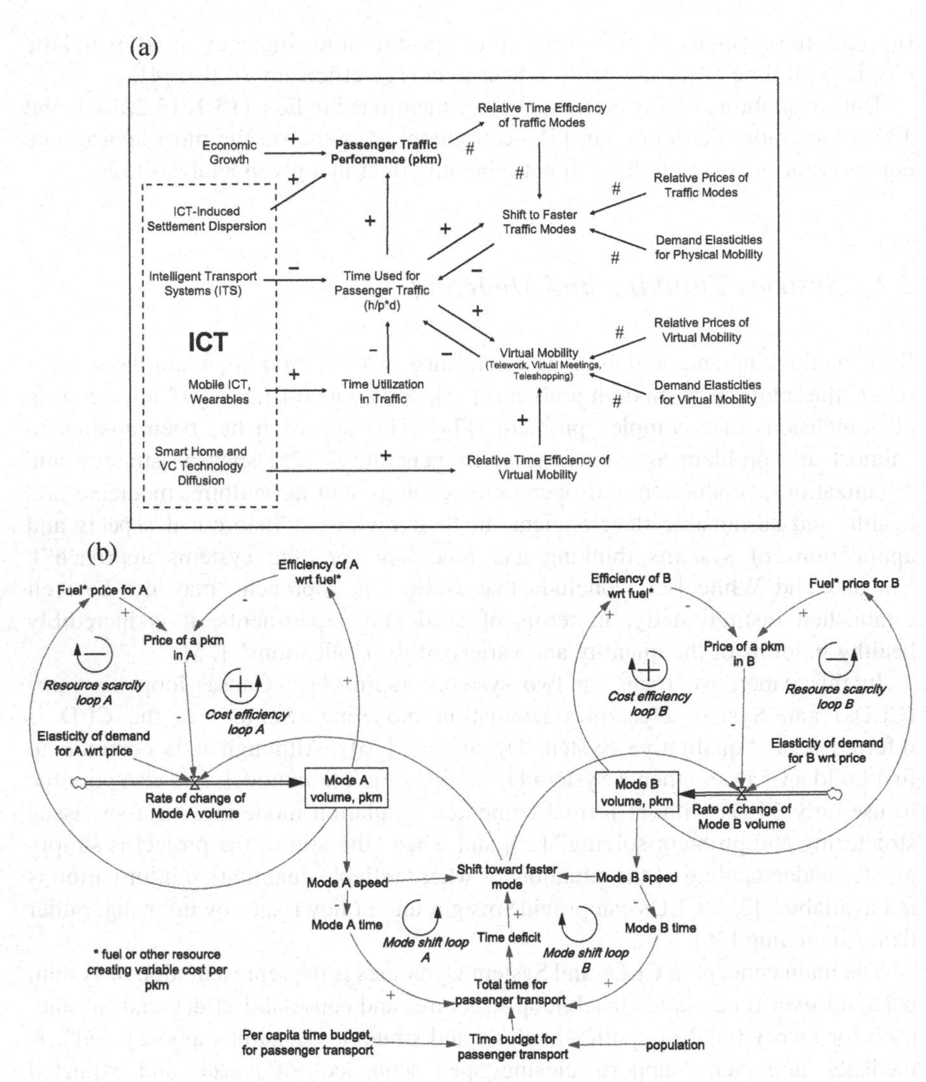

Fig. 13.1 Causal loop diagram for Model 1: (**a**) more abstract diagram for the development of passenger transport performance, taken from the IPTS interim report [31]: "ICT has second-order effects when applied passenger traffic (all applications subsumed under intelligent transport systems) and third-order effects in the long term via settlement dispersion, time use in traffic, smart home and videoconferencing technology. The '#' sign is used where the multidimensional variables are involved, leading to complex causal relationships." (**b**) less abstract diagram focused on main feedback loops

implies that "if a cause increases, the effect increases above what it would otherwise have been" and vice versa [10]. Similarly, a negative link polarity "means that if the cause increases, the effect decreases below what it would otherwise have been" and vice versa [10]. A CLD (as a qualitative technique) can be translated into

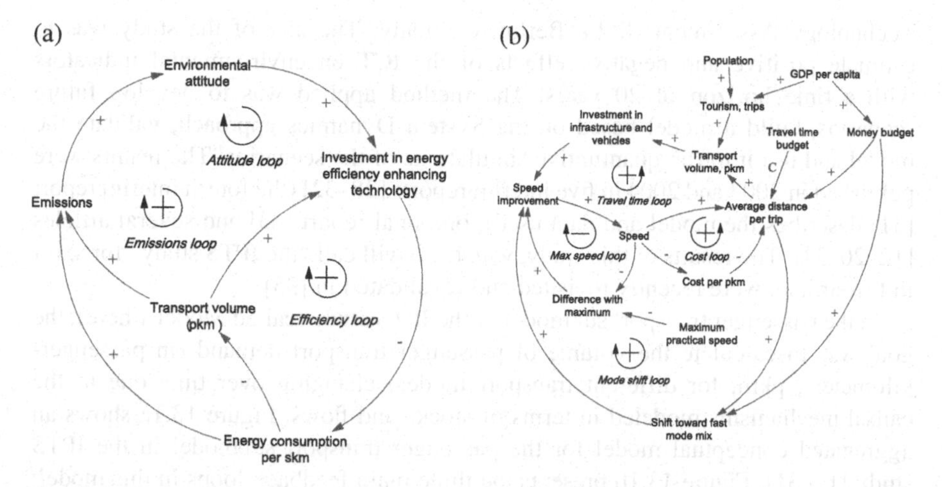

Fig. 13.2 Causal loop diagram for Model 2 [14]: (**a**) pollution-saving loops (**b**) basic forces in transport systems

stocks (accumulations or levels) in the system and their inflows and outflows (rates) [10, Chap. 6]. Mathematically, a system of difference equations is used to define computational System Dynamics models [10, 11].

3 Selected Models

Two models from literature have been chosen to illustrate how rebound effects are modeled using System Dynamics feedback loops. The first example (Model 1) models the dynamics of how ICT positively or negatively affects the passenger transport demand and modal split. The second example (Model 2) models the dynamics of how pollution-saving technologies positively or negatively affect the tourist transport demand and GHG emissions.

Feedback loops related to Model 1 and Model 2 (as shown in Figs. 13.1 and 13.2, respectively), which are used to conceptualize the rebound effects, are presented and discussed in the following sections.

4 Model 1: Future Impacts of ICT on Environmental Sustainability: Submodel Passenger Transport (IPTS Study)

In 2002, the European Commission's Institute for Prospective Technological Studies (IPTS) commissioned a study to explore the current and future environmental effects of ICT to a consortium led by the Institute for Futures Studies and

Technology Assessment (IZT), Berlin, Germany. The aim of the study was to estimate positive and negative effects of the ICT on environmental indicators with a time horizon of 20 years. The method applied was to develop future scenarios, build a model based on the System Dynamics approach, validate the model and use it to run quantitative simulations of the scenarios. The results were published in 2003 and 2004 in five interim reports [28–32] (the fourth interim report [31] describes the model and data used), one final report [33] and several articles [12, 20, 34]. The results of this study, which we will call "the IPTS study" for short in this article, were recently revisited and revalidated in [35].

In the passenger transport submodel of the IPTS study (called Model 1 here), the goal was to calculate the volume of passenger transport demand (in passenger-kilometers, pkm, for different transport modes) changing over time due to the causal mechanisms modeled in terms of stocks and flows. Figure 13.1a shows an aggregated conceptual model for the passenger transport submodel in the IPTS study [12, 31]. Figure 13.1b presents the three main feedback loops in this model: cost efficiency loop, resource scarcity loop, and mode shift loop. For simplicity, the model has been reduced to two competing modes of transport, here called "A" and "B" (which can be thought of, e.g. private car and public bus), whereas the original model differentiates among five modes of physical transport and is more complex. Each loop exists twice (symmetrically) because of the two modes chosen here for illustration, i.e., mode A and B.

Figure 13.1b is not a pure CLD, since it includes parts of the stock-and-flow diagram to better describe the central mechanism of shifting transport demand between traffic modes. In the IPTS study, this mechanism was generalized to *n* modes based on multimodal passenger transport models developed by Hilty [36, 37]. Five traffic modes were differentiated in Model 1: Private car, bus and coach, tram and metro, train, and air transport. In addition, three modes of "virtual mobility" including home-based telework, virtual meetings, and teleshopping were represented, which was a new feature developed for the IPTS study.

Two types of rebound effect were modeled in Model 1 with regard to passenger transport [12]: direct economic rebound effects and time rebound effects. The following subsections present how these two categories of rebound effects were addressed in Model 1.

4.1 Direct Economic Rebound Effects in Model 1

The direct economic rebound effects in Model 1 are represented via demand elasticities for passenger transport. The IPTS study considered the rebound induced by the price level of each passenger transport mode: besides changes of market prices (e.g. the oil price), which are external to the mode, higher efficiency (e.g. fuel efficiency of vehicles) can lead to lower prices per pkm (direct rebound effect), which will create more demand according to empirical elasticity parameters. As shown in Fig. 13.1b the elasticity parameter of each mode together with per-pkm price of the mode are controlling the inflow rate of transport volume associated with the mode.

Elasticity of demand with regard to price is defined as a relative change in demand divided by a relative change in price. It will not be realized immediately, i.e., when the price changes, in that moment no change occurs in demand, but gradually over the years. Empirical studies of elasticity of demand therefore usually distinguish between "short term" and "long term" elasticity. Model 1 expresses the temporal aspect of elasticity by adding a time constant to each elasticity value. Adaptations in Model 1 including the elasticity-based adaptation of demand (also the shift between modes based on relative speed and time deficit, which will be discussed later) are not immediate in Model 1, but controlled by time constants.

Direct economic rebound effects in Model 1 are represented via the following feedback loops (see Fig. 13.1): cost efficiency loop and resource scarcity loop.

(a) *Cost efficiency loop*

Traffic volume (pkm) for each mode—modeled as a stock—is controlled by an inflow rate depending on the elasticity parameter and the per-pkm fuel price associated with the mode. (For simplicity, fuel is used here a *pars pro toto* for the sum of all resources needed to produce a pkm which cause variable cost; these resources may vary depending on the mode of transport.)

The elasticity parameter represents "classical" elasticity of demand with regard to price (in Model 1: the "PED" submodel included for each mode). Because the fuel price per pkm does not only depend on the fuel price per liter but also on the efficiency with which the fuel is used ("Efficiency of A with regard to fuel"), the price of 1 pkm is affected by efficiency and will, depending on the elasticity, influence the demand (traffic volume). The efficiency can increase by technical measures (e.g. more efficient vehicles) or by better utilization of vehicles (more people in the vehicle means more pkm per vehicle-km). It is possible that more volume increases efficiency for several reasons (the "(+)"). However, for each concrete transport mode, one has to account for the specific causal link between volume and fuel efficiency and how fuel efficiency affects the price the user finally has to pay.

(b) *Resource scarcity loop*

Fuel (or any other resource needed to produce a pkm) may change in price if the total demand for this resource changes, depending on how supply reacts to demand in the market. Besides fuels, we may think of road pricing, which reflects the resource "infrastructure capacity" that is used to produce transport. Increasing use of any limited resource will at some point lead to an increase in price, which is reflected in this feedback loop. Again, it depends on the mode how this causal relationship is modeled in detail.

4.2 Time Rebound Effect in Model 1

In addition to direct economic rebound, the IPTS study (Model 1) included time rebound, another type of rebound effect based on time efficiency in consumption. Especially in passenger transport, time is a scarce resource and may affect behavior more than money. Model 1 (like Model 2) belongs to a class of models which abstain from converting time to money (which would be a straightforward approach in economic modeling) and keep financial budgets and time budgets of users separate. The time rebound effect was considered crucial in the IPTS study, because a core characteristic of ICT is the potential to accelerate processes.

Time rebound effects in Model 1 are modeled via the following mechanisms, as shown in Fig. 13.1:

- Travel time budget mechanism
- Mode shift loop

These mechanisms work with time (not money as it is the case for cost efficiency loop and resource scarcity loop); a central variable is the speed of transport of each mode.

(a) *Travel time budget mechanism*

For the transport submodel the time rebound was considered via the so-called constant travel time hypothesis, assuming that the average daily time spent in transport over the whole population is more or less stable [38] (a critique of this hypothesis will be addressed later in this paper). At any point in time, the given travel volumes of all modes and their current speed levels make it possible to compare actual travel time with this time budget. If there is a deficit, this will cause a shift of the modal split from slower to faster modes. If Mode A is currently slower than Mode B, then traffic volume will shift to mode B, with some time constants similar to the ones mentioned for economic elasticity, and also with some limitations of the substitution potential. In the full IPTS model with five modes, this can for example mean that people having to commute over a higher distance will then maybe use a private car instead of the public bus, or that car drivers faced with increasing congestion will switch to the train or metro.

(b) *Mode shift loop*

As shown in Fig. 13.1, the mode-shift loop includes a causal link between the volume of each mode and the speed of this mode. This reflects the fact that utilization of each mode has an effect on time. It is important to see that this relationship can be different for each mode. For example, in public transport higher volume can lead to a better service (increased density in time and space) such as a higher frequency and more bus lines, which increases door-to-door speed for the passenger. Whereas in the private car mode, increased volume usually means that speed goes down, especially when congestion occurs. Model 1 makes this difference between "self-accelerating" and "self-limiting" transport modes and can therefore account for complex changes in demand, in particular when also the virtual modes and other effects of ICT come into play.

One of these effects is called the "time utilization effect" in the IPTS study (not represented in Fig. 13.2b, but shown in Fig. 13.2a): Because of mobile work that is possible to some limited, but increasing degree due to ICT, the time spent in traffic is not fully counted as transport time, i.e., a part of it is not charged from the travel time budget. Of course, the degree of time utilization is different from mode to mode (higher in public modes than in the private car mode) and changes over time with progress in mobile ICT devices and infrastructures. This is a core feature of the IPTS study. Time utilization effects can create more transport demand and influence the modal split towards public transport.

Two features of Model 1 could not be shown in Fig. 13.1. First, different modes of transport can share infrastructure, which means that their speeds are coupled to a certain degree (e.g. public buses may be slowed down by congestion caused by private car traffic). This can be expressed in Model 1 by so-called coupling factors for each pair of transport modes. Second, there is an overall reinforcing feedback loop of passenger transport demand which works via settlement dispersion: more traffic volume slowly increases the level of dispersion. It is the level of dispersion which decides how a time deficit is corrected; the correction is in fact a mix between the two possibilities of shifting to a faster mode or reducing the distance covered.

5 Model 2: Tourism Transport, Efficiency, and GHG Emissions

The second model (Model 2) is taken from a study by Peeters [14] on modeling tourism transport demand considering rebound effects of technological efficiency improvement. In a similar way to Model 1, Model 2 has also addressed two types of rebound effects with regard to tourism transport [14]: direct economic rebound effects and time rebound effects (based on travel time budget).

The following subsections present how these two categories of rebound effects were addressed in Model 2.

5.1 Direct Economic Rebound Effect in Model 2

As shown in Fig. 13.2a, the direct economic rebound effects are represented via two reinforcing feedback loops in Model 2: Efficiency enhancing loop and emissions loop.

(a) *Efficiency enhancing loop*

This is the main reinforcing loop, which starts with investment in efficiency enhancing technology. The efficiency reduces energy consumption per seat-kilometer (skm), and thus it reduces cost per skm, which in turn can induce increases in transport volume (pkm) depending on the price elasticity of transport demand (although the economic elasticity is not clearly presented in [14]). More transport generates funds that can be used as more investment in technology improvement, creating a reinforcing loop that improves efficiency.

(b) *Emissions loop*

The reinforcing loop of efficiency improvement transport volume does not necessarily reduce total emissions due to the increase in transport volume in the reinforcing loop. Which of the two loops of efficiency and emissions has the most impact depends on the specifics of the transport system described by the model [14].

(c) *Attitude loop*

A third relevant loop in this system is the attitude loop, a balancing loop because an increase in environmental pressure will tend to increase the willingness to invest in pollution-saving technology, which also improves efficiency [14].

5.2 Time Rebound Effects in Model 2

Model 2, as shown in Fig. 13.2b, contains three reinforcing feedback loops—travel time loop, cost loop, and mode shift loop—and one balancing loop, i.e., max speed loop. The causal loop diagram in Fig. 13.2b is based on three basic assumptions drawn from literature [14]:

- Tendency to travel longer distances (a significant part of a population has the aspiration to increase their range),
- Travel time budget (on a population level the total amount of time spent for actually traveling from home to destinations and back is more or less constant)
- Constant share of income (the average amount of money spent on transport per year on a population level is a constant share of income).

(a) *Travel time loop*

Assuming a constant travel time budget, if people have more money they will be able to travel more kilometers within the constant time budget (This is valid for the whole population, but not for the individual as they can temporarily change the amount of time and money spent on travel.) From "average distance" a reinforcing loop boosts the distances traveled [14].

(b) *Mode shift loop*

As shown in Fig. 13.2b, an increase in money budget and in average travel distance will increase the share of faster modes. Faster transport modes are used over longer distances [14].

(c) *Cost loop*

This reinforcing loop runs through cost of transport. With an increase in speed, operational costs generally reduce because productivity is increased faster than per hour operational costs, allowing for a higher number of kilometers to be sold [14].

6 Main Findings and Discussion

6.1 Main Findings from Reviewing Model 1 and Model 2

Both Model 1 and Model 2 have employed feedback loops to explore the dynamics of transport volume (in pkm). A better understanding and estimation of the demand for transport volume is important because energy demand and GHG emissions are associated with transport volume.

Both models represent the same types of rebound effects in passenger transport, using feedback loops including cost efficiency and resource scarcity loops (for direct economic rebound), and a mode shift loop with travel time budget (for time rebound).

Both models showed that efficiency cannot necessarily reduce total emissions if the transport volume increases because of time rebound and direct economic rebound. These conclusions derived from the models are rooted in empirical evidence as follows. Model 1 is an implemented simulation model the results of which have been evaluated using empirical data (see Ref. [35]). Model 2 is a purely conceptual qualitative model without simulation. The validity of model 2 is justified in the original publication by reference to historical studies on the US and EU showing the existence of the feedback loops [14].

The two models included similar external variables such as population and the economic growth as drivers of transport demand. (See the upper right part in Fig. 13.2b and upper left part in Fig. 13.2a). The efficiency loop modeled in Model 2 (Fig. 13.2a) includes investment in efficiency enhancing technology. However, investments are not explicitly represented in Model 1.

The concept of economic elasticity of demand with regard to price is used in both models. However, Model 1 addressed this in a more explicit way by presenting elasticity parameters for different transport modes.

Both models use the constant travel time budget assumption in a similar way to show the dynamics of speed versus demand; higher speed implies using the unutilized travel time to cover more distance.

6.2 *Time Rebound*

The two models highlight an important aspect of rebound effects which is related to more efficient use of time. Sorrell and Dimitropoulos [4] note that time efficiency, a parallel to direct rebound from energy efficiency, has not been included in many studies that seek to quantify direct rebound effects of efficiency improvements. Time rebound occurs in two ways: first, through the rebound effects with respect to time, e.g. choosing to travel longer; second, through "trading off energy efficiency for time efficiency" [6], e.g. choosing to travel by air rather than by train.

6.3 *Critique of the Travel Time Budget Approach*

Both models assume that at a population level the total amount of time spent for actually traveling from home to destinations and back is more or less constant. The idea of such a travel time budget—which has been developed since 1970s in the field of transport research and supported by empirical studies (e.g. see Ref. [39])—has encountered critiques. Höjer and Mattsson [40] briefly review this mechanism, its advantages and weaknesses and find it "hardly reasonable to presuppose that travel time is constant when planning for future transport systems and urban structures." They critically review some explanations of this hypothesis discussed in the literature regarding biological reasons (e.g. that "people like a certain amount of stability in both habit and behavior" may be related to evolutionary processes and cave-period humans tending to spend the same average time for daily travels) as well as economic and social explanations (see Ref. [40] for details).

It is worth further investigating the advantages and weaknesses of employing the hypothesis of constant travel time compared to other alternatives. Two points regarding the critique can be considered.

First, the travel time budget approach provides a mechanism to model the scarcity of the resource time. Without accounting for time scarcity, a model could possibly predict that someone who can afford it would travel for 24 h per day.

Second, Model 1 already showed a way to relax the constant travel time hypothesis without loosing its advantages: the concept of (travel) time utilization, or dual use of time, mitigates some of the problems of this approach. As shown in Fig. 13.1a, the IPTS included the variable of time utilization in traffic (This variable is not presented in Fig. 13.1b to make the diagram as simple as possible for the purpose of this paper), which means that if passengers can do something else while traveling (e.g., using mobile ICT), this "something else" makes travel less "time consuming". The IPTS study included several factors regarding time utilization. For example, an hour on the train while reading is not a full travel time hour. As shown in Fig. 13.1a, time utilization can create more transport demand and it can influence the modal split via the mechanisms already explained. It has roughly the same effect as an increase in speed.

An alternative approach (not employed in Model 1 or 2) is to convert time into money, leading to the question of the subjective economic value of time spent on travel. The economic value of travel time has been investigated in empirical studies since the 1970s. As an example, if drivers have the choice to pay a fee to cross a bridge or to accept detour for crossing a bridge without paying a fee, these choices can be related to their income (monetary value of their time). It is known from such studies that the value people assign to the time spent while driving a car is between 1/3 and 1/2 of their net hourly income [37]. So it is not the same as working, but it is related to income. The advantage of this approach is that time cost could be added to fuel cost and other variable costs, yielding one price for a pkm. It would then be easier to apply demand elasticity data to determine the size of the rebound effect. However, one problem with monetizing time is that the marginal value may increase dramatically; e.g. the second hour per day spent in traffic might be much more expensive than the first one. Considering this makes the approach less different from the constant travel time budget approach than it may look like.

6.4 *Quantification of Rebound Effects*

Economic studies have presented the calculation of the magnitude of rebound effects. For example, Borenstein [19], in his microeconomic framework or evaluating energy efficiency rebound, provided illustrative calculations for improved auto fuel economy and lighting efficiency and showed that rebound likely reduced the net savings from these energy efficiency improvements by roughly 10–40 %.

How could such a quantitative analysis be conducted using the models discussed in this paper? Each of the models would have to be run in two versions (so-called competitive models, [41]), an original version and a version with those feedback loops cut which are responsible for rebound effects. The model outputs, such as total energy consumption of passenger transport or total passenger transport volume, could then be compared quantitatively among the two versions. Such a simulation experiment could also be refined to a larger number of model versions by disabling only one type of rebound effect at a time.

6.5 *System Archetypes for Rebound Effects*

As we saw in Models 1 and 2, there are particular patterns of feedback for direct economic rebound and time rebound that occur in situations of efficiency improvement and generate particular patterns of behavior. Such generic structures are termed as "system archetypes" [42] which often explain certain situations, in which competing feedback loops determine the behavior of the systems and assist [25]. Future work on system archetypes for rebound effects would facilitate and standardize the future uses of systems thinking and modeling instruments in environmental policy. In such efforts, it is also useful to clarify the connections between rigorous definitions of direct rebound effects (e.g., see Eqs. (13.1, 13.2, 13.3, and 13.4)) and the patterns of feedback modeled in the system archetypes.

6.6 Contribution of Systems Thinking and Modeling to Rebound Analysis

Many economic, sociological and psychological studies have addressed different aspects, both qualitative and quantitative, of rebound effects [5–8]. See Ref. [18] for an overview of sociological, ecological economics, and environmental systems analysis perspectives on rebound analysis. What is the contribution of systems thinking and System Dynamics modeling to the analysis of rebound effects? Can we generalize insights from our case study on rebound effects in passenger transport to rebound effects in other sectors?

First, we should note that most of existing studies in economics and sociology seek to provide empirical evidence on rebound effects and their magnitude. For example, econometric studies use private household surveys and data on elasticity of demand for useful work or energy consumption to estimate the direct rebound effect [43]; use input output models to estimate the indirect (income) rebound effect [44]; and use Computable General Equilibrium (CGE) models to address economy-wide effects [45]. Further, sociological studies seek to explain various social factors and dynamics behind the increase in consumption—factors such as search for identity, status competition, advertising, lock-in within institutional structures like the work-and-spend cycle, and individualization [46].

What systems thinking and modeling instruments provide is a set of tools such as CLDs and simulation combined with collaborative approaches to "offer a holistic way of appreciating all dimensions of a complex problem" [24] and to better understand the causal relations (circular causality) between the factors influencing a certain effect observed in a real-world context. Given the current knowledge about a system (in which a rebound effect occurs) and knowledge about historically and socially constructed influencing factors, a systems thinking and modeling approach assists in structuring a policy issue regarding rebound effects, conceptualizing and linking socio-economic causal relations, and communicating and enhancing such understanding of the system and its behavior over time in collaboration with stakeholders. Thereby, systems modeling provides a platform for testing various policy options and system intervention scenarios to mitigate rebound effects—For example, see the policy discussions in studies employing a systems approach and also addressing rebound effects: Hilty et al. [12], Stepp et al. [13], Peeters [14], and Dace et al. [15].

7 Conclusions

The two models we discussed represent the same types of rebound effects in passenger transport. Both are multi-modal transport models, considering the dynamic change of modal split as well. Feedback loops (closed causal chains) are an obvious concept to model rebound effects at a macro-economic level as it is done in System Dynamics (as opposed to use behavior rules at the micro-economic level in agent-based simulation).

Despite the similarities, the comparison of the two models showed that there can be much variety in the details of modeling rebound effects in passenger transport. Model 1 puts greater emphasis on the different characteristics of transport modes and how they interact, on time utilization and virtual modes, whereas Model 2 explicitly considers investment in technology and environmental attitude as variables in the main feedback loops.

The contribution of the systems thinking and modeling approach to rebound analysis originates from its holistic approach, its capability to be built upon our empirical knowledge on rebound effects; its capability to represent various policy options and intervention scenarios; and its support for integration of rigorous simulation tools with high-level diagramming tools easy to be used by a variety of stakeholders in a collaborative modeling environment.

Future work may employ systems thinking and modeling in designing policy instruments, addressing both efficiency and rebound effects in a holistic perspective.

Acknowledgements The authors would like to thank Empa (Technology and Society Lab), KTH (Centre for Sustainable Communications), and Vinnova, which made this work possible as a part of the first author's Ph.D. project.

References

1. O. Edenhofer, R. Pichs-Madruga, Y. Sokona, S. Agrawala, I. A. Bashmakov, G. Blanco, J. Broome, and others, *Climate Change 2014: Mitigation of Climate Change: Working Group III Contribution to the Fifth Assessment Report of the Intergovernmental Panel on Climate Change*. Cambridge University Press, 2014.
2. P. H. G. Berkhout, J. C. Muskens, and J. W. Velthuijsen, "Defining the rebound effect," *Energy Policy*, vol. 28, no. 6–7, pp. 425–432, Jun. 2000.
3. A. Arvesen, R. M. Bright, and E. G. Hertwich, "Considering only first-order effects? How simplifications lead to unrealistic technology optimism in climate change mitigation," *Energy Policy*, vol. 39, no. 11, pp. 7448–7454, Nov. 2011.
4. S. Sorrell and J. Dimitropoulos, "The rebound effect: Microeconomic definitions, limitations and extensions," *Ecol. Econ.*, vol. 65, no. 3, pp. 636–649, Apr. 2008.
5. L. A. Greening, D. L. Greene, and C. Difiglio, "Energy efficiency and consumption — the rebound effect — a survey," *Energy Policy*, vol. 28, no. 6–7, pp. 389–401, Jun. 2000.
6. S. Sorrell, J. Dimitropoulos, and M. Sommerville, "Empirical estimates of the direct rebound effect: A review," *Energy Policy*, vol. 37, no. 4, pp. 1356–1371, Apr. 2009.

7. I. Røpke and T. H. Christensen, "Energy impacts of ICT – Insights from an everyday life perspective," *Telemat. Inform.*, vol. 29, no. 4, pp. 348–361, Nov. 2012.
8. W. Abrahamse, L. Steg, C. Vlek, and T. Rothengatter, "A review of intervention studies aimed at household energy conservation," *J. Environ. Psychol.*, vol. 25, no. 3, pp. 273–291, Sep. 2005.
9. J. W. Forrester, *Industrial Dynamics*. Cambridge Massachusetts: The MIT Press, 1961.
10. J. D. Sterman, *Business dynamics: systems thinking and modeling for a complex world*, vol. 19. Irwin/McGraw-Hill Boston, 2000.
11. G. P. Richardson, "System dynamics, the basic elements of," in *Complex Systems in Finance and Econometrics*, Springer, 2011, pp. 856–862.
12. L. M. Hilty, P. Arnfalk, L. Erdmann, J. Goodman, M. Lehmann, and P. A. Wäger, "The relevance of information and communication technologies for environmental sustainability – A prospective simulation study," *Environ. Model. Softw.*, vol. 21, no. 11, pp. 1618–1629, Nov. 2006.
13. M. D. Stepp, J. J. Winebrake, J. S. Hawker, and S. J. Skerlos, "Greenhouse gas mitigation policies and the transportation sector: The role of feedback effects on policy effectiveness," *Energy Policy*, vol. 37, no. 7, pp. 2774–2787, Jul. 2009.
14. P. Peeters, "Chapter 4 Tourism transport, technology, and carbon dioxide emissions," in *Tourism and the Implications of Climate Change: Issues and Actions*, vol. 3, 0 vols., Emerald Group Publishing Limited, 2010, pp. 67–90.
15. E. Dace, G. Bazbauers, A. Berzina, and P. I. Davidsen, "System dynamics model for analyzing effects of eco-design policy on packaging waste management system," *Resour. Conserv. Recycl.*, vol. 87, pp. 175–190, Jun. 2014.
16. S. Sorrell, "Jevons' Paradox revisited: The evidence for backfire from improved energy efficiency," *Energy Policy*, vol. 37, no. 4, pp. 1456–1469, Apr. 2009.
17. C. Gossart, "Rebound Effects and ICT: A Review of the Literature," in *ICT Innovations for Sustainability*, L. M. Hilty and B. Aebischer, Eds. Springer International Publishing, 2015, pp. 435–448.
18. M. Börjesson Rivera, C. Håkansson, Å. Svenfelt, and G. Finnveden, "Including second order effects in environmental assessments of ICT," *Environ. Model. Softw.*, vol. 56, pp. 105–115, Jun. 2014.
19. S. Borenstein, "A microeconomic framework for evaluating energy efficiency rebound and some implications," National Bureau of Economic Research, 2013.
20. L. Erdmann and L. M. Hilty, "Scenario Analysis: Exploring the Macroeconomic Impacts of Information and Communication Technologies on Greenhouse Gas Emissions," *J. Ind. Ecol.*, vol. 14, no. 5, pp. 826–843, 2010.
21. J. D. Khazzoom, "Economic Implications of Mandated Efficiency in Standards for Household Appliances," *Energy J.*, vol. 1, no. 4, pp. 21–40, Oct. 1980.
22. T. H. Oum, W. G. Waters, and J.-S. Yong, "Concepts of price elasticities of transport demand and recent empirical estimates: an interpretative survey," *J. Transp. Econ. Policy*, pp. 139–154, 1992.
23. P. Hjorth and A. Bagheri, "Navigating towards sustainable development: A system dynamics approach," *Futures*, vol. 38, no. 1, pp. 74–92, Feb. 2006.
24. N. C. Nguyen, D. Graham, H. Ross, K. Maani, and O. Bosch, "Educating systems thinking for sustainability: experience with a developing country," *Syst. Res. Behav. Sci.*, vol. 29, no. 1, pp. 14–29, 2012.
25. J. Mingers and L. White, "A review of the recent contribution of systems thinking to operational research and management science," *Eur. J. Oper. Res.*, vol. 207, no. 3, pp. 1147–1161, Dec. 2010.
26. E. F. Wolstenholme, "Qualitative vs quantitative modelling: the evolving balance," *J. Oper. Res. Soc.*, pp. 422–428, 1999.
27. N. Videira, F. Schneider, F. Sekulova, and G. Kallis, "Improving understanding on degrowth pathways: An exploratory study using collaborative causal models," *Futures*, vol. 55, pp. 58–77, 2014.

28. L. Erdmann and F. Wurtenberger, "The future impact ICT on environmental sustainability. First Interim Report. Identification and global description of economic sectors," Institute for Prospective Technology Studies (IPTS), Sevilla, 2003.
29. L. Erdmann and S. Behrendt, "The future impact ICT on environmental sustainability. Second Interim Report," Institute for Prospective Technology Studies (IPTS), Sevilla, 2003.
30. J. Goodman and V. Alakeson, "The future impact ICT on environmental sustainability. Third Interim Report. Scenarios," Institute for Prospective Technology Studies (IPTS), Sevilla, 2003.
31. L. M. Hilty, P. Wäger, M. Lehmann, R. Hischier, T. F. Ruddy, and M. Binswanger, "The future impact of ICT on environmental sustainability. Fourth Interim Report. Refinement and quantification," Institute for Prospective Technological Studies (IPTS), Sevilla, 2004.
32. P. Arnfalk, "The future impact ICT on environmental sustainability. Fifth Interim Report. Evaluation and Recommendations," Institute for Prospective Technology Studies (IPTS), Sevilla, 2004.
33. L. Erdmann, L. M. Hilty, J. Goodman, and P. Arnfalk, "The future impact ICT on environmental sustainability. Synthesis Report," Institute for Prospective Technology Studies (IPTS), Sevilla, 2004.
34. P. Wäger, L. M. Hilty, P. Arnfalk, L. Erdmann, and J. Goodman, "Experience with a System Dynamics model in a prospective study on the future impact of ICT on environmental sustainability," in *IEMSs 3rd Biennial Meeting Summit on Environmental Modeling and Software*, Burlington, USA, 2006.
35. M. A. Achachlouei and L. M. Hilty, "Modeling the Effects of ICT on Environmental Sustainability: Revisiting a System Dynamics Model Developed for the European Commission," in *ICT Innovations for Sustainability*, L. M. Hilty and B. Aebischer, Eds. Springer International Publishing, 2015, pp. 449–474.
36. L. M. Hilty, *Ökologische Bewertung von Verkehrs-und Logistiksystemen: Ökobilanzen und Computersimulation*. IWÖ, 1994.
37. L. M. Hilty, "Umweltbezogene Informationsverarbeitung–Beiträge der Informatik zu einer nachhaltigen Entwicklung," *Habil Hambg.*, 1997.
38. G. Hupkes, "The law of constant travel time and trip-rates," *Futures*, vol. 14, no. 1, pp. 38–46, Feb. 1982.
39. D. Metz, "The Myth of Travel Time Saving," *Transp. Rev.*, vol. 28, no. 3, pp. 321–336, May 2008.
40. M. Höjer and L.-G. Mattsson, "Determinism and backcasting in future studies," *Futures*, vol. 32, no. 7, pp. 613–634, Sep. 2000.
41. L. M. Hilty, R. Meyer, and T. F. Ruddy, "A general modelling and simulation system for sustainability impact assessment in the field of traffic and logistics," *Environ. Inf. Syst. Ind. Public Adm. Idea Group Publ.*, pp. 167–185, 2001.
42. E. Wolstenholme, "Using generic system archetypes to support thinking and modelling," *Syst. Dyn. Rev.*, vol. 20, no. 4, pp. 341–356, 2004.
43. J. Schleich, B. Mills, and E. Dütschke, "A brighter future? Quantifying the rebound effect in energy efficient lighting," *Energy Policy*, vol. 72, pp. 35–42, Sep. 2014.
44. R. Kok, R. M. J. Benders, and H. C. Moll, "Measuring the environmental load of household consumption using some methods based on input–output energy analysis: A comparison of methods and a discussion of results," *Energy Policy*, vol. 34, no. 17, pp. 2744–2761, Nov. 2006.
45. S. Grepperud and I. Rasmussen, "A general equilibrium assessment of rebound effects," *Energy Econ.*, vol. 26, no. 2, pp. 261–282, Mar. 2004.
46. I. Røpke, "Theories of practice — New inspiration for ecological economic studies on consumption," *Ecol. Econ.*, vol. 68, no. 10, pp. 2490–2497, Aug. 2009.

Part IV
Sustainable Enterprises and Management

Chapter 14
Software and Web-Based Tools for Sustainability Management in Micro-, Small- and Medium-Sized Enterprises

Matthew Johnson, Jantje Halberstadt, Stefan Schaltegger, and Tobias Viere

Abstract Recently, new approaches to organizational level sustainability management and reporting have emerged in the form of software and web-based applications. At first glance, it appears that such software and web-tools are applicable in small and medium-sized enterprises (SMEs), as they offer user-friendly and cost-effective alternatives to assess, manage and report on company-wide sustainability activities. Nevertheless, it remains academically and practically uncertain if such technological advancements will be adopted by a great number of SMEs. Using the Individual-Technology-Organization-Environment (ITOE) model as a theoretical framework and empirical data from a recent survey with 1,250 German SMEs, this paper investigates various firm-internal and external factors that might influence managers' decisions to adopt or reject this new technology. This paper reveals which factors might play a role in the adoption of such web-tools in SMEs. In addition, this paper proposes a conceptual framework of an IT-assisted sustainability analysis and reporting scheme for micro-enterprises and startups. Based on existing software for larger enterprises, the paper describes the main content and potential layout of such a web-based tool.

Keywords Small and medium-sized enterprises • Sustainability management • Individual-technology-organization-environment framework • Quick-check • Web-based tools

M. Johnson (✉) • J. Halberstadt • S. Schaltegger
Centre for Sustainability Management, Leuphana University Lüneburg, Lüneburg, Germany
e-mail: johnson@uni.leuphana.de; jantje.halberstadt@uni.leuphana.de; schaltegger@uni.leuphana.de

T. Viere
Institute for Industrial Ecology, Pforzheim University, Pforzheim, Germany
e-mail: tobias.viere@hs-pforzheim.de

J. Marx Gómez et al. (eds.), *Advances and New Trends in Environmental and Energy Informatics*, Progress in IS, DOI 10.1007/978-3-319-23455-7_14

1 Introduction

Sustainability management integrates economical, ecological and social aspects into the core business activities while finding business links to remain competitive and economically viable [5]. A company should steer its activities in such a manner to reduce its negative effects and achieve positive outcomes for its social and environmental aspects related to business operations, while contributing to the sustainable development of society and the economy [29]. Visions and strategies of corporate sustainability can help to integrate all these activities into the core business of a company.

To support this strategic integration, companies are provided with a vast set of operational approaches, including management software applications to manage sustainability-driven goals and strategies. A wide range of software programs facilitate various managerial tasks with many areas of application, including accounting, research and development, procurement and production, supply chain management as well as cross-functional activities [33]. However, an all-embracing software integrative all aspects of sustainability is currently unavailable [22]. Even so, few SMEs would likely adopt such a software program due to high implementation costs. There also is an apparent inhibition from SMEs to allow a software application to manage sustainability aspects without the additional consultation of experts [13, 17].

With limited exceptions [1, 10], research has not yet investigated the adoption of sustainability management software and web-tools in SMEs. Furthermore, a research gap has emerged on organizational-level factors that influence the decision to adopt or reject such software and web-tools. This chapter begins by filling the knowledge gap on the reasons for adoption based on two conference papers at EnviroInfo 2014 [11, 18]. By doing so, it provides initial discernments on the main influential factors that might affect the adoption of software and web-tools [18]. In addition, this chapter incorporates a second paper from EnviroInfo [11] proposing a conceptual framework for even the smallest of enterprises to analyze and report on their sustainability activities. The findings in these combined papers provide new insights on the adoption and functionality of sustainability software in SMEs.

2 Web-Based Tools for German SMEs

In recent years, innovative approaches to sustainability management have emerged in the form of software and web-based applications to support companies of all sizes with their sustainability activities [21, 31]. Organizational based software and web-tools have been designed to facilitate various management tasks related to sustainability, such as self-assessment and strategy formation on sustainability aspects, sustainability control and benchmarking (e.g. EPM-Kompass) [10]; and

Table 14.1 An overview of web-based sustainability management tools

Product	Sustainability aspects	Application area
360 Report	Integrative	Reporting
CR Kompass	Integrative	Assessment, reporting
EcoEnterprise	Ecological	Control, management
EcoWebDesk	Ecological	Control, management
EPM-Kompass	Ecological	Control
EffiCheck	Ecological	Assessment
Green Software	Integrative	Assessment, management
N-Kompass	Integrative	Assessment, management
Quick-Scan	Ecological	Assessment
Verso	Integrative	Assessment, reporting

sustainability reporting [15, 31] as well as administration of occupational safety and environmental management (e.g. EcoTra) [20].

Software and web-based applications supporting the implementation of sustainability management can facilitate various management tasks including the assessment, control, management and communication (i.e. reporting) of sustainability activities. Commercialized software applications are increasingly emerging, promising to enable the overall coordination and communication of sustainability-related tasks shared between various functions and employees within the company. While it is understood that software is not a substitute for the human factor – from strategic visions and planning to the manual input and co-ordination of data – it appears that software can offer many promising ad-vantages once the strategies and responsibilities have been properly assigned. Table 14.1 below presents an overview of the available web-based tools for SMEs:

At first glance, it appears that sustainability management software and web-tools are applicable to SMEs. These applications offer a cost-effective approach to introducing sustainability management in the company and allowing managers to deal with sustainability activities in an organized manner. They can be tailored to an enterprise's particular structure and provide user-friendly features, such as a multi-user function allowing multiple persons to work simultaneously on one project as well as offering a manageable step-by-by instructions, so that additional training is not required to input and retrieve the necessary data. While several authors promote the applicability of such software [10], there is a lack of empirical evidence on the adoption of such software in SMEs. It re-mains unclear if firm-level software and web-based tools for sustainability management will be applied by a great number of SMEs. Previous research has yet not investigated which internal and external factors play a role in decision-making to adopt such technologies. Therefore, these practical and scientific uncertainties have lead us to propose the following research question:

> *Which firm-internal and external factors influence the adoption of software and web-based tools for sustainability management in SMEs?*

Instead of examining the current success and failure rates of individual software application and web-tools, this paper investigates various organizational and external factors that might influence adoption rates from a wider perspective. Using the *Individual-Technology-Organization-Environment* (ITOE) framework [32], it is possible to quantify which particular factors influence the rate of adoption for these new technologies for an enterprise's sustainability management. The next section will explain how the research question was addressed using this framework.

3 Theoretical Background

With the ITOE framework [32] this paper examines various firm-internal and external factors that might influence decision-making for new technologies in SMEs. The ITOE framework was selected over other frameworks, such as the diffusion of innovation framework and task-technology-fit, as it can simultaneously examine adoption of new technologies in four different contexts, including individual (or personal) factors, technological factors, organizational (or internal) factors, and environmental (or external) factors.

The ITOE framework has been frequently applied to research on the adoption of new software and web-based solutions in SMEs, particularly with Enterprise Resource Planning (ERP) software [4, 27, 28] and e-business solutions [23, 35]. These papers reveal which and how various factors, such as prior IT-knowledge, attitude towards new software, top management support and external IT-support, play a role in firm-wide decision making to adopt such software. For example, Ramdani et al. [27] illustrate how the adoption of ERP software in SMEs is mostly influenced by top management support since the primary decision-maker in SMEs is typically the owner-manager.

However, no account has been found for the ITOE framework in context of environmental or sustainability software. Therefore, this paper applies an adaptation of the ITOE framework to assess what exactly influences SMEs to adopt sustainability management software. Figure 14.1 below shows the overall research model as well as the various factors among the four contexts that were taken into consideration for this paper.

Within the individual context, three factors were selected, including prior IT-knowledge, innovativeness and attitude. Prior IT-knowledge explains an individual's beliefs about level of competency with IT, which in this case is the perceived ability to use the computer and related software applications. Innovativeness refers to the managers' willingness to take risks and try something new through experimentation. Attitude refers to a managers' positive or negative feelings about a new technology. In this context, attitude towards sustainability management web-tools are being assessed.

From the technological context, five factors were provided, including relative advantage, compatibility, complexity, trialability and observability. Relative advantage refers to the degree in which a manager perceives the software or

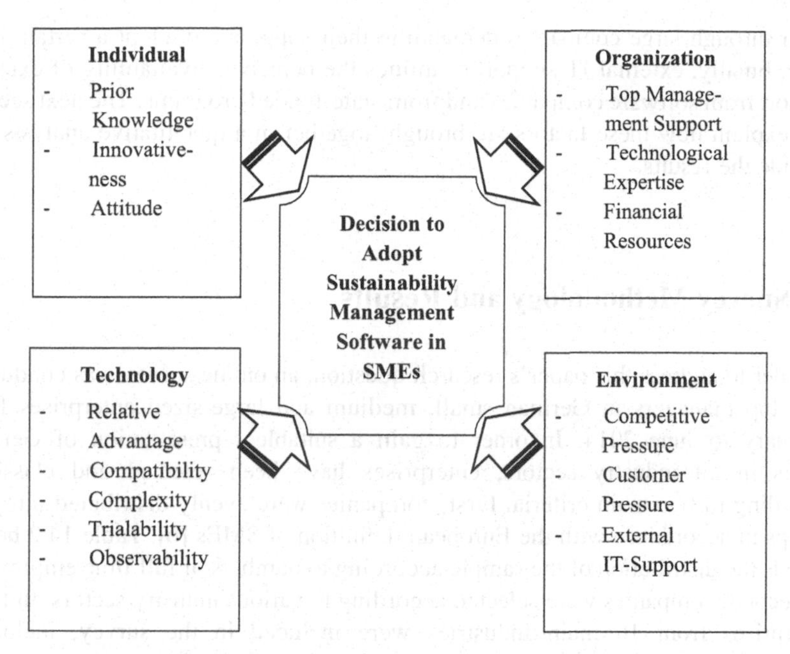

Fig. 14.1 Adaptation of ITOE framework for survey

web-tool to be superior to the previous method of operation. This factor is considered a key factor in improving the rate of new technology adoption to the extent that the innovation is perceived as [12]. However, it might not be as relevant in the case of sustainability management software and web-tools as most SMEs have not previously have had a formal approach to sustainability up till now [7]. Compatibility explains the degree in which software is perceived to be well-matched with existing organizational structure and software usage. Complexity is the perceived extent to which a new technology is difficult to understand and use. This would be reflected as a negative value in comparison to rate of adoption. Trialability and observability focus on the degree in which software can be experimented on a limited basis and can be visible to others.

In the organizational context, four factors were included in the model – top management support, technological expertise, financial resources and firm size. For starters, support from top management can highly affect if such software will be implemented [12, 28]. Furthermore, the availability of in-house software support (technological expertise) and ample financial resources may play a role in decision-making. Company size, measured by full-time employees, has been revealed as a major determinant for the rate of new technology adoption [12].

From the environmental context, three factors were selected, including competitive pressure, customer pressure and external IT-support. Competitive pressure measures the perceived intensity level of competition and resulting pressure to adopt new technologies to remain competitive. Customer pressure is the perceived feeling of demands from customers to adopt software. In the case of SMEs, this may occur through large companies demanding their suppliers to adopt a certain software. Finally, external IT-support examines the perceived availability of external support from software companies and from state-funded programs. The next section will explain how these factors are brought together in a quantitative analysis and provide the results.

4 Survey Methodology and Results

In order to address this paper's research question, an online survey was conducted with top managers in German small, medium and large-sized enterprises from February to June 2014. In order to gain a suitable representation of German SMEs in all industry sectors, enterprises have been selected and classified according to two main criteria. First, companies were evenly distributed into five groups in accordance with the European definition of SMEs [6]. Table 14.2 below reveals the distribution of the sample according to numbers of full-time employees.

Second, companies were selected according to various industry sectors. In total, enterprises from 10 main industries were included in the survey, including manufacturing, energy utilities, construction, wholesaler and retailers, transportation, gastronomy, and various service sectors. The number of companies selected from each industry was based on percentages of enterprises in each sector.

A total of 1250 enterprises were sent an e-mail invitation to the online survey. However, 96 of these invitations were sent back as *not deliverable*. In total, the survey produced 145 usable questionnaires from the 1154 e-mails delivered. The response rate is 12.6 %, which is comparable with other surveys with similar focus of sustainability management in SMEs [17].

The online survey consisted of questions with mostly closed-form responses using a seven-point Likert scale. The dependent variable is the adoption of sustainability management software with the question, "Does your company currently use or plan to adopt sustainability management software within the next 2 years?" Questions on the relevant factors were organized according to the four contexts – individual, technological, organizational and environmental. For each individual factor within each context (e.g. "top management support"), three to eight questions

Table 14.2 Distribution of sampled companies

Company size (full-time employees)	Number of contacted companies	Number of companies completing the survey
10–49	250	34
50–99	250	29
100–249	250	31
250–499	250	26
500+	250	25
Totals	1,250	145

Table 14.3 Averages and differences between adopters and non-adopters

Factors	Adopters		Non-adopters		Differences
Individual factors Avg. S.D. Avg. S.D. Avg					
Prior IT-knowledge	4.43	1.06	4.55	1.39	−0,12
Innovativeness	6.01	0.85	5.85	0.91	0.16
Attitude	4.80	1.41	2.93	1.35	1.87
Technological factors					
Relative advantage	4.59	1.23	3.91	1.27	0.68
Compatibility	4.47	1.29	3.67	1.24	0.80
Complexity	4.18	1.42	3.86	1.21	0.32
Trialability	4.02	1.61	2.33	1.48	1.69
Observability	4.90	1.61	2.02	1.52	2.88
Organizational factors					
Top management	4.36	1.44	2.92	1.51	1.44
Financial resources	4.93	1.67	4.31	1.79	0.62
Expertise	5.47	1.20	4.54	1.65	0.93
Environmental factors					
Competition	3.87	1.52	2.80	1.37	1.07
Customer pressure	3.23	1.51	2.95	1.56	0.28
External IT-support	3.65	1.29	2.76	1.18	0.89

were provided, and these questions were later averaged to represent the factor in the analysis stage. For example, the question was asked: "Does your company's top management support the implementation of sustainability software?"

An initial evaluation of the results examined the descriptive statistics of the data including averages (Avg.) and standard deviations (S.D.) of studied factors of the TOE framework. These factors were investigated based on the answer to the lead question on adoption, that is either "decision to adopt" or "decision to reject". Adopters are managers who currently use sustainability management software and/or who intend to adopt such software within the next 2 years. Non-adopters are managers who neither use nor plan to adopt such software. As expected, the group *decision to reject* was much greater (110 enterprises) than the group *decision to adopt* (35 enterprises). Table 14.3 below shows the descriptive statistics of averages and standard deviations from the various influential factors between the two groups of respondents.

From Table 14.3 we observe significant differences between both groups, adopters and non-adopters, with the factors personal attitude, trialability, observability, top management support and competitive pressures. From these preliminary results, we can deduce that managers' perceived awareness of commercialized software for sustainability management is a major determinant for adoption, where they can also test it on a limited basis (trialability) and see others using it (observability). Furthermore, the overall positive attitude towards software combined with added support from top management also positively influences the chances that such software will be used.

Other factors had also similar results, including top management support, technological expertise, competitive pressure and external IT-support. Even though the differences of two factors in the environmental context (i.e. competitive pressure and external IT-support) are substantial, the overall averages are moderate and even below average even for adopters. This means that external factors play a marginal role in the decision to adopt sustainability management software. Only one factor, namely prior IT-knowledge, was stronger for non-adopters than adopters; however, the difference is so small that it is difficult to argue that commercialized software for sustainability management might be able to increase such IT-knowledge.

In a second step, a multi-logit regression analysis was conducted on those variables that had the greatest difference in mean values between the two groups (adopters and non-adopters). These selected variables include attitude, trialability, top management support, competitive pressure and external IT-support. Company size according to employee amounts was also included as a control variable. After a preliminary reliability screening, the variable 'observability' was removed because it too strongly predicts adoption, which makes all the other factors insignificant. In addition, the problem of multicollinearity arose for the factor observability in the regression model, as the variance inflation factor (VIF) was above 4. While observability is clearly the strongest variable predicting adoption, other factors also play a role in the adoption of sustainability management software. From another point of view, it could be argued that the other variables first influence observability and then the latter strongly influences adoption. Table 14.4 below shows the results of the regression analysis.

Besides the strong influence of company size, the results show that the decision to adopt sustainability management software mainly depends on the observability – the awareness that sustainability management software exists – and trialability – SME managers have been able to try it out. Furthermore, SMEs managers will likely adopt sustainability management software if they have an overall positive attitude towards the software. It is also important that top management supports the decision to adopt it. Future research could further investigate these influential factors in qualitative interviews to better understand why companies should adopt or reject such software.

While these results provide new insights on influential factors for the adoption of sustainability management software, several concerns remain. On one hand, it remains uncertain if companies with existing environmental and sustainability management systems have less of a need for commercialized software, as they have probably some IT-solution already, for example self-made Excel spreadsheets and Word documents. On the other hand, companies that are not interested in sustainability management in the first place will not perceive any benefit for related software.

While this study provides good insights for SMEs with more than 10 employees, the results did not include an important sub-category of SMEs – micro-sized enterprises. From the overview of the available sustainability management software and a conceptual framework based on the business model canvas by Osterwalder et al. [24], this paper now presents a novel concept for an online tool for micro-enterprises to assess and report on sustainability impacts of their business.

5 Web-Based Tool for Micro-enterprises

With few exceptions [36], previous research has not proposed company-level tools for sustainability management in micro-enterprises and particularly start-ups. While some research does exist on sustainable business models and plans [2], these models are more focused on sustainable innovations versus the core business itself. Such business models are difficult for most start-ups to implement because they mainly revolve around new business units than the core message of a new company. These business models also do not provide tools for a comprehensive sustainability evaluation and reporting system [3, 25].

Furthermore, IT-solutions have not considered the early stages of business creation from the actual start-up of a company to its further development as a micro-enterprise. In fact, it appears that literature has overlooked certain category of businesses in the sustainability management context. According to the European Commission this category includes micro-enterprises with less than 10 employees and no more than two million Euro annual revenue.

However, mounting evidence suggests that start-ups and micro-enterprises should be considered in light of sustainable development for several reasons. First of all, sustainability is relevant for all companies in every industry of every economy [29]. Secondly, sustainability will never be achieved if the smallest companies do not get involved [13]. Not only do micro-enterprises constitute a majority of all registered businesses, e.g. 2.8 million enterprises (ca. 80 %) in Germany fall into the micro-enterprise range, they also feed many products and services into the larger companies as suppliers and service-vendors. Thirdly, while it could be argued that individual micro-enterprises transmit an insignificant burden on the environment, it is their collective impact and spill-over into larger enterprises that raises major concerns.

Fourthly, besides the direct burdens placed on society and the environment, indirect effects can be attributed to the exemplary roles that entrepreneurs and owner-managers of small businesses hold in economies and societies that desperately look for heroes to right the wrongs of environmental degradation and intra-generational injustices through sustainability-driven goals and measures. When considering the good examples set by social entrepreneurs, such as Muhammad Yunus, and ecopreneurs, such as Klaus Hipp, new business founders need not just inspirational stories, but effective operational means and devices to steer their business endeavors into future-oriented sustainability enterprises [30].

Last but not least, start-ups generally do not remain small but rather are growth-oriented [8, 34]. As the size of the enterprise increases, so too does the relevance and knowledge for sustainability management [13]. In addition to well-known management problems of fast-growing enterprises [16], small business managers must be informed about the increasing environmental and social demands that rise with increasing size. For example, in the future it is plausible that medium-sized enterprises starting with 100 employees will be mandated by corporate law to state their environmental and social impacts through annual sustainability reports [19]. Those owner-managers that have addressed with sustainability issues from the beginning might achieve a competitive advantage over those that decide to wait it out. The challenges of sustainability management in start-ups and fast growing companies should be integrated so to avoid a lengthy, costly period of playing catch-up.

Thus, the questions are raised: why should a start-up or micro-sized enterprise wait to reach a certain size in order to measure, manage and report on its sustainability activities? How could such a sustainability management program be conceptualized? What benefits would it bring the enterprise? Lastly, how might IT-solutions provide simple yet effective means to accomplishing these goals?

The aim of this section of the paper is, therefore, not only to close the theoretical gap on appropriate sustainability measures for start-ups and micro-enterprises, but also to propose a conceptual framework for an IT-supported application that allows a company to easily access and report its sustainability activities. This conceptual model will hopefully set the foundation for further practical developments. Based on previous research on sustainability management tools in SMEs [14, 17], this conceptual model proposes the contents and step-wise process of an IT-support tool for both start-ups and micro-enterprises. This IT-supported tool, called the "Sustainability Quick-Check" (SQC) model, will be explained in the next section.

6 Sustainability Quick-Check

Many of the existing processes for the preparation of sustainability assessments and reports are complex and contain a variety of indicators and metrics. In turn, this provides no clear path or structure for intuitive handling. One possible reason may be attribute to the fact that software applications were intended to be sold with additional consulting services. The aim of this paper is to develop a manageable and straightforward tool with a clear structure and based on understandable steps for a start-up and micro-enterprise.

The development of the SQC model is broken down into three complementary and sequential stages. In the first stage, a systematic analysis of the existing sustainability management tools and software and tools were examined. Based on Johnson [18], it is established that not all management tools are applicable even in small and medium-sized enterprises (SMEs) with 10 or more employees. The most applicable tools for small businesses are those that correspond with well-established management practices, such as a quality management system, training and education on sustainability management, risk analysis, supply chain management and even an environmental management system.

In the second stage of analysis, several SME-adequate software and web-based applications were closely examined (refer to Table 14.1). These web-based tools not only offer user-friendly, cost effective ways to analyze and report on sustainability management in SMEs, combined they provide a good overview of what criteria and indicators should be considered for sustainability management in small businesses. While these various applications offer great insights applicable topics and indicators for SMEs, it is still uncertain if these software packages and web-applications will be adopted by very small enterprises and start-up companies.

In the third stage, a grid was developed that allows a structured overview of sustainability topics and corresponding indicators for start-ups and micro-enterprises. The idea behind this structure was to combine the results from both the first stage of analysis with the ideas from business model canvas [24] and value chain [26]. Suitable sustainability key performance indicators and metrics were classified into various SQC-categories, such as production, supply chain management, sales and marketing and administration and supporting business functions (including strategy and human resources), and further broken down into key activities, key resources and key partners from both environmental and social perspectives. Figure 14.2 below depicts example of possible categories, fields and aspects for the SQC model.

The SQC model is based on some of the components of the aforementioned software, the value chain according to Porter [26] and the business model canvas by Osterwalder et al. [24]. The value chain is the presentation and analysis of the primary (e.g. logistics, production or operations, sales and marketing) and supporting secondary activities (e.g. administration, human resources, research and development) that together bring value to a company's products and services. Similarly, this value chain has been used to assess environmental and social sustainability aspects along all these business activities [29]. Therefore, the value chain served as the basis for our selection of the four SQC categories, including production, supply chain, market and internal firm structure. Primary activities can be located in the first, second and third categories. For example, inbound and outbound logistics are combined with supply chain management and procurement into one category. The supporting activities provide an indirect but still supporting role in the production of products and/or services, and these are mostly located in the fourth category.

The business model canvas is a method of visualization of business models [24]. Business models describe the basic principles by which organizations create value, with the distinctions made between three aspects: the product-market combination, the configuration of value chains and main revenue mechanisms. For the SQC, the configuration of value chains is considered to be particularly important, since this the area where sustainability-related decision are made. Also, this part of the business model fits well with Porter's [26] value chain. A brief description of the product-market combination should precede the initial analysis, but it is actually not a part of the SQC since it is tailored for all kinds of startups and micro-enterprises. The environmental and social aspects of companies are already a part

Basic Structure of the *Sustainability Quick Check (SQC)*		Sustainability	
		Ecological Aspects	Social Aspects
SQC-Category	Assessment field	Example Criteria	
Production of Product or Service	Key Activities	Energy and Water consumption in production (G4-EN3/ EN8)	Adherence to working hours and guarantee of workplace safety (G4-LA5, LA6)
	Key Resources	Use of non-toxic and recycling materials and packaging (G4-EN1, EN28)	Use of fair trade materials, incl. free from forced and child labor
	Key Partners	Selection of regional, sustainable production partners, (G4-EN17)	Support of the disadvantaged, e.g. collaboration with disabled persons
Supply Chain Management, incl. Logistics and Procurement	Key Activities	Shortening transport routes	Supply chain code of conduct and enforcement (audits); supplier training
	Key Resources	Environmentally conscious procurement for environmentally safe materials (G4-EN2)	Purchasing requirements for fair products
	Key Partners	Selection of regional, environmentally friendly partners	Supplier selection and negotiations for fair and safe working practices
Firm Structure, Administration and Human Resources	Key Activities	Training and support on the ecological performance of employees	Pay attention to equality in the workplace; Guidelines for recruitment (G4-LA1)
	Key Resources	Energy efficient Administration building (G4-EN3)	Employees with fair wages (G4-EC5)
	Key Partners	Employee participation in environmental activities	Employee participation in firm-internal decisions as well as firm-external community engagement projects

Fig. 14.2 Sustainability quick-check example criteria

the business model, and they will be described separately in the product-market combination.

The business model canvas depicts a total of nine areas of a business model. The fields deemed particularly relevant for the SQC are the key activities, key resources and key partners. Key activities are those actions that are particularly important for a particular area of a business (in this case for each category, such as production of products and services). Key resources can be physical, intellectual, human or financial resources. In addition, a sustainability management tool that carries information about desired sustainable processes, such as guidelines for environmentally conscious procurement and supply chains, can also be considered a key resource. Key partners consider essential partnerships into order to fulfill the key activities. Examples of partners are buyer–supplier relationships, and also strategic alliances with competitors and additional support organizations. This area ensure that sustainability issues are at the heart of cooperation, but partners must also be audited and consulted for conformity to an enterprises' sustainability goals.

These aspects should be monitored within each of the SQC categories to ensure that sustainability-related targets are met, and that he enterprise has the proper resources and partnerships to fulfill these actions. The analyzed sustainability reports can then account on the three pillars of sustainability: economic, environmental and social aspects. Since the development and description of business models and the development of business plans – economic criteria are already involved with every environmental and social aspect of the SQC. Therefore, the economic aspects are not given an own column in the model.

By associating environmental and social areas of action in the SQC categories, each key area can be seen as an individual aspect that provides the basis for an overall combinative effort for sustainability in a start-up or micro-enterprise. These aspects are also related to core indicators found in the GRI reporting scheme. These indicators can thus be assessed within the framework of a software application as bullet points to cover or as questions that must be answered within a project to establish sustainability criteria within a very small business. These core indicators selected were mostly confirmed through an overview of the new G4-criteria [9].

7 Conclusions

Overall, this paper was able to gain greater insights on the factors that influence the adoption of sustainability software in SMEs. It opens the discussion and offers new find pathways to consider in the adoption by highlighting the main factors that might encourage further adoption in SMEs. From a practical standpoint, it should help software developers understand their target market and position the product more effectively toward the end-user. In this way, the results can make a considerable contribution for future research to build from as well as support the further development of software in SMEs.

The results of this paper also provide both academic and practical implications. From an academic standpoint, the paper provides numerous points of departure for further interdisciplinary research. In the context of startup-related research, for instance, the IT-supported tool can be used as a basis for sustainability-centered business plans. From a practical perspective, this conceptual tool can encourage consultants of startups and software developers to include sustainability criteria in the creation of new software and further services. Based on this conceptual framework, mini-sustainability quick-checks and reports can be created as complementary parts of business plans and marketing-related activities.

References

1. Àlvarez, I. "Branchenorientierte und IT-gestützte Energieeffizienz und Benchmarking in KMU-Netzwerken," in: Marx Gómez, J., Lang, C. and Wohgemuth, V. (eds.) *IT-gestütztes Ressourcen-und Energiemanagement*, pp. 21–33, Berlin: Springer, 2013.
2. Boons, F., and Lüdeke-Freund, F. "Business models for sustainable innovation: State-of-the-art and steps towards a research agenda," *Journal of Cleaner Production*, vol. 45, pp. 9–19, 2013.
3. Borga, F., Citterio; A., Noci, G., and Pizzurno, E. "Sustainability report in small enterprises: Case studies in Italian furniture companies," *Business Strategy and the Environment*, vol. 18, no. 3, pp. 162–176, 2008.
4. Buonanno, G., Faverio, P., Pigni, F., Ravarini, A., Sciuto, D., and Tagliavini, M. "Factors affecting ERP system adoption: A comparative analysis between SMEs and large companies," *Journal of Enterprise Information Management*, vol. 18, no. 4, pp. 384–426, 2005.
5. Dyllick, T., and Hockerts, K. "Beyond the business case for corporate sustainability," *Business Strategy and the Environment*, vol. 11, no. 2, pp. 130–141, 2002.
6. European Commission. *The New SME Definition: User Guide and Model Declarations.* Available at: http://ec.europa.eu/enterprise/policies/sme/files/sme_definition/sme_user_guide_en.pdf. (2005). Accessed 29 November 2013.
7. Graafland, J., van de Ven, B., and Stoffele, N. "Strategies and Instruments for Organising CSR by Small and Large Businesses in the Netherlands," *Journal of Business Ethics*, vol. 47, pp. 45–60, 2003.
8. Gregory, B. T., Rutherford, M. W., Oswald, S., and Gardiner, L. "An empirical investigation of the growth cycle theory of small firm financing," *Journal of Small Business Management*, vol. 43, no. 4, pp. 382–392, 2005.
9. GRI, Global Reporting Initiative. G4 Sustainability Reporting Guidelines, Reporting Principles and Standard Disclosures. https://www.globalreporting.org/reporting/g4/Pages/default.aspx. (2014). Accessed 19 June 2014.
10. Günther, E., and Kaulich, S. "The EPM-KOMPAS: An instrument to control the environmental performance in small and medium-sized enterprises," *Business Strategy and the Environment*, vol. 14, no. 6, pp. 361–371. 2005.
11. Halberstadt, J., and Johnson, M. "Sustainability Management for Start-ups and Micro-Enterprises: Development of a Sustainability Quick-Check and Reporting Scheme," in: Marx Gómez, J., Sonnenschein, M., Vogel, U., Winter, A., Rapp, B., and Giesen, N. (eds.), *EnviroInfo 2014 – 28th International Conference on Informatics for Environmental Protection*, pp. 17–23, Olderburg: BIS-Verlag, 2014.
12. Hashem, G., and Tan, J. "The adoption of ISO 9000 standards with the Egyptian context: A diffusion of innovation approach," *Total Quality Management*, vol. 18, no. 6, pp 631–652, 2007.

13. Hillary, R. Small and Medium-Sized Enterprises and the Environment. Sheffield: Greenleaf Publishing, 2000.
14. Hörisch, J.; Johnson, M. P. and Schaltegger, S. "Implementation of Sustainability Management and Company Size: A Knowledge-Based View," *Business Strategy and the Environment*, Early Online View, 2014.
15. Isenmann, R. "Internet-based sustainability reporting," *International Journal of Environment and Sustainable Development*, vol. 3, no. 2, pp. 145–167, 2004.
16. Jarillo, J. C. "Entrepreneurship and growth: The strategic use of external resources," *Journal of Business Venturing*, vol. 4, no. 2, pp. 133–147, 1989.
17. Johnson, M. P. "Sustainability Management and Small and Medium-Sized Enterprises: Managers' Awareness and Implementation of Innovative Tools," *Corporate Social Responsibility and Environmental Management*, Early Online View, 2013.
18. Johnson, M., Viere, T., Schaltegger, S., and Halberstadt, J. "Application of Software and Web-Based Tools for Sustainability Management in Small and Medium-Sized Enterprises," in: Marx Gómez, J., Sonnenschein, M., Vogel, U., Winter, A., Rapp, B., and Giesen, N. (eds.), *EnviroInfo 2014 – 28th International Conference on Informatics for Environmental Protection*, pp. 17–23, Olderburg: BIS-Verlag, 2014.
19. Kolk, A. "A decade of sustainability reporting: developments and significance," *International Journal of Environment and Sustainable Development*, vol. 3, no. 1, pp. 51–64, 2004.
20. Maijala, A., and Pohjola, T. "Web-based environmental management systems for SMEs: Enhancing the diffusion of environmental management in the transportation sector," in: Schaltegger, S., Bennett, M. and Burritt, R. (eds.) *Sustainability Accounting and Reporting*, pp. 655–677, London: Springer, 2006.
21. Marx Gómez, J. M., Lang, C., and Wohlgemuth, V. IT-gestütztes Ressourcen-und Energiemanagement. Konferenzband zu den 5. BUIS-Tagen. Berlin: Springer Verlag, 2013.
22. Muuß, K. and Conrad, C. *Nachhaltigkeit managen. Softwaresysteme für das Nachhaltigkeitsmanagement*, Bremen: Brands and Values, 2012.
23. Oliveira, T. and Martins, M. F. "Literature review of information technology adoption models at firm level," *The Electronic Journal Information Systems Evaluation*, vol. 14, no. 1, pp. 110–121, 2011.
24. Osterwalder, A., & Pigneur, Y. Business Model Generation: A Handbook for Visionaries, Game Changers, and Challengers, 2010.
25. Perrini, F. and Tencati, A. "Sustainability and stakeholder management: the need for new corporate performance evaluation and reporting systems," *Business Strategy and the Environment*, vol. 15, no. 5, pp. 296–308, 2006.
26. Porter, M. E. Competitive advantage: creating and sustaining superior performance. Nova, 1985.
27. Ramdani, B, Kawalek, P., and Lorenzo, O. "Predicting SMEs' adoption of enterprise systems," *Journal of Enterprise Information Management*, vol. 22, no. 1, pp. 10–24, 2009.
28. Ramdani, B., Chevers, D., and Williams, D. A. "SMEs' adoption of enterprise applications: A technology-organisation-environment model," *Journal of Small Business and Enterprise Development*, vol. 20, no. 4, pp. 735–753, 2013.
29. Schaltegger, S., and Burritt, R. "Corporate sustainability," in: Folmer, H. and Tietenberg, T. (eds.) *The international yearbook of environmental and resource economics 2005/2006: A Survey of current issues*, pp. 185–222, Northampton, MA: Edward Elgar Publishing, 2005.
30. Schaltegger, S. and Wagner, M. "Sustainable Entrepreneurship and Sustainability Innovation: Categories and Interactions," *Business Strategy and the Environment*, vol. 20, no. 4, pp. 222–237, 2011.
31. Süpke, D., Marx Gómez, J., and Isenmann, R. "Web 2.0 sustainability reporting: Approach to refining communication on sustainability," in Wohlgemuth, V., Page, B., and Voigt, K. (eds.) *Environmental Informatics and Industrial Environmental Protection: Concepts, Methods and Tools*, pp. 235–243, Aachen: Springer, 2009.

32. Tornatzky, L.G. and Fleischer, M. *The Processes of Technological Innovation*, 1975. Lexington, MA: Lexington Books.
33. Windolph, S. E., Harms, D., and Schaltegger, S. "Motivations for Corporate Sustainability Management: Contrasting Survey Results and Implementation," *Corporate Social Responsibility and Environmental Management*, Early Online View, 2013.
34. Yim, Hyung Rok. "Quality shock vs. market shock: Lessons from recently established rapidly growing US startups," Journal of Business Venturing, vol. 23, no. 2, pp. 141–164, 2008.
35. Zhu, K., Kraemer, K., and Xu, S. "Electronic business adoption by European firms: A cross-country assessment of the facilitators and inhibitors," *European Journal of Information Systems*, vol. 12, no. 4, pp. 251–268, 2003.
36. Zorpas A. "Environmental management systems as sustainable tools in the way of life for SMEs and VSMEs," *Bioresource Technology*, vol. 101, pp. 1544–1557, 2010.

Chapter 15
Towards Collaborative Green Business Process Management as a Conceptual Framework

Timo Jakobi, Nico Castelli, Alexander Nolte, Niko Schönau, and Gunnar Stevens

Abstract Organizational strategies for saving energy are currently largely defined by three main courses of action: From a process organization perspective, efforts are being made to optimize processes and invest into more energy efficient infrastructure; from a behavioral perspective, one-time interventions such as energy campaigns or feedback mechanisms are common means to reduce environmental impact. However, both approaches face limitations concerning the scope of intervention. Researching organizational needs in the wild, we conducted action-based research regarding energy optimization practices. We discovered a lack of integrated approaches as regards fostering sustainability in organizations and deriving strategies for bridging the gap between strategic planning and everyday work in order to manage sustainability strategies more effectively and efficiently. We conclude by laying out a research agenda, which we seek to address in course of the ongoing research project in order to gain more sophisticated understanding of how to conduct collaborative green business process management in the wild.

Keywords Sustainability • Business process management • Collaboration • Organizations

---

T. Jakobi (✉) • N. Castelli • G. Stevens
Human Computer Interaction, University of Siegen, 57068 Siegen, Germany
e-mail: timo.jakobi@uni-siegen.de; nico.castelli@uni-siegen.de; gunnar.stevens@uni-siegen.de

A. Nolte
Information and Technology Management, Ruhr-University of Bochum, 44780 Bochum, Germany
e-mail: nolte@iaw.rub.de

N. Schönau
Information Science and New Media, University of Siegen, 57068 Siegen, Germany
e-mail: niko.schoenau@uni-siegen.de

J. Marx Gómez et al. (eds.), *Advances and New Trends in Environmental and Energy Informatics*, Progress in IS, DOI 10.1007/978-3-319-23455-7_15

1 Introduction

Energy is increasingly gaining importance as a critical factor to many businesses' success. Traditionally, due to its low costs and also out of the sheer inability to account for energy consumption costs of single processes or products, energy costs have always been perceived as overhead costs. There was neither the urgent need nor any suitable tools for making energy consumption costs accountable. However, with increasing energy prices, businesses are seeking for the tools to change this situation and optimize resource consumption systematically.

Organizations' typical countermeasures are closely connected to the advent of ubiquitous computing technologies and the rise of smart, networked organizations. In providing affordable technological means for tracking and making energy consumption accountable, digital measurement is key to introducing energy monitoring into controlling. Traditional approaches for the strategic optimization of organizations, such as Business Process Management (BPM) [22], were quick to adopt the new parameter into so-called GreenBPM approaches [5, 13, 16, 17].

Classic BPM is typically characterized by a top-down approach and is often driven by external specialists. Therefore, criticism towards such conduct addresses the approaches' inherent structure, as measures are in danger of failing practicability for routines at operative level. In order to better address local workers as sources of energy consumption, one popular suggestion is to foster more sustainable work practices. Their effectiveness, however, is largely limited by organizationally defined processes. As a result, neither strategy makes use of the full potential of rendering an organization sustainable in terms of energy consumption (Fig. 15.1). It is, therefore, an open research question as to how far there is potential to integrate both methods into a collaborative green business process management approach. Successful integration of local workers' expertise is expected not only to better motivate people involved in processes to alter their behavior with respect to energy consumption. It is also expected to uncover additional potentials to save energy in making use of workers' specific process knowledge. However, the suitable tools to inform stakeholders within a collaborative workshop on green business process management are unknown. Motivated by an ongoing case study, in this paper we outline possible strategies and tools for raising collective awareness and supporting the direct involvement of all stakeholders in the analysis and rearrangement of organizational work by introducing **Collaborative GreenBPM**. We therefore show potentials of extending existing models which already take into account environmental data for business process management with a collaborative approach. Referring to an ongoing case study, we outline prototypical tools and methods of moving from GreenBPM to **Collaborative GreenBPM** and outline a research agenda for supporting participants' views on energy consumption data when conducting collaborative workshops for business process management based on environmental data.

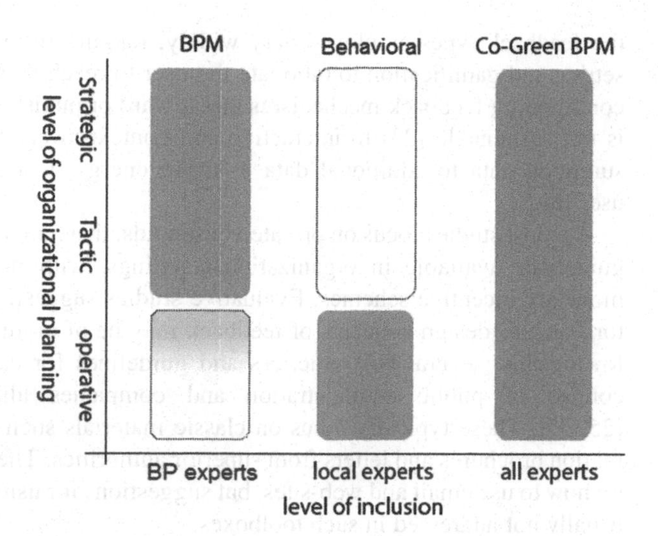

Fig. 15.1 Comparing the two dominant approaches increasing organizational sustainability and introducing Co-GreenBPM

2 Sustainability in BPM and Organizational Development

2.1 Energy Feedback for Behavioral Change

In the course of the oil crisis in the 1970s, research of environmental psychology started to take an interest in the influence of behavior on energy consumption and investigated consumption feedback as a means of encouraging energy conservation [4]. At the same time, a body of theoretical approaches emerged within environmental research, seeking to understand individual's (un-) sustainable behaviors. Early and most common approaches adopted rational choice theory [19], arguing that energy-relevant behavior is conceptualized as an act of informed decision-making by consumers. Over time, other theories (like Stern's Value Belief Norm Theory [37]) emerged, considering e.g. subjective norms, beliefs and the influence of social surrounding. Both concepts of norms and rational behavior provide theoretical ground for persuasion and feedback campaigns, which nowadays are the most widespread methods of trying to implement changes in behavioral energy consumption [8, 11].

In recent years, several design concepts and energy feedback systems inspired by these theories have emerged in the domestic context [1]. In addition to the differentiation between the theoretical foundations, approaches mainly differ on the levels of data gathering, data processing and data visualization. Due to the developments in smart metering technology, it has become possible to capture real-time consumption data disaggregated at device- or room-level to gain detailed understanding of the consumption of specific devices [12]. Instead of visualizing raw consumption data, more and more data is processed and intelligently analyzed with smart algorithms to support the user with detailed feedback and treatment suggestions (e.g. automatic evaluation of consumption) [40]. The visualization of data and the feedback types employed vary widely, ranging from approaches using goal-setting and gamification to motivate the user to reach a specific goal [20] through conditioning feedback mechanisms that reward or punish users if the consumption is not sustainable [23] to interactive and context-aware feedback that links consumption data to additional data to make energy data more meaningful to the user [6].

As most studies focus on private households, there are only few experiences and guidelines available in organizational settings, with most of them relying on monetary incentive schemes. Evaluative studies suggest, however, that other factors such as design concepts of feedback may be of more relevance [14, 18]. Following this, several best practices and guidelines for campaigns, largely in the context of public administration and companies alike have been created [25, 35]. These typically focus on classic materials such as posters, flyers, information brochures and letters from superior authorities. They also offer some advice on how to use email and web-sites, but suggestions for using smart technologies are usually not addressed in such toolboxes.

More recent research tries to make use of such existing ubiquitous sensing technologies in developing feedback solutions for organizations [2, 24]. First, general design guidelines and wireframe sketches were developed by Foster et al. [10] using focus group sessions. Based on a literature review about techniques of intervention appropriate for the workplace, Yun et al. [41] implemented a first functional prototype of an energy-dashboard.

Looking at studies investigating and evaluating eco-feedback in organizations, there are fewer examples showing mixed results. According to Carrico and Riemer [8], even monthly feedback with a motivating message can provoke energy savings in a case study with university workers. Also in the context of a university, Murtagh et al. [26]'s study tested eco-feedback applications on employees' desktops, finding significant reductions of consumption. However, due to manifold work- and context-dependent restrictions, many studies notice a complex nexus of dependencies between feedback, organizational constraints and behavior, resulting in numerous reasons not to switch off.

Another problem which was identified is the long-term impact of measures, typically appearing as a one-time intervention. Until now, this phenomenon has only been addressed by very few studies. As one of the first to observe this, Schwartz et al. [36] used smart metering technology in a research institute, evoking significant positive short-term effects, with conservation fading successively over time. This finding is backed by a more recent study by Jakobi et al. [21], who investigated office energy consumption and the possibility to provide tailored feedback as a one-time intervention based on design guidelines from energy feedback for households. Jakobi et al. also point out how processes and formal organizational restrictions limit the effectiveness of feedback. Piccolo et al. [30] took a closer look at this, equipping an office environment with a debating tool, smart monitors and a haptic feedback tree in order to raise awareness and provide a collaborative discussion platform. On analyzing discussion input, it became obvious how technical, formal and informal constraints exist within and around the group, shaping and limiting behavioral change.

In summary, while some feedback mechanisms from research in domestic contexts can successfully be adapted to organizational eco-feedback solutions, there are open challenges: Firstly, feedback systems effective in the long run are needed to embed more sustainable behavior into organizational routines. This in turn could reduce the costs of one-time interventions, at the same time supporting sustaining effects and learning about energy efficiency. Secondly, feedback fails to overcome formal organizational hurdles, thus leaving behind strategic inefficiencies. It is an open research question, how far combining the benefits of behavioral energy feedback for workers with common change management methods thus might hold the potential to increase the added value of measuring energy consumption in organizations when allowing behavioral change to further affect organizational process management.

2.2 *BPM for Organizational Sustainability*

The optimization of organizational practices often is managed through business process management approaches. Literature contains very diverse approaches in context of business process management methods. These approaches are mostly organized in lifecycles that are repeated iteratively, to achieve a continuous improvement process (CIP).

One of the most prominent approaches is the lifecycle by van der Aalst [7] which iterates through four phases (c.f. Fig. 15.2). In the process design phase, business processes are identified and designed/redesigned. The configuration phase includes configuration and selection of the system and implementation of first-phase designs. In the enactment phase, the configured systems are used to execute and monitor the operational business processes. Finally, during the diagnosis and evaluation phase, monitored information is analyzed to identify problems and to detect potential room for improvements. While this approach clearly focuses on the execution and monitoring of processes supported by a workflow engine, there are other approaches that place stronger emphasis on the phases of process analysis and process design.

One of these lifecycles is the one recently published by Dumas et al. [9]. In contrast to the aforementioned lifecycle by van der Aalst et al. there is a clear distinction between phases in which processes are visualized and analyzed (c.f. Fig. 15.3). However, before analyzing a process, the lifecycle starts with a phase termed "process identification". The main goal of this phase is to identify the start and end of the process in question as well as the factors potentially influencing it. This phase also aims to identify relevant stakeholders which are then used as a source of information for the process in the next phase (c.f. process discovery in Fig. 15.3), ultimately leading to a visualization of the process within a process model that describes the current state of the process (AS-IS-model). This model is

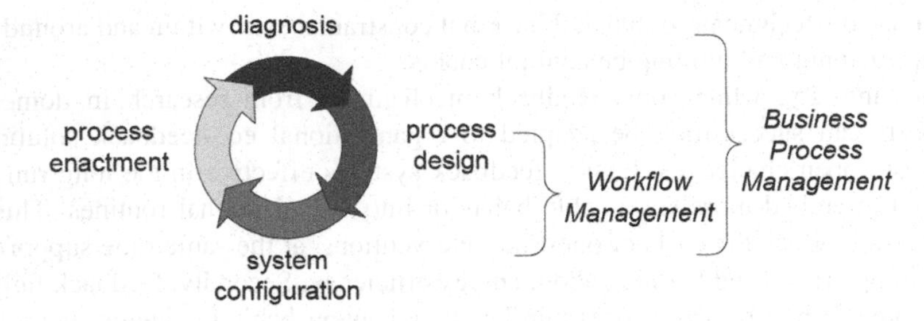

Fig. 15.2 BPM lifecycle by van der Aalst et al. [7]

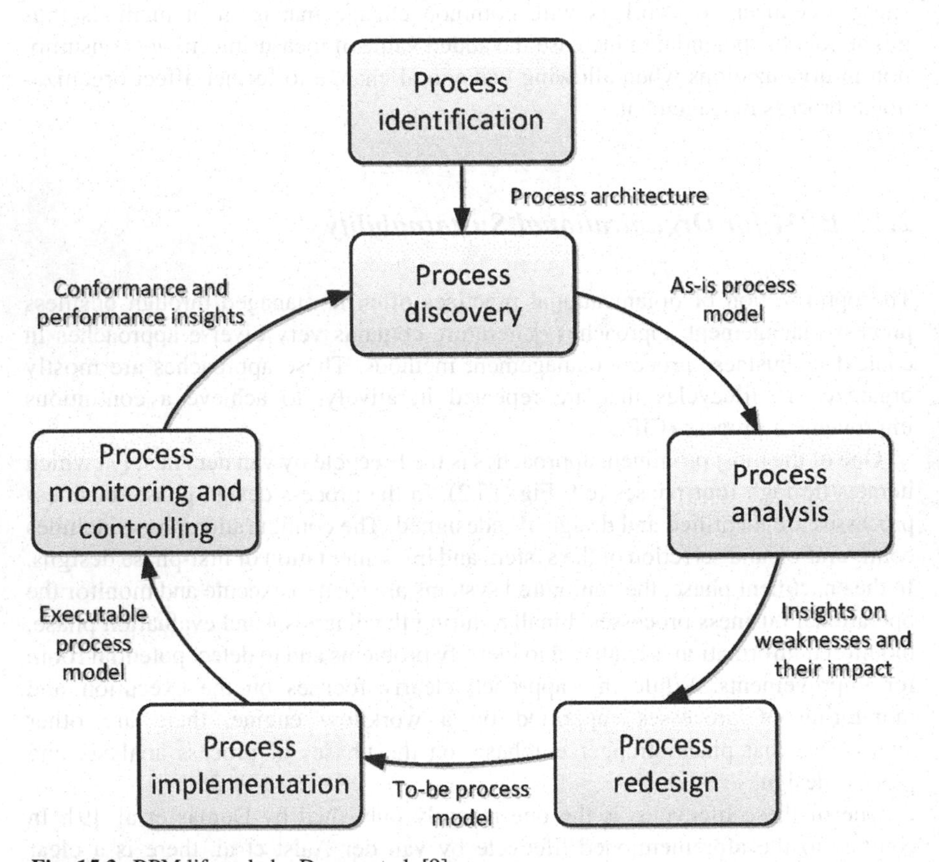

Fig. 15.3 BPM lifecycle by Dumas et al. [9]

then used to analyze the process in question, to identify problems and flaws. Afterwards, the process is re-designed in order to tackle the identified flaws before it is implemented. During implementation, it is monitored and the lifecycle is run again if necessary.

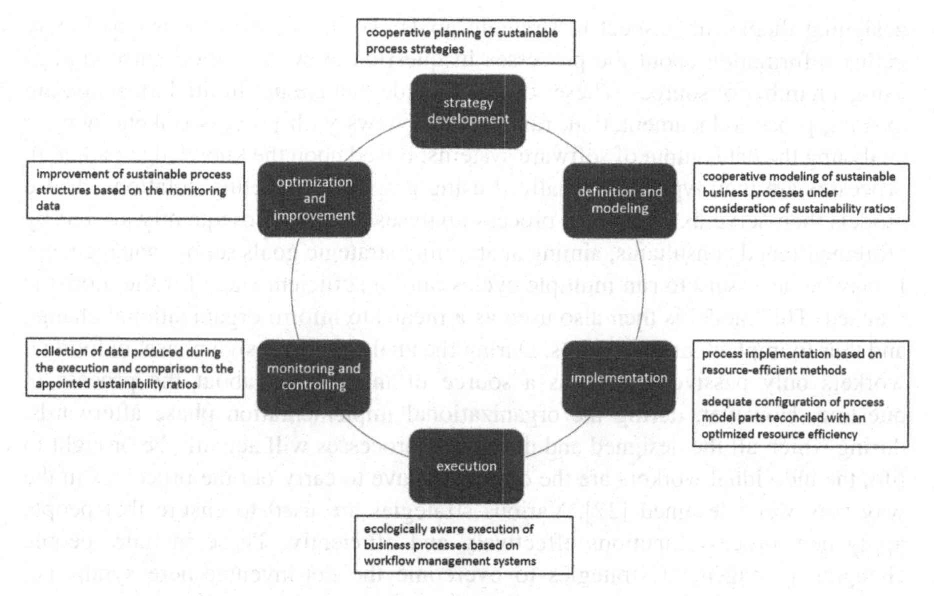

Fig. 15.4 GreenBPM model by vom Brocke et al. [5]

Both the aforementioned lifecycles – as well as others – are based on the Deming cycle (Shewhart cycle) of plan, do, check and act [39]. Approaches within GreenBPM commonly refer to this model as well. Therefore, in case of a holistic infrastructure for collecting and distributing environmental context data from processes, GreenBPM can adopt and adapt to existing BPM tools, instead of reinventing the wheel.

This is reflected by the GreenBPM model of vom Brocke et al. [5] (c.f. Fig. 15.4), which expands the dimensions of the BPM model by Becker and Kugeler [3]. The model includes six phases (description of processes, workflow definition, workflow instance execution, monitoring of workflow instance execution, workflow reporting and entire process reporting) evaluated by dimensions of cost, quality, flexibility and time. Vom Brocke et al. add "sustainability" as a dimension of decision making, arguing for the consideration of sustainability objectives in workflows. Deriving a GreenBPM approach from van der Aalst [7], Nowak et al. [28] demonstrate how integrating environmental data into BPM has effects on both general conduct (inclusion of further stakeholders) and design of specific phases (including new key figures). Therefore, sustainability as an issue influences the whole BPM process; yet this does not necessarily imply adding another stream of data, but sometimes just processing existing data in a different way.

One major downside of all of the aforementioned lifecycles is that they are usually management-driven as a result of strategic planning. This means that they are mostly driven by a top-down approach in which stakeholders are only partially involved, if at all. This effect is reinforced by the fact that corresponding initiatives are often conducted by external consultants analyzing existing processes and (re-)

designing them with respect to strategic goals. To do so, these consultants may gather information about the processes in question as well as their surroundings using a number of sources. These sources include, but are not limited to, analyzing existing process-documentation, running interviews with process stakeholders or analyzing the data output of software systems. Based upon the knowledge gathered, processes are then typically visualized using graphical modeling notations. These models then serve as a basis for process analyses and are subsequently altered by aforementioned consultants, aiming at attaining strategic goals set by management. It may be necessary to run multiple cycles until a sufficient stage for the model is reached. This model is then also used as a means to inform organizational change and thus to implement the process. During the analysis and design phase, individual workers only passively serve as a source of information about the process in question. However, during the organizational implementation phase afterwards, during which all the designed and developed processes will actually be brought to life, the individual workers are the ones who have to carry out the processes in the way they were designed [22]. Various strategies are used to ensure that people apply new process functions effectively and efficiently. These include: people changing management strategies to overcome the not-invented-here syndrome; set up training strategies to overcome knowledge deficits; and management and controlling strategies to verify whether the actual process is in line with the process designed.

Approaches like the ones described above can be considered common practice within organizations. However, this conduct limits the influence of people directly involved in or affected by the processes at stake to the provision of data. Thus they are not allowed to directly participate in design, which potentially limits people's motivation to adapt to the newly designed processes. Furthermore, by gathering data from single sources and putting it together afterwards, knowledge about the process is only presented from a single perspective, thus leaving out important information. This may result in inefficient processes, if they are seen to be designed against routines and informal workflows made use of by local experts.

In order to address the aforementioned limitations, different approaches have been created in recent years, which can be subsumed as collaborative modelling [27, 29, 34]. At the center of these approaches are workshop concepts, in which stakeholders – together with consultants – jointly develop models of processes, analyze them and discuss possible changes. These approaches allow for stakeholders to directly participate in process design, thus potentially increasing their motivation to adapt to process modifications afterwards and limiting possible misguided designs. Furthermore, these approaches also allow people to exchange perspectives within workshops, discuss alternatives and come up with a more sophisticated solution on which all stakeholders can agree. So far, however, they lack a discussion of how to appropriately support workers as well as management and external business process managers to take environmental data, such as energy consumption data, into account.

3 Collaboration in Green Process Management

In this chapter we describe the first impressions we gained from an ongoing case study, in which we are investigating and accompanying the implementation of a holistic energy management system. During our interviews with several stakeholders, ranging from operative to strategic management, the need for tailored feedback according to individual roles and specific process knowledge as well as ways to make energy accountable become apparent. We then outline the challenges and benefits of Collaborative GreenBPM and present a preliminary set of tools for supporting collaborative process analysis and modeling.

3.1 Understanding Needs in Practice

In an ongoing case study within a metal working business, we are exploring possible support for the organizations' style of tracking, evaluating and using energy consumption data for more sustainable processes. While the former are themselves highly complex tasks and out of the scope of this contribution, we here focus on different workers' needs in terms of energy consumption data in order to gain the highest possible added value from the data available.

The company is operational in the area of fastening technology. It employs more than 2300 people in 30 subsidiaries worldwide, and thus can be characterized as largely decentralized. In contrast, the company has a centralized energy management department, which is part of a shared service center and is responsible for managing all subsidiaries in this matter.

Based on an Action Research Methodology [38], we conducted 19 interviews with different stakeholders within the companies. These included workers at operative level, employees explicitly concerned with existing energy management and parts of the companies' management.

Our status quo analysis was aimed at understanding general organizational energy-management activities, as well as individual and role-based preconditions regarding the existing infrastructure, in addition to goals, needs and possible benefits for the internal energy management. The interviews lasted between 30 min and 1.5 h and were recorded on tape for later transcriptions and analysis.

As a result of our research and in close cooperation with the company, we developed an integrated energy management concept that consisted of the major areas hardware, software and processes. The key of the concept includes a measuring sensor system which allows the energy consumption of e.g. the machinery to be monitored. This is already partially realized in practice (e.g. one subsidiary was equipped with sensor technology) Data is stored in a central real-time database system to allow the precise evaluation of the energy consumed in the production process. We additionally plan to roll out our concept to other sites and partners both from the retail industry and from financial institutes to make the approach more robust.

Throughout the process of accompanying the organization on their way to making choices regarding hardware, software and infrastructure, we also gained insights on the data preferred to better understand energy consumption from each individual's perspective. For example, an energy manager told us quite explicitly:

> Well you have to understand that my information need is quite different from the need of my CEO. Whereas I need a more specific view on the cause-effect relationship, he needs a more aggregated view

This depicts a representative statement regarding our general impression of individual and role-dependent needs of energy consumption feedback. Furthermore, during our interviews we were repeatedly confronted with workers' ideas of how to make use of energy feedback data, or how to improve work procedures for more sustainability.

3.2 *Envisioning Co-GreenBPM*

Taking the aforementioned empirical findings and existing approaches for process improvement into account, we envision Co-GreenBPM as being an approach that ties together individual energy feedback with mechanisms of collaborative modeling. This approach should make use of existing business process management tools, which have proven effective to change organizational structures.

Based on the identified gap regarding potentials of a joint approach integrating both process management and behavioral approaches for optimizing process sustainability, we envision a Co-GreenBPM approach. Our concept of Co-GreenBPM stands in the tradition of collaborative approaches as laid out above. It aims towards a more holistic approach that allows for stakeholders to actively trigger and participate in all phases of the BPM lifecycle. Referring back to the aforementioned lifecycle by vom Brocke et al. [5], we thus do not aim to alter the lifecycle as a whole. Rather our aim is to strengthen the participation of stakeholders throughout the lifecycle (c.f. Fig. 15.4). This in turn allows us to tap into the full potential of energy saving by making use of local expertise while at the same time fostering acceptance of new processes. We therefore propose to add means of collaboration such as collaborative modeling during process analysis and re-design to the all phases of the lifecycle. We also propose to allow process stakeholders to actively intervene in GreenBPM, thus aiming towards a more bottom-up and people-centered strategy that affects all phases of the lifecycle. Finally, we propose energy feedback to be linked to process models as those models can be seen as the central artifact for process analysis and (re-)design. Linking process steps to energy data using interfaces such as the ones described in the following section might provide a hint for tying abstract representations such as process models to real world data. This in turn might help process stakeholders who are not trained in using models to tie actual energy consumption to abstract representations of process steps in process models.

In order to increase such tools' efficiency and acceptance, however, we aim to add the workers' perspective and local expertise to guide the definition of new workflows by making use of their knowledge on tweaks of everyday working activities. Workshops are at the center of this approach, in which people involved in or affected by processes can discuss "their" respective energy consumption, identify potentials and alter processes with respect to tapping into these potentials. By bringing people together from multiple teams and potentially multiple departments, it is expected that the identification of energy saving potentials that go beyond individual workplace adjustments will be fostered. We furthermore envision Co-GreenBPM as being a bottom-up rather than top-down approach. Using energy feedback systems that allow process stakeholders to view their current and past energy consumption in a manner suitable for them and not only at an individual level but also with respect to the processes they perform, could enable them to identify space for optimization. It should then be possible for them to trigger the aforementioned workshops. Subsequently, the energy feedback allows the impact of the changes that they made to a process to be assessed and may further trigger another round of workshops thus at best resulting in a continuous process of improvement.

However, we additionally do not neglect the potential triggering of Co-GreenBPM by management. Instead, we argue that even if it is a top-down initiative, the worker-level should be involved, thus integrating multiple perspectives and increasing motivation among participants to subsequently actively alter their behavior.

Our concept therefore relies on a bottom-up management approach, which brings changes to the cycles previously used in BPM and considering several issues.

This brings changes to the cycles previously used in BPM with respect to the following issues (c.f. Fig. 15.5):

1. While BPM initiatives are mostly triggered by management, we aim to allow for process stakeholders and especially process participants to actively start (Green) BPM initiatives. Moreover, if polluting practices are not caused by a lack of motivation, but are due to current process design, workers should be empowered to become aware of this and contribute this knowledge to a continuous improvement process, thus sparking a BPM initiative.
2. In contrast to a top-down change management, in a Co-GreenBPM, next to BPM-specialists, all relevant stakeholders should be asked to participate. Process participants should especially be involved as they are experts of the demands of the situated work practices and opportunities provided by knowledge at operational level. In general, the diversity of the stakeholders makes the process more complex, which has a negative impact on the efficiency of process design phase. Yet including a wide array of knowledge generally has a positive impact with respect to perspectives covered when analyzing a process. To overcome the asymmetries of knowledge among the stakeholders in particular,

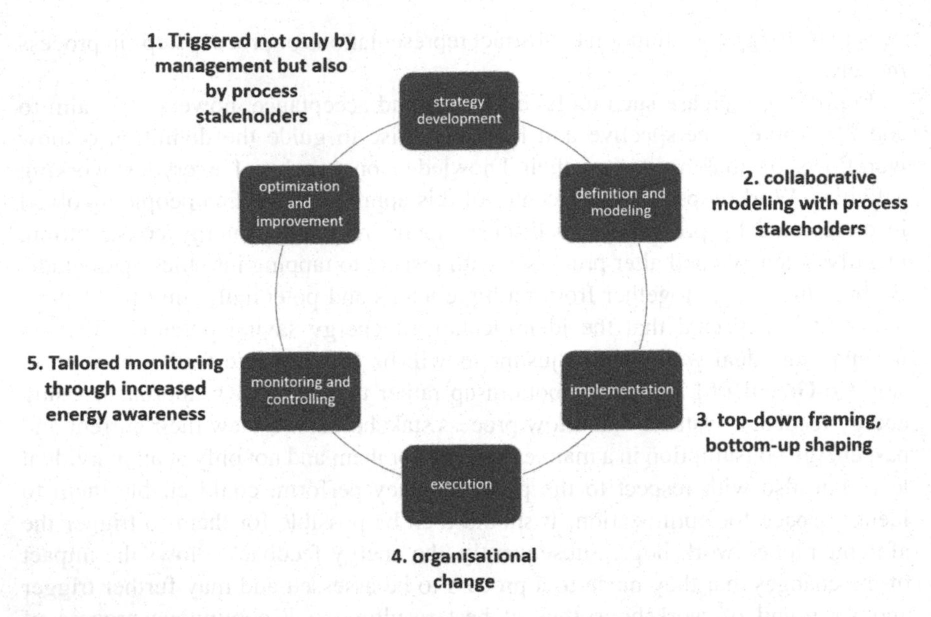

Fig. 15.5 Co-GreenBPM lifecycle based on classic understanding of vom Brocke [5]

modelling should take place in collaborative workshops where all participants are valued as experts for their particular domain. Furthermore these workshops provide an environment where abstract visualizations of processes in models can be combined with energy data within those process models, thus creating a possibility for stakeholders of relating models to their individual real world practices.

3. When a process is designed and implemented, the overall circumstances it is conducted in have to be taken into account, e.g. the strategy that the company follows. While this strategy is set by management, it should leave enough room for process participants to shape a process in a form that they can imagine would fit their requirements. Processes stakeholders should thus not only become involved with respect to carrying out a process but also with respect to shaping it, while taking the overall strategy into account.
4. While a lot of BPM initiatives aim towards processes that can later be transformed into a workflow and potentially even carried out by a workflow engine, our approach also aims towards processes that might not be carried out with respect to a predefined workflow. We rather also intend to allow Co-GreenBPM to be useful for e.g. knowledge-intensive processes. This in turn means that execution has to focus on an organizational change project, with stakeholders being strongly involved in order to ensure its success.
5. Traditional eco-feedback systems mainly focus on direct feedback to motivate people to act sustainably in the current situation [8, 11]. In the context of Co-GreenBPM eco-feedback, people should additionally be encouraged to reflect on their work practice with regard to green processes (and vice versa).

Furthermore, in a green process, environmental data is needed by various stakeholders (e.g. the management, the worker, the controller, etc.). Hence, data visualization needs to be tailored to the individual demands of different stakeholders. Furthermore feedback has to be tied to process steps in a visual way, thus building the bridge between abstract representations of process steps and real life data. Providing individualized views on process and energy consumption feedback and tying that feedback to process steps will enable stakeholders to use their tacit knowledge to reflect how processes and (!) situated work could be designed in more sustainable ways. Such tailored feedback for included stakeholders needs to provide a basis for decision making when designing sustainable processes. Taking into account energy's complex nature [31], this, however, raises the question, which kind of data workers need to reflect on their energy consumption behavior in terms of both their own workplace and the organizations processes.

3.3 Envisioning Tools and Methods

In order to include the heterogeneous group of stakeholders and their individual expertise, we envision a highly customizable set of tools for understanding and evaluating energy consumption and thus enabling a change process. At a methodological level, our approach is based on the established concept of collaborative modeling [32–34]. This concept focuses on a series of workshops in which stakeholders jointly analyze processes they are involved in by visualizing them using process models. During these workshops, they are supported by a facilitator who manages the communication and also translates the contributions by the process stakeholders into elements of a modeling notation [15]. It has also been observed that process stakeholders are capable of analyzing parts of processes they are involved in on their own [27]. Due to this possibility, process analysis may also be extended across the boundaries of workshops and may thus ultimately also occur while the process is being executed. To analyze the sustainability of processes, we extend on these approaches by suggesting appropriate tools of making energy consumption accountable. We are currently developing prototypic solutions in close collaboration with employees of the company and plan to evaluate them in practice to increase our understanding of designing for respective stakeholders.

To address the significantly varying requirements, the first software tool we designed provides the different stakeholders with individually tailored and highly customizable feedback. As a direct consequence of our empirical work, employees should be enabled to create their own, meaningful key performance indicators and visualizations for their area of expertise using this software (Fig. 15.6).

To support communication across hierarchical boundaries, we are implementing a sustainability-reporting tool that supports semi-automatic creation of sustainability reports with a drag-and-drop mechanism. Templates for reporting standards (e.g. GRI v4) are already implemented in the tool, but employees also have the

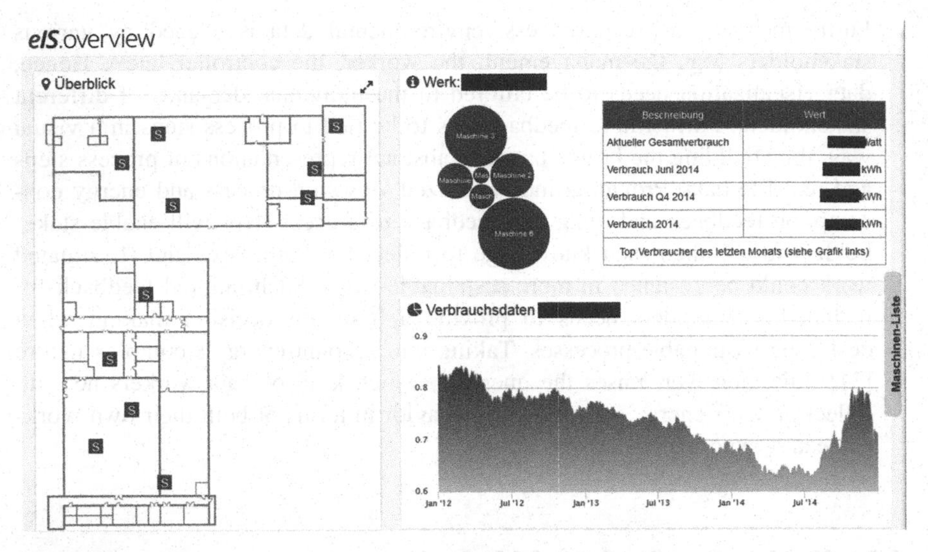

Fig. 15.6 Prototype of energy awareness board for flexible feedback allowing for individual tailoring

opportunity to create their own templates for custom reporting. It is possible, for example, to create an in-house process-reporting template, which employees can build individually or collaboratively; and they are empowered to customize the reports to fit their individual information need, e.g. by incorporating KPIs that are compiled by the KPI editor. The intention is to support communication across organizational structures and to raise energy awareness for colleagues' or other departments' understanding of energy consumption. To foster general awareness about processes and their energy consumption intensity, we are currently building a tool for digital representation of organizational sites.

This tool allows e.g. executive producers, maintenance, or employees to monitor real time consumption of the plant and its (measured) appliances and workers. Furthermore, we will implement real-time algorithms that allow users to specify ranges of energy consumption, to allow proactive maintenance. Although the tool does not aim specifically to support a green bpm process, it gives the relevant stakeholders the feedback which is both desired and needed to engage in a CoGreenBPM.

In addition to this overview, we prepared an organizational-role-based dashboard for the company to support the need for information even further. By virtue of the role-based construction, each employee automatically receives the energy feedback data of the relevant processes, machines, equipment and environment which are relevant to him. The data is presented in a meaningful form. In addition, the dashboard can be customized by the employees to show more information if necessary. Both the KPI-editor and the dashboard widgets are an integral part of the sustainability-reporting tool, since the KPIs and the widgets can be integrated into the reports.

Fig. 15.7 Tailorable dashboard for understanding and managing energy consumption for different levels of accountability

Within the next few months, our prototypes will be rolled out and evaluated in practice in order to develop them further and to gain a deeper understanding of organizational needs and the potentials and challenges when introducing a Co-GreenBPM (Fig. 15.7).

4 Discussion and Research Agenda

We have addressed the gap between green processes and green practices by combining environmental, psychological and organizational theories. We presented Co-GreenBPM as a conceptual framework to bridge the gap, taking into account work practices as well as strategic process improvement. We further outlined the key challenges of how environmental data should be included into a collaborative GreenBPM in order to enable stakeholders to make sense of data, provide group awareness for energy consumption and actively include them into process management, thus tapping the full sustainability potential in organizations. While we do acknowledge that there is work on collaborative modeling as well as on Green BPM still to be done, we think that the combination of both approaches in Co-GreenBPM the way is novel and extends existing methods in multiple ways:

- CoGreemBPM is not limited to processes that are carried out and controlled by using workflow engines. It rather explicitly aims at any process in which collaboration takes place, thus explicitly taking organizational change into account.
- In contrast to other frameworks Co-Green BPM places emphasis on process stakeholders as triggers, drivers and active participants in process analysis and (re-)design, focusing on reducing energy consumption and can thus be considered a bottom-up approach. While not diminishing the importance of management support (e.g. the provision of a frame for process change), our emphasis clearly lies on active (!) stakeholder involvement. This is expected to not only allow for identifying potentials for saving energy but also to increase the motivation of stakeholders to actively pursue an energy saving approach.
- Co-Green BPM includes continuous awareness and feedback mechanisms which do not only serve as a means to assess the current state of energy consumption but that may also be used to trigger process assessments as mentioned before.
- Technical as well as conceptual combination of process models with energy awareness links abstract representations of processes (models) to real world phenomena thus again strengthening the combination of organizational change and individual behavior.

As Co-GreenBPM does not aim towards reinventing BPM but rather at altering existing approaches in order to tap their full potential, organizations may even build upon established strategies.

Our next step will be to aim towards enriching our theoretical considerations with further empirical work. We will investigate in more detail effective ways of including environmental data to identify green process improvements and set up collaborative modeling workshops accordingly. This covers multiple areas and may serve as a future research agenda:

- Investigating various views, interests and motivations on green processes, including questions such as what kind of environmental data is needed, both every day and at strategic work level. This also includes ways of feeding back such data effectively and efficiently to stakeholders in a collaborative BPM process.
- Identifying ways of combining energy feedback with process models thus connecting process steps in models directly with energy data using interfaces as described before. This explicitly takes into account whether or not the combination of models and energy feedback actually leads to a better understanding of processes and ultimately to better processes with respect to energy consumption.
- Assessing the impact of environmental data on decision-making, process modelling and process adoption. Concerning this, we are currently approaching a variety of organizations including manufacturing, trading sector and office work.

- Assessing the impact of the Co-Green BPM on the motivation of stakeholders for behavioral change. This includes not only the question if stakeholders are willing to participate in organizational change based on energy feedback and a corresponding analysis of processes not only in Co-Green BPM; but also aims to understand factors that lead participants to start Co-Green BPM themselves.

While the approach is firmly grounded in literature and backed up by first empirical findings, it is still limited as it is a conceptual approach that has to undergo a more thorough assessment, thus answering the aforementioned research questions.

We will continue our action research and plan to conduct a first beta version of Co Green BPM within the afore-mentioned organization by bringing together strategic and operative planning and working. We will further foster collaborative work of the stakeholders on making their organization more sustainable, on both levels: everyday behavior, infrastructure and processes in order to tap into the full potential of smart energy measurements in organizations.

References

1. Abrahamse, W., Steg, L., Vlek, C., and Rothengatter, T. A review of intervention studies aimed at household energy conservation. *Journal of Environmental Psychology 25*, 3 (2005), 273–291.
2. Azar, E. and Menassa, C.C. A decision Framework for energy use reduction initiatives in commercial buildings. *WSC '11 Proceedings of the Winter Simulation Conference*, (2011), 816–827.
3. Becker, J. and Kugeler, M. *Prozessmanagement – Ein Leitfaden zur prozessorientierten Organisationsgestaltung*.
4. Brandon, G. and Lewis, A. Reducing household energy consumption: a qualitative and quantitative field study. *Journal of Environmental Psychology 19*, 1 (1999), 75–85.
5. Vom Brocke, J., Seidel, S., and Recker, J. *Green business process management: towards the sustainable enterprise*. Springer, 2012.
6. Castelli, N., Stevens, G., Jakobi, T., and Schönau, N. Switch off the light in the living room, please! –Making eco-feedback meaningful through room context information. *BIS-Verlag*, BIS-Verlag (2014), 589–596.
7. Van Der Aalst, W.M., Ter Hofstede, A.H., and Weske, M. Business process management: A survey. In *Business process management*. Springer, 2003, 1–12.
8. Dourish, P. Points of Persuasion: Strategic Essentialism and Environmental Sustainability. *Persuasive Pervasive Technology and Environmental Sustainability, Workshop at Pervasive*, Citeseer (2008).
9. Dumas, M., La Rosa, M., Mendling, J., and Reijers, H.A. *Fundamentals of business process management*. Springer, 2013.
10. Foster, D., Lawson, S., Linehan, C., Wardman, J., and Blythe, M. 'Watts in it for me ?' Design Implications for Implementing Effective Energy Interventions in Organisations. (2012).
11. Froehlich, J., Findlater, L., Landay, J., and Science, C. The design of eco-feedback technology. *Proceedings of the 28th international conference on Human factors in computing systems – CHI'10*, ACM Press (2010), 1999–2008.

12. Gamberini, L., Corradi, N., Zamboni, L., et al. Saving is fun: designing a persuasive game for power conservation. *Proceedings of the 8th International Conference on Advances in Computer Entertainment Technology*, ACM (2011), 16.
13. Ghose, A., Hoesch-Klohe, K., Hinsche, L., and Le, L.-S. Green business process management: A research agenda. *Australasian Journal of Information Systems 16*, 2 (2010).
14. Handgraaf, M.J.J., Van Lidth de Jeude, M.A., and Appelt, K.C. Public praise vs. private pay: Effects of rewards on energy conservation in the workplace. *Ecological Economics 86*, (2013), 86–92.
15. Herrmann, T. Systems design with the socio-technical walkthrough. *Handbook of research on socio-technical design and social networking systems. Information Science Reference*, (2009).
16. Hoesch-Klohe, K., Ghose, A., and Lê, L.-S. Towards green business process management. *Services Computing (SCC), 2010 I.E. International Conference on*, IEEE (2010), 386–393.
17. Houy, C., Reiter, M., Fettke, P., and Loos, P. Towards Green BPM–Sustainability and resource efficiency through business process management. *business process management workshops*, Springer (2011), 501–510.
18. ifeu-Institut Heidelberg. *Auswertung der Budget- und Anreizsysteme zur Energieeinsparung an hessischen Schulen*. Hessisches Ministerium für Wirtschaft, Verkehr und Landesentwicklung, Heidelberg, 2004.
19. Jackson, T. *Motivating sustainable consumption*. Sustainable Development Research Network, Surrey: Centre for Environmental Strategies, Guildford, Surrey, 2005.
20. Jacucci, G., Spagnolli, A., Gamberini, L., et al. Designing Effective Feedback of Electricity Consumption for Mobile User Interfaces. *PsychNology Journal 7*, 3 (2009), 265–289.
21. Jakobi, T., Stevens, G., and Schwartz, T. Verhaltensbasiertes Energiesparen am Arbeitsplatz: Eine vergleichende Studie. In D. Kundisch, L. Suhl and L. Beckmann, eds., *MKWI 2014 – Multikonferenz Wirtschaftsinformatik : 26. – 28. Februar 2014 in Paderborn*. Unversitätsverlag Paderborn, Paderborn, 2014, 76–88.
22. Jeston, J. and Nelis, J. *Business Process Management: Practical Guidelines to Successful Implementations*. Elsevier/Butterworth-Heinemann, 2008.
23. Kirman, B., Linehan, C., Lawson, S., Foster, D., and Doughty, M. There's a monster in my kitchen: using aversive feedback to motivate behaviour change. *CHI'10 Extended Abstracts on Human Factors in Computing Systems*, ACM (2010), 2685–2694.
24. Lehrer, D. and Vasudev, J. Evaluating a social media application for sustainability in the workplace. *Proceedings of the 2011 annual conference extended abstracts on Human factors in computing systems – CHI EA'11*, (2011), 2161.
25. Matthies, E., Kastner, I., Klesse, A., and Wagner, H.-J. High reduction potentials for energy user behavior in public buildings: how much can psychology-based interventions achieve? *Journal of Environmental Studies and Sciences 1*, 3 (2011), 241–255.
26. Murtagh, N., Nati, M., Headley, W.R., et al. Individual energy use and feedback in an office setting: A field trial. *Energy Policy*, 62 (2013), 717–728.
27. Nolte, A. and Prilla, M. Anyone can use Models: Potentials, Requirements and Support for Non-Expert Model Interaction. *International Journal of e-Collaboration (IJeC) 9*, 4 (2013), 45–60.
28. Nowak, A., Leymann, F., and Schumm, D. The Differences and Commonalities between Green and Conventional Business Process Management. IEEE (2011), 569–576.
29. Persson, A. *Enterprise modelling in practice: situational factors and their influence on adopting a participative approach*. Stockholm University, 2001.
30. Piccolo, L., Baranauskas, C., Fernández, M., Alani, H., and De Liddo, A. Energy consumption awareness in the workplace: technical artefacts and practices. (2014).
31. Pierce, J. and Paulos, E. Materializing Energy. *Proceedings of the 8th ACM Conference on Designing Interactive Systems*, ACM (2010), 113–122.
32. Prilla, M. and Nolte, A. Integrating Ordinary Users into Process Management: Towards Implementing Bottom-Up, People-Centric BPM. In *Enterprise, Business-Process and Information Systems Modeling*. Springer, 2012, 182–194.

33. Renger, M., Kolfschoten, G.L., and De Vreede, G.-J. Challenges in collaborative modelling: a literature review and research agenda. *International Journal of Simulation and Process Modelling 4*, 3 (2008), 248–263.
34. Rittgen, P. Collaborative modeling–a design science approach. *Proceedings of the 42nd Hawaii International Conference on System Sciences (HICSS-42)*, (2009).
35. Schlomann, B., Fleiter, T., Hirzel, S., Arens, M., and Rohde, C. *Möglichkeiten, Potenziale, Hemmnisse und Instrumente zur Senkung des Energieverbrauchs und der CO2-Emissionen von industriellen Branchentechnologien durch Prozessoptimierung und Einführung n, 2011 / IREES GmbH: Institut für Ressourceneffizienz und Energiestrategien.* Fraunhofer ISI, 2011.
36. Schwartz, T., Betz, M., Ramirez, L., and Stevens, G. Sustainable energy practices at work: understanding the role of workers in energy conservation. *Proceedings of the 6th Nordic Conference on Human-Computer Interaction: Extending Boundaries*, (2010), 452–462.
37. Stern, P.C. and Aronson, E. *Energy Use: The Human Dimension.* W.H. Freeman & Company, 1984.
38. Stevens, G. and Nett, B. Business Ethnography as a research method to support evolutionary design. *Schnitte durch das Hier und Jetzt, Zeitschrift Navigationen 2*, 09 (2009).
39. Walton, M. *The Deming Management Method.* Penguin, 1986.
40. Weiss, M., Staake, T., Mattern, F., and Fleisch, E. PowerPedia: changing energy usage with the help of a community-based smartphone application. *Personal and Ubiquitous Computing 16*, 6 (2011), 655–664.
41. Yun, R., Lasternas, B., Aziz, A., et al. Toward the design of a dashboard to promote environmentally sustainable behavior among office workers. *Proceeding PERSUASIVE'13 Proceedings of the 8th international conference on Persuasive Technology*, (2013), 246–252.

Part V
Environmental Decision Support

Chapter 16
A Generic Decision Support System for Environmental Policy Making: Attributes, Initial Findings and Challenges

Asmaa Mourhir, Tajjeeddine Rachidi, and Mohammed Karim

Abstract This chapter presents the methodology and preliminary findings related to the development and roll-out of a Decision Support System for the Environment (D2S4E), designed to assist the Moroccan national decision makers in the evaluation of the state of the environment, and in their approach to sustainable development and environmental policy-making. D2S4E exhibits a number of salient features and attributes that help with modeling environmental issues and framing the decision making process in an integrated and holistic approach. Initial pilot deployment in the region of *Grand Casablanca* led to the identification of nine development initiatives that were fixed following the results of the selection of the significant issues in the region by the different stakeholders based on proactive scenarios. We also discuss some of the major challenges we faced during the different project phases in spite of the perceived importance of D2S4E for environmental management and evaluation at the highest levels.

Keywords Environmental decision support • DPSIR framework • Cognitive maps • Fuzzy logic • Policy simulations

1 Introduction

During the last decades, intensive agricultural production, uncontrolled irrigation schemes, population growth, industrialization, and urbanization incurred Morocco enormous environmental degradation: The annual cost of environmental damage has been estimated at nearly 3.7 % of Morocco's GDP ([1]). In response to the

A. Mourhir (✉) • T. Rachidi
Computer Science Department, School of Science and Engineering, Al Akhawayn University, Ifrane, Morocco
e-mail: A.Mourhir@aui.ma; T.Rachidi@aui.ma

M. Karim
Physics Department, Faculté des Sciences Dhar Mahraz, Université Sidi Mohamed Ben Abdellah, Fès, Morocco
e-mail: mohammed.karim@usmba.ac.ma

J. Marx Gómez et al. (eds.), *Advances and New Trends in Environmental and Energy Informatics*, Progress in IS, DOI 10.1007/978-3-319-23455-7_16

alarming environmental situation, the main approach adopted by Morocco to achieve the targets of sustainable development is environmental. One of the key initiatives in this regard is the establishment of the Moroccan National Environmental Observatory (ONEM), which aims at improving knowledge related to the Moroccan environmental systems, and performing deep analyses of interactions between the environment and development activities. As such, ONEM's mission consists in: (i) collecting, analyzing and disseminating environmental information; (ii) producing statistical data and indicators on sustainable development; (iii) developing a network of decision makers for environmental monitoring; and (iv) contributing to the elaboration of public policies in terms of sustainable development and environmental protection. The approach adopted by ONEM to achieve its mission is based on Regional Observatories of the Environment and Sustainable Development (OREDDs).

The main objectives of the different environmental initiatives raise a number of primary challenges to the modeler. Any method that attempts to support modelling of complex environmental issues must include techniques beyond frameworks that tend to merely make an inventory of important indicators. First, it needs to include stakeholders' opinions into formal policy design, and consider tradeoffs across the different disciplines. It should also support impact assessment and proactive scenario simulation as well as ranking of alternatives and multi-criteria decision support. The majority of environmental solutions are theme specific with a focus on one or another aspect, i.e., either the emphasis is on modelling of ecological indicators for a specific theme, on data and environmental reporting, on Geographical Information Systems (GIS), or on communication and dissemination of environmental information using knowledge portals [2–8]. Fortunately, integrated Environmental Decision Support Systems (EDSSs) offer a number of tools and activities that help with structuring the decision making process. Integration of processes helps overcome the limitations of fragmented solutions which tend to be loosely related and have an emphasis on sectorial problems, with decisions made using limited problem perceptions [8, 9].

In this chapter we describe the approach and findings related to the development and implementation of an integrated, full-fledged Decision Support System for the Environment (D2S4E). The adopted methodology promotes multiple dimensions of integration that help with framing the decision making process. Integration is exhibited at the discipline level (social, economic, ecology, policy), at the participatory level using a variety of stakeholders (experts, decision makers, industry groups, community, etc.), and the data and model levels (Fuzzy Cognitive Maps, Multiple Attribute Decision Making (MADM) and Fuzzy Logic).

The methodology is based on five key pillars, namely:

- A participative and collaborative approach; the project was conducted in collaboration with a number of institutions for data collection and exchange as key stakeholders in order to provide a solution that caters for the desired needs and aspirations.

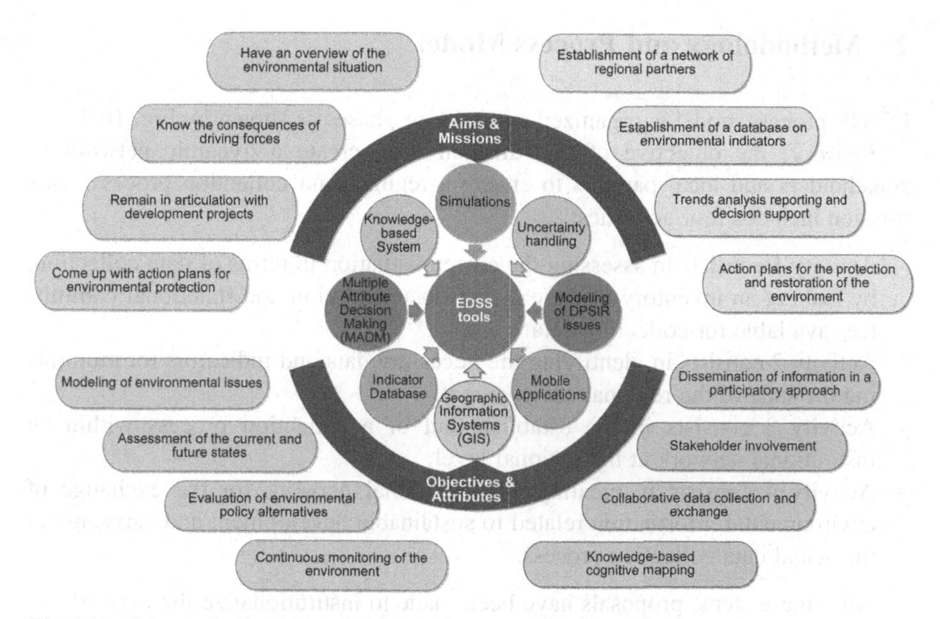

Fig. 16.1 Schematic description of D2S4E aims, missions and tools

- An integrated perspective that encapsulates all aspects of environmental degradation including socio-economic pressures and possible policy alternatives using the DPSIR (Driving forces – Pressures – State – Impacts – Responses) model.
- Use of a knowledge-based approach to capture expert input about key decision factors in complex environmental DPSIR loops.
- Favor of interdisciplinary group modeling through cognitive mapping sessions and scenario development.
- Adoption of an iterative and adaptive development and evaluation approach based on successively augmented prototypes.
- Acknowledgement of ambiguities and uncertainties using Fuzzy Logic, and
- Communication with the society to disseminate environmental information, using a GIS Web Mapping application.

Figure 16.1 shows a schematic description of the D2S4E aims, missions and tools.

The remainder of the chapter is organized as follows. We present the D2S4E methodology and process model in Sect. 2. We discuss the major system attributes in a relationship to the tools and sub-systems used as instruments for decision making in Sect. 3. We provide a summary of preliminary findings and results from a pilot deployment in the region of *Grand Casablanca* in Sect. 4. Section 5 discusses some of the challenges faced during the different project phases, and we highlight some conclusions along with ideas for future work in Sect. 6. The chapter extends the conference paper [10] in Sects. 2 and 4.

2 Methodology and Process Model

D2S4E process model is organized around four phases as shown by Fig. 16.2.

Phase 1: the objective of this mission is to create a dynamic network of stakeholders and local partners to ensure a regular data collection process. This mission includes four activities:

- Activity 1 consists in assessing the current situation in terms of data collection, by making an inventory of the existing documentation and functional committees available for collecting data.
- Activity 2 consists in identifying the necessary data and indicators for monitoring the state of the regional environment.
- Activity 3 consists in the establishment of a collection process within an institutional network at the regional level.
- Activity 4 consists in creating an operational network for the exchange of environmental information related to sustainable development and carrying out the actual data collection process.

After these steps, proposals have been made to institutionalize the network for the collection of environmental data and information, the network is controlled by the OREDDs in collaboration with the ONEM as a National focal point.

Phase 2 is essentially a mission articulated around two main axes: (1) development of an information system in the region, and (2) bring the system developed in

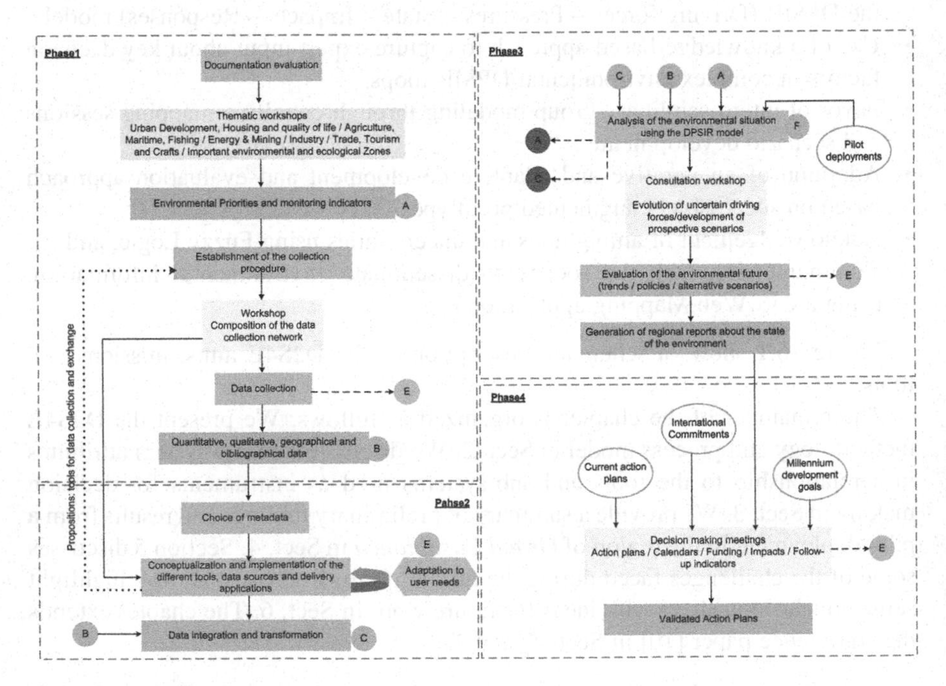

Fig. 16.2 D2S4E process model

(1) towards a comprehensive system with a spatial dimension, regionally able to assist in decision making with full knowledge of the territory it covers.

Phase 3 consists in performing an integrated assessment of the environment at the regional level and the development of the regional report on the state of the environment. The objectives of the mission are: (i) to analyze the state of the regional environment using the DPSIR model, (ii) to evaluate and analyze environmental trends at the regional level, (iii) to analyze environmental scenarios and their impacts regionally, and (iv) to develop future of the regional environment.

The objective of *Phase 4* is to offer a number of action plans for the protection and restoration of the regional environment, with the aim to achieve especially the following goals: (i) the reduction of all types of pollution from domestic and industrial wastes and reaching compliance with the environmental laws and regulations, (ii) achieving the millennium development goals, in particular on drinking water and sanitation, and (iii) to meet the international commitments, knowing that Morocco is actually a party to over 100 multilateral agreements (MEAs) on environmental protection and sustainable development. As part of this mission will also be established a timetable for the implementation of action plans; an evaluation of funding opportunities for all stakeholders; the study of the action plans' impact on the economic and social environment of the region; and the proposed monitoring and performance indicators required for the implementation of action plans.

3 D2S4E Major Attributes

The detailed design decisions and technology choices of D2S4E have been discussed in [11], and are summarized in Fig. 16.3. The D2S4E building block architecture gives a number of evident attributes, like the ability to scale up, flexibility and simplicity of process reusability and duplication when there is a need to. The modular and generic design allows D2S4E to follow the need to build complex environmental services to be offered at different scales and open to a wide audience, ranging from NGOs, experts, decision makers, and the public. The system allows composing different functionalities into higher level entities, thus allowing the creation of different deployment scenarios, possibly distributed at different administrative regional and national levels.

At the heart of D2S4E lies an *Information System* (EIMS), it includes a database, together with activities to manage indicators and associated data; its goal is to assist decision makers access pertinent and consistent information on the state of the environment at various spatiotemporal scales. The EIMS also stores data provider information, environmental themes, geographical and administrative scopes, environmental thematic issues, maps, a bibliography, and a metadata catalogue.

A *Network Component and a Web Application* have been established in D2S4E to ensure regular collection and dynamic exchange of historical data. Stakeholders and partners can actively contribute to the production of environmental

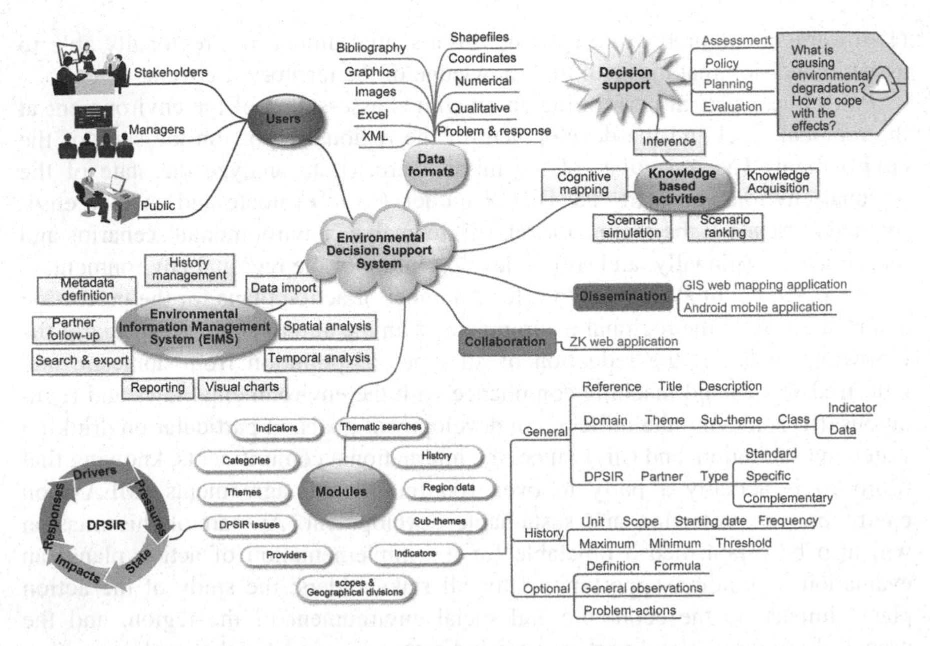

Fig. 16.3 Summary of D2S4E sub-systems and delivery applications

information, through a web portal for the upward transmission and circulation of the information to the OREDDs without further moderation.

Since public participation and the right to have access to environmental information were instructed by the Moroccan Environmental Charter [12], a dynamic *GIS Web Mapping Application* is used as an inexpensive means for disseminating environmental information with a wide audience of users. Users can query the database of indicators and choose layers of statistics in different geographical divisions that are cross-referenced to point to *ESRI Shapefiles*.

The role of the *Knowledge-based System* is to capture and use environmental expert knowledge and process data acquired from the EIMS and the GIS sub-systems against the knowledge base, and generate advice or recommendations as support for managers and decision makers.

The *Decision Support Interface* offers a number of activities that help with structuring the decision making process. It allows managers and decision makers build and explore future paths, evaluate different policies or regulations, and plan for effective responses. It also helps easing the choice among relevant alternatives using multiple attribute decision making.

3.1 Integrated Evaluation of Environmental Issues

3.1.1 Integration of Disciplines

Integrated Assessment (IA) is a 'meta-discipline' that has emerged to support sustainability. The goal if IA is to guarantee that scoping goes beyond sectorial interests and limited problem perceptions, focusing on finding the balance between socio-economic needs and the protection of ecosystems in a holistic approach, which is an essential prerequisite for achieving the targets of sustainable development and helps reducing the occurrence of unexpected negative futures [13]. Therefore, the framework employed to perform an integrated evaluation of environmental issues in D2S4E is the DPSIR model. The model was adopted by the European Environment Agency (EEA) in 1995, and has since been widely used in the study of environmental problems for many applications with the aim to support decision making at different scales [14–18]. According to the DPSIR framework there is a causal chain from "driving forces" including anthropogenic and industrial activities that cause "pressures" on the environment, like emissions and waste, which lead to changes in the environmental, physical, chemical or biological "state", which will induce negative "impacts" on ecosystems, assets and human health, eventually leading to "responses" that might relate to the four other elements: preventive measures directed towards driving forces and pressures, remedial measures towards state, and mitigating measures towards impacts [19, 20]. A case study, illustrated in Fig. 16.4, demonstrates how the DPSIR framework was applied to assist decision makers with the identification of key pressures on water quality in the *Bouregreg-Chaouia* basin.

3.1.2 Integration with Stakeholders

According to the authors of [21], one of the key features of IA is that it's a process with disciplinary equilibration, enriched by stakeholder involvement with team-shared objectives and norms. If policy actions are mandated without considering stakeholders' input, they might result in the different actors perusing their strategies regardless of the developed policies [22, 23]. D2S4E was developed in collaboration with the Regional Environmental and Sustainable Development Observatories (OREDDs) and their partners, the National Environmental and Sustainable Development Observatory (ONEM), and the Environment Department (DE), as well as academics. Potential areas of environmental concern along with the set of indicators to be used were identified by performing a close study through a number of consultations and discussions with specialized environmentalists.

The collaborative and iterative approach yielded a highly adaptive and flexible decision support system. Stakeholder participation was maintained all through the different project phases, they were engaged in selecting indicators, collecting data, identifying the needs and suggesting steps for system development. They had an

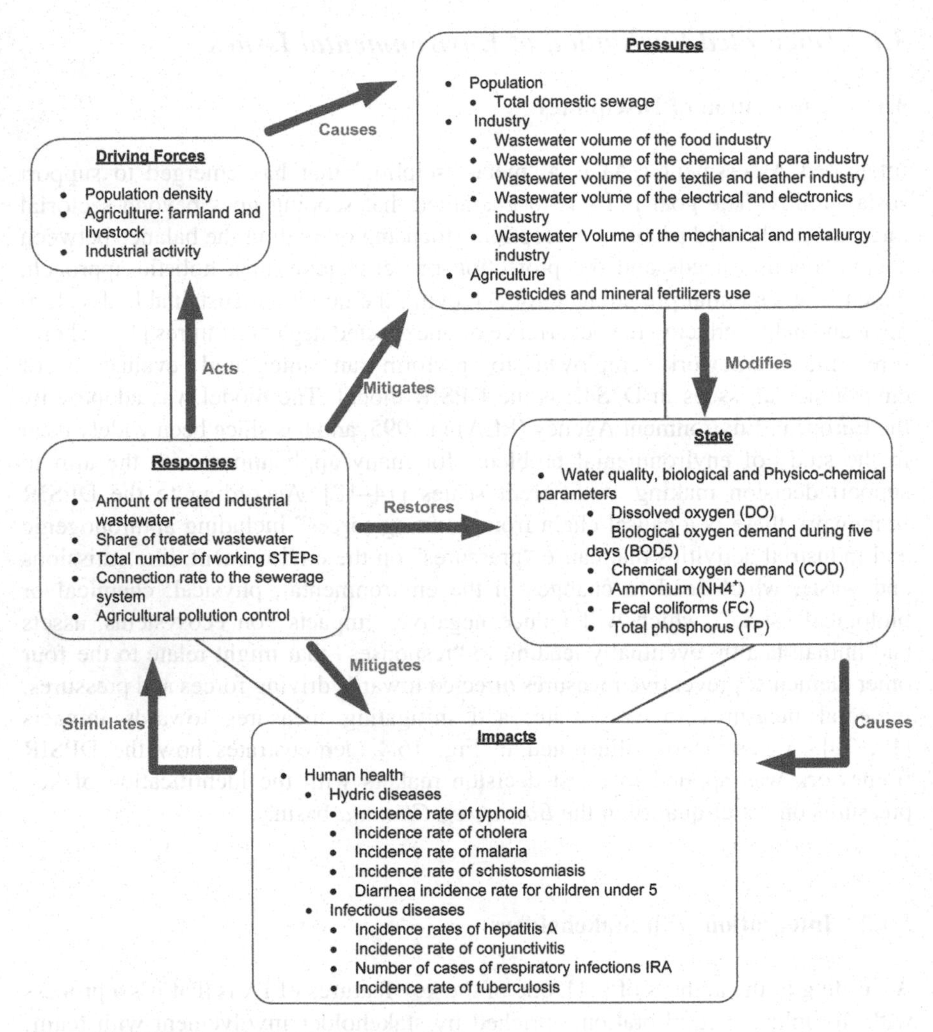

Fig. 16.4 DPSIR application to water quality

input as well in the modelling process and application, the development of policy alternatives through scenario simulation and analysis, as well as generation of the regional reports. The continuous participation of stakeholders was very important in orchestrating the tasks and giving process guidance. Stakeholder involvement was also undoubtedly helpful in tailoring the services and models with the actual needs, in assessing the outcomes and in constantly feeding D2S4E with feedback. It should be also noted that the DPSIR allowed OREDDs, apart from performing environmental diagnostic and evaluation by ecosystem, to perform a complementary analysis of the specific issues, related to the different activities, and stakeholders, who interfere in fact more or less in a direct and deliberate manner with their environment. Stakeholder involvement in this manner not only improves understanding of the studied ecosystem functioning, but also helps actors adopt the appropriate profile so as to contribute to the achievement of a common set of environmental goals considering the cross discipline tradeoffs.

3.2 A System Dynamics Approach to Model Complex Environmental Issues

Although the DPSIR framework helps with modeling environmental issues, there is a gap between the design of the DPSIR loops and their effective use with regards to support of environmental management and policy formulation. Moreover, the requirement to include stakeholders' opinions on environmental issues, from a variety of disciplines, into formal policy analysis raises some main challenges for the modeler. First, it must advocate for group modeling, and be capable of aggregating the elicited knowledge from a variety of participants with possible conflicts in a meaningful manner. Second, there are many sources of uncertainty that can be faced in modelling environmental problems. Uncertainty can be exhibited at the data level, derived from empirical observations and measurements; or at the model level, because experts usually present ambiguous or conflicting knowledge due to the lack of standards, diversity of modelers' beliefs and experiences. Clearly, any method that attempts to support evaluation and analysis of environmental issues must include techniques to assist with the abovementioned complexities. Namely: (1) it must be capable of capturing the different sectorial conflicts, (2) it must be capable of dealing with any possible feedback loops between the elements of the studied system, actually, socio-economic drivers will trigger management measures to decrease the pressures on a particular ecosystem or mitigate the impacts of pressures; therefore there is a need for a modelling approach that simulates feedback loops (3) it must also be able to apprehend the somewhat abstract expert mental images, with rather subjective opinions by emulating natural language.

Fortunately, soft computing techniques like Fuzzy Cognitive Maps (FCMs) offer an approach to bridge the gap and help with analyzing the decision alternatives by achieving the aforementioned aims. They also overcome the limitations of other reasoning methods like Bayesian Networks or Rule-based systems [24]. Cognitive maps were initiated by the political scientist Robert Axelrod [25] and were extended by Kosko [26] to include the concept of fuzziness. A FCM is a system model, that is on the structural level represented in the form of a graph whose nodes represent the concepts of the system being studied and edges represent causal relationships between these concepts. This graph structure facilitates the representation of causal reasoning to study the system dynamics. FCMs have shown a good promise in modelling quantitative and qualitative scenarios and in establishing a social learning process between stakeholders and modelers [27–32].

The advantage of using FCMs in policy development is twofold. Firstly, at the systemic level, FCMs allow group modeling, and have the ability to elicit and combine individuals' knowledge in an easy and meaningful way. Secondly, a FCM is not a static representation of the world; a FCM is an inference system in which calculations will be made to perform an assessment of the consequences of a specific system state by propagating dynamically the causalities in a network of concepts. Execution of FCM causal reasoning allows hence simulation of "what-if" scenarios for a given system from different initial states. Once a FCM is fed with an initial stimulus, it is possible to improve our understanding of a system's dynamics by studying the final stable state. One of the major activities in environmental policy management is the examination of alternative solutions prior to decision making, based on strategic objectives. In D2S4E, FCM simulations help decision makers evaluate the impact of the different policy alternatives on key variables, in a given environmental problem, according to multiple expert perceptions; or to discover what should be the paths that a set of indicators should take in order to achieve a specific goal. Simulations offer a description of projections of the studied system in the future, based on a consistent set of conditions, and assuming that the rules which define the model do not change. For example, what would be the impact on populations, ecosystems and activities if the actions against climate change are not effectively implemented by 2030 in the region? We are thus seeking to gain insight about the projections of the forced scenario, which are based on assumptions that may or may not be realized, and are therefore subject to substantial uncertainty.

The use of Fuzzy Logic is particularly more suitable for modeling vague qualitative knowledge in cognitive maps, because of its linguistic nature. Fuzzy Logic is based on fuzzy sets theory introduced by Zadeh [33]; it has known an increased interest during the last few decades in ecological and environmental modeling [34–36]. Fuzzy Logic includes the so-called classical Boolean Logic. Unlike conventional logic that only considers two properties of a given item, true or false, Fuzzy Logic allows partial truths. Using membership functions [37], that are valued in the real interval [0 1], fuzzy sets theory allows the evaluation of the gradual membership of elements in a specific fuzzy set. One element can more or less belong to a subset which allows modeling ambiguities and more flexibility of reasoning compared to Boolean Logic. In spite of the underlying knowledge elicitation complexity when Fuzzy Logic is used, it allowed us, besides the handling of uncertainty, to establish a more natural user interface since knowledge input and results can be communicated in numerical formats or in natural language terms.

4 Preliminary Findings from Pilot Deployment

D2S4E has been deployed in the pilot regions of *Meknès-Tafilalt* and *Grand Casablanca*. The cross sectorial analysis and environmental evaluation helped discovering a number of challenges that environmental policies shall be confronted to in the coming years. Table 16.1 presents a summary of the main factors and

Table 16.1 Major environmental degradation factors, their causes and effects, future trends as identified in the region of Grand Casablanca

Theme	Factors	Causes	Effects	Trends
Water	Domestic and industrial waste	Population growth and socio-economic development, shortcomings in liquid sanitation, lack of water treatment plants (STEPs) and dilapidated sewerage network in some neighborhoods, uncontrolled autonomous sanitation, raw sewage discharged into the natural environment, inadequate sanitation in the province of Mohamadia, low connection rate: between 50 and 90 %	Waste deposits, high levels of pollutants (BOD_5, TP, FC), degradation of groundwater and surface water, multiplicity of nuisances (bad odors, water pollution, diseases)	
	Use of fertilizers and pesticides	Intensive farm operation, high levels of nitrates	Water pollution by nitrogen and phosphorus	
	Over exploitation of water resources	Increase in population of the Grand Casablanca region, evolution of industrial and tourism sectors, increased water demand in agriculture	Continuous deficit observed of groundwater level in the area of Bouregreg	Trends by 2030: drinking and industrial water demand will increase by 46 % in urban area and 68 % in the rural area; The region will depend on the water resources of other regions

(continued)

Table 16.1 (continued)

Theme	Factors	Causes	Effects	Trends
Biodiversity	Deforestation	Waste discharges recreation and soil compaction by vehicles	Forest species affected by degradation, loss of ecological niches and wildlife, degradation and eutrophication of wetlands, fish mortality, loss of habitats of wild birds, endangered marbled teal birds	
	Material exploitation sites	Exploitation of various materials, abandoned sites without reclamation, non respect of requirement specifications	Loss of vegetation structures, disturbance and pollution contribute to the loss of biodiversity, negative effects on groundwater	
Coastline	Liquid discharges and urbanization	Industrial wastewater discharged directly into the marine environment, urbanization of the coastal area, tourism development	Costal pollution, poor quality of swimming water, degradation of marine resources	
	Sand extraction	Illegal sand extraction	Erosion of sand dunes along the coast	
Air	Air emissions	Pollution due to vehicle fleet, exploitation sites, industrial units, incineration of industrial wastes by the informal sector	Air quality substandard in some areas, health impact, bad odors and fumes	Trends for 2015: SO_2 and NOx exceed the limit values; SO_2, NOx, CH_4 and MPS decrease from the 2004 to the favorable scenario
Waste	Contamination by waste	Uncontrolled landfills, black waste points, plastic, direct drain without treatment, increased production of waste, poor collection and waste management	Water and soil contamination by leachate, proliferation of insects, bad odor emanation, accumulation of waste on the banks of rivers	Trends by 2030: production of household waste will increase by 20.85 % compared to 2015; Waste collection rate is expected to reach 100 % by 2020 in urban areas and won't exceed 50 % in rural areas

(continued)

Table 16.1 (continued)

Theme	Factors	Causes	Effects	Trends
Climate change (risk)	Global warming	Climate conditions	Degradation of the natural plant environment, losses in agricultural production, coastal submersion	Under the baseline scenario, the region will experience a worsening of extreme weather events linked to global climate change (cold/heat waves, desertification) accompanied by an exhibition of more activities related to population growth and economic development of the region
Floods (risk)	Precipitations with lack of infrastructure	Development of human activities in areas with a flood hazard	Risk of pollution of surface waters, infrastructure damage, human damage (homeless, injured, dead), crisis during episodes of flooding/severe flooding	Trends by 2020: prevention by stream recalibration will reduce the risk by approximately 80 % in the 24 priority sites

causes of the socio-economic pressures that knows the region of *Grand Casablanca* considered by environmental theme. The environmental problems of the *Grand Casablanca* region are particularly complex and interdependent. It can be noticed that the demographic development coupled with the lack of governance and coordination between stakeholders gives more breadth to the identified problems. The liquid and solid waste infrastructure's deficit also represents a significant source of pressure involved in the degradation of the living environment in general, and the quality of water and soils in particular. This deficit is aggravated by an inadequate legal framework, lack of control and lack of standards. The development of the industry sector is responsible for a substantial degradation of soil and air due to dusts. The significant air pollution of the *Grand Casablanca* region is also generated by population growth and development of the transport sector. As for the coastline of the *Grand Casablanca*, it is a complex vulnerable area, highly coveted and subject to conflicts, resulting in a high exploitation of the coasts with large population and urbanization pressures.

Given the specific nature of the adopted environmental approach, it was proposed to focus prospective analysis on two scenarios: a baseline scenario and an alternative proactive scenario. The *baseline scenario* is based primarily on a likely continuation of current trends. The organization and policies continue to be organized around the current dynamics growth, mandated policies and regulations. This scenario may have two variants:

- An aggravated trend scenario (STA) in which one does not take into account the various planned programs; and
- A contextual baseline scenario (STC) that addresses the different programs planned by the local regulatory institutions.

On the other hand, the *voluntary alternative scenario* (SAV) is based on strengthening these programs to ensure a better environmental future. The evolution of trends for a number of themes based on proactive scenarios is shown in Table 16.1.

The analysis and comparison of scenario planning and natural resource management, has led to the development of an action plan named "*Future of the Environment*". Nine development initiatives were fixed following the results of the selection of priority issues in the region by the different stakeholders: (1) improving sewerage; (2) control of air pollution; (3) improving waste management; (4) risk management; (5) coastal management; (6) water resources preservation; (7) soil and biodiversity protection; (8) governance; (9) education and awareness.

5 Challenges Faced

In spite of the recognized importance of D2S4E in environmental management and evaluation at the highest levels, there have been a number of challenges that hindered the advancement of the different project phases. We shall discuss the perhaps most problematic ones across both development and implementation of the pilots, namely elicitation of needs, scarcity of data and knowledge, and the lack of criteria to frame the selection of pertinent indicators.

5.1 Communication with End Users

One of the challenges we faced, that was also discussed in [38] is that the stakeholders' needs, initially, were not clearly detailed or even understood by the end users (OREDDs and ONEM) themselves. What complicated the situation further is that the EIMS sub-system of D2S4E was funded by the Environment Department (DE), but subcontracted to private consulting companies and implemented by academic researchers independently of the originally intended end users, which resulted, quite naturally, in creating a problematic communication setting for the developers who had to cope with the increasing and continuously changing requirements. To mitigate somehow the problem, we adopted an iterative and interactive approach, conducted in a collaborative spirit with just the right amount of formality. We built a number of prototypes that served initially to validate the requirements and provide suggestions, taking into account the changing needs, however the process was very costly and time consuming.

5.2 *Data and Knowledge Quality Versus Scope of Analysis*

Applying the DPSIR framework to an environmental issue should be strengthened by a solid knowledge base. As with any decision support system, the obtained results highly depend on the quality of data and knowledge availability. Unfortunately, in D2S4E, the data collection step was performed at the early stages independently from the identification of environmental issues and the application of the DPSIR model. The data collection process turned out not to have sufficient historical data to support the selected loops. As a result, the loops were revisited numerous times, leading to scope reduction and considerable loss of time. Moreover, in an integrated evaluation approach, availability of knowledge in the different areas of the DPSIR model cannot be assumed. Learning trends and knowledge extraction was quite hard too because of scarcity of data. For example, in the pilot region of *Meknès-Tafilalt*, we noted that 85 % of indicators have no history, and that only 10 % of indicators contribute to the available data with a filling percentage lower than 40 %, which means that there are missing data for many indicators and for many periods. Another factor that contributed to the problem is the underlying acquisition cost of non free data. This situation has pushed DE to put pressure on the different stakeholders in order to maintain the level of interest and to make available the funds required to collect the necessary data.

5.3 *Environmental Data and Indicator Framework*

Given the complexity and the interlacing nature of environmental issues, the list of indicators to be used in the integrated environmental evaluation took considerable time to finalize. The project didn't fully succeed to handle the actual complexity of environmental issues, because of the lack of scientific consistency in defining and adopting a clear framework or criteria for choosing the indicators that best describe a specific ecological problem. It is felt that a procedure for environmental indicator selection should be adopted. The procedure should have clear criteria and also present a method to evaluate the indicators against the chosen criteria. The set of indicators need to capture the complexities of an ecosystem but yet be simple, easily monitored and measured [39]. They should also be relevant to the policy life cycle and help decision makers take informed decisions. The result of this process should then be used as a baseline for data collection and environmental monitoring.

5.4 *Lack of Synergy Between Regional Project Implementations*

The project was designed to be implemented simultaneously in the different regions of Morocco, with no best practices derived or reused from one region into another one. There is significant duplication of effort at the regional scale, in the areas of indicator selection and modeling of environmental issues, and there are many instances of 'best practice' which could be shared. Despite the substantial resources going into these different implementations, some regions are significantly behind schedule in designing DPSIR loops for regional environmental concerns, and there is little integration of the regional information systems at the national level. There are some gaps in the built systems; for instance, air quality and biodiversity issues are poorly covered; topics which are relatively well covered in most regions include water resources and solid waste management.

6 Conclusion

In this chapter we described the background, methodology, attributes and challenges of an Environmental Decision Support System developed to support the Moroccan national strategy for environmental protection and sustainable development.

One of the major assets of the environmental decision support system is that it was designed and deployed within a framework that includes stakeholder participation and expert knowledge. This linkage was maintained within the frame of a national project conducted by the Moroccan Environment Department. Applying the DPSIR framework by integrating ecological and socio-economic indicators is very important in an integrated evaluation of the environmental situation and in reaching the targets of sustainable development.

However, it remains very challenging to mobilize enough expertise pertaining to inter-related domains of a given environmental issue. The knowledge-based approach was proposed to clearly define the network of causal links between indicators and in associating the qualitative and quantitative data, which are all necessary activities in the modeling process but lacking in the DPSIR framework.

In order to ensure successful implementation of environmental decision support systems, it is of a tremendous importance to adopt a formal framework, with clear criteria for defining the scopes and selecting indicators in order to obtain the ones that are very representative of a specific ecological problem, and yet can be easily monitored and measured. It is also very important that the focus, in environmental decision support system development, get shifted from functionality to adoption of proven technologies to capture elaborate usability and user oriented process models, in order to ensure correct elicitation of the end user needs and their involvement in the different stages of the project.

Future research will focus on exploiting the potential of the environmental decision support system. Two important research topics are considered. First, we would like to allow scientists to develop and integrate domain-specific models in a seamless and transparent manner. Second, we would like to build a platform to combine, analyze and exploit heterogeneous data from various sources, including near real time data streams generated by sensors. Research towards the exploitation of the data generated by such devices may lead to better trend analysis and to more innovative services.

References

1. Sarraf, M., Belhaj, M., Jorio, A.: Report No 25992-MOR: Kingdom of Morocco Cost Assessment of Environmental Degradation (2003).
2. Mayer-Föll, R., Keitel, A., Hofmann, C., Lukács, G., Briesen, M., Otterstätter, A.: Integrated Analysis and Reporting of Environmental Data Through Cooperation and Technical Innovations – Despite Organisational Changes, Increasing Requirements and Decreasing Budgets. Informatics for Environmental Protection – Networking Environmental Information (2005).
3. Booty, W.G., Lam, D.C.L., Wong, I.W.S., Siconolfi, P.: Design and implementation of an environmental decision support system. Environmental Modelling & Software 16(5), 453–458 (2001). doi:10.1016/S1364-8152(01)00016-0.
4. Vögele, h., Klenke, M., Kruse, F., Groschupf, S.: A New and Flexible Architecture for the German Environmental Information Network. Informatics for Environmental Protection – Networking Environmental Information (2005).
5. Madari, U.: Supporting Strategic Decision Making on Climate Change Through Environmental Information Systems: The Case of ENVIS. In. (2012).
6. Evans, S.: A Prototype Environmental Information System for London (London Environment Online, Leo). bartlett (1999).
7. Pillmann, W., Geiger, W., Voigt, K.: Survey of environmental informatics in Europe. Environmental Modelling & Software 21(11), 1519–1527 (2006). doi:10.1016/j.envsoft.2006.05.008.
8. Rizzoli, A.E., Young, W.J.: Delivering environmental decision support systems: software tools and techniques. Environmental Modelling & Software 12(2–3), 237–249 (1997). doi:10.1016/S1364-8152(97)00016-9.
9. Guariso, G., Werthner, H.: Environmental decision support systems. E. Horwood, (1989).
10. Mourhir, A., Rachidi, T., Karim, M.: Design and Implementation of an Environmental Decision Support System: tools, attributes and challenges. Proceedings of the 28th Conference on Environmental Informatics – Informatics for Environmental Protection, Sustainable Development and Risk Management (2014).
11. Mourhir, A., Rachidi, T., Karim, M.: A Decision Support System for the Environment: Application to Morocco. J. Int. Environmental Application & Science 9(1), 10–23 (2014).
12. Ministry of Energy Mines Water and Environment: Morocco's National Charter for Environment and Sustainable Development. In: Green Compass Research. (2010).
13. Robinson, J.: Defining a Sustainable Society: Values, Principles and Definitions. Sustainable Society Project, Department of Environment and Resource Studies, University of Waterloo, (1990).
14. Lorenzoni, I., Jordan, A., Hulme, M., Kerry Turner, R., O'Riordan, T.: A co-evolutionary approach to climate change impact assessment: Part I. Integrating socio-economic and climate change scenarios. Global Environmental Change 10(1), 57–68 (2000). doi:10.1016/S0959-3780(00)00012-1.

15. Bidone, E.D., Lacerda, L.D.: The use of DPSIR framework to evaluate sustainability in coastal areas. Case study: Guanabara Bay basin, Rio de Janeiro, Brazil. Reg Environ Change 4(1), 5–16 (2004). doi:10.1007/s10113-003-0059-2.
16. Holman, I.P., Rounsevell, M.D.A., Shackley, S., Harrison, P.A., Nicholls, R.J., Berry, P.M., Audsley, E.: A Regional, Multi-Sectoral And Integrated Assessment Of The Impacts Of Climate And Socio-Economic Change In The Uk. Climatic Change 71(1–2), 9–41 (2005). doi:10.1007/s10584-005-5927-y.
17. Karageorgis, A., Kapsimalis, V., Kontogianni, A., Skourtos, M., Turner, K., Salomons, W.: Impact of 100-Year Human Interventions on the Deltaic Coastal Zone of the Inner Thermaikos Gulf (Greece): A DPSIR Framework Analysis. Environmental Management 38(2), 304–315 (2006). doi:10.1007/s00267-004-0290-8.
18. Borja, Á., Galparsoro, I., Solaun, O., Muxika, I., Tello, E.M., Uriarte, A., Valencia, V.: The European Water Framework Directive and the DPSIR, a methodological approach to assess the risk of failing to achieve good ecological status. Estuarine, Coastal and Shelf Science 66(1–2), 84–96 (2006). doi:10.1016/j.ecss.2005.07.021.
19. Smeets, E., Weterings, R.: Environmental indicators: Typology and overview. In. European Environment Agency, (1999).
20. Gabrielsen, P., Bosch, P.: Environmental indicators: typology and use in reporting. In. - European Environment Agency, (2003).
21. Jakeman, A.J., Letcher, R.A.: Integrated assessment and modelling: features, principles and examples for catchment management. Environmental Modelling & Software 18(6), 491–501 (2003). doi:10.1016/S1364-8152(03)00024-0.
22. Voinov, A., Bousquet, F.: Modelling with stakeholders. Environmental Modelling & Software 25(11), 1268–1281 (2010).doi:10.1016/j.envsoft.2010.03.007.
23. Korfmacher, K.S.: The Politics of Participation in Watershed Modeling. Environmental Management 27(2), 161–176 (2001). doi:10.1007/s002670010141.
24. Taber, R.: Knowledge processing with Fuzzy Cognitive Maps. Expert Systems with Applications 2, 83–87 (1991). doi:10.1016/0957-4174(91)90136-3.
25. Axelrod, R.: Structure of Decision: The Cognitive Maps of Political Elites. Princeton University, (1976).
26. Kosko, B.: Fuzzy Cognitive Maps. International Journal of Man-machine Studies 24, 65–75 (1986).
27. Kontogianni, A.D., Papageorgiou, E.I., Tourkolias, C.: How do you perceive environmental change? Fuzzy Cognitive Mapping informing stakeholder analysis for environmental policy making and non-market valuation. Applied Soft Computing 12(12), 3725–3735 (2012). doi:10.1016/j.asoc.2012.05.003.
28. Henly-Shepard, S., Gray, S.A., Cox, L.J.: The use of participatory modeling to promote social learning and facilitate community disaster planning. Environmental Science & Policy 45(0), 109–122 (2015). doi:10.1016/j.envsci.2014.10.004.
29. van Vliet, M., Kok, K., Veldkamp, T.: Linking stakeholders and modellers in scenario studies: The use of Fuzzy Cognitive Maps as a communication and learning tool. Futures 42, 1–14 (2010). doi:10.1016/j.futures.2009.08.005.
30. Kok, K.: The potential of Fuzzy Cognitive Maps for semi-quantitative scenario development, with an example from Brazil. Global Environmental Change 19, 122–133 (2009). doi:10.1016/j.gloenvcha.2008.08.003.
31. Kafetzis, A., McRoberts, N., Mouratiadou, I.: Using Fuzzy Cognitive Maps to Support the Analysis of Stakeholders' Views of Water Resource Use and Water Quality Policy. In: Glykas, M. (ed.) Fuzzy Cognitive Maps, vol. 247. pp. 383–402. Springer Berlin Heidelberg, (2010).
32. Jetter, A., Schweinfort, W.: Building scenarios with Fuzzy Cognitive Maps: An exploratory study of solar energy. Futures 43, 52–66 (2011). doi:10.1016/j.futures.2010.05.002.
33. Zadeh, L.A.: Fuzzy sets. Inform. Control, vol.8, 338–353 (1965).
34. Shepard, R.B.: Quantifying Environmental Impact Assessments Using Fuzzy Logic. Springer, (2005).

35. Silvert, W.: Ecological impact classification with fuzzy sets. Ecological Modelling 96(1–3), 1–10 (1997). doi:10.1016/S0304-3800(96)00051-8.
36. Salski, A.: Ecological Applications of Fuzzy Logic. In: Recknagel, F. (ed.) Ecological Informatics. pp. 3–14. Springer Berlin Heidelberg, (2006).
37. Ross, T.J.: Fuzzy Logic with Engineering Applications. Wiley, (2009).
38. McIntosh, B.S., Ascough Ii, J.C., Twery, M., Chew, J., Elmahdi, A., Haase, D., Harou, J.J., Hepting, D., Cuddy, S., Jakeman, A.J., Chen, S., Kassahun, A., Lautenbach, S., Matthews, K., Merritt, W., Quinn, N.W.T., Rodriguez-Roda, I., Sieber, S., Stavenga, M., Sulis, A., Ticehurst, J., Volk, M., Wrobel, M., van Delden, H., El-Sawah, S., Rizzoli, A., Voinov, A.: Environmental decision support systems (EDSS) development – Challenges and best practices. Environmental Modelling & Software 26(12), 1389–1402 (2011). doi:10.1016/j.envsoft.2011.09.009.
39. Dale, V.H., Beyeler, S.C.: Challenges in the development and use of ecological indicators. Ecological Indicators 1(1), 3–10 (2001). doi:10.1016/S1470-160X(01)00003-6.

Chapter 17
Towards an Environmental Decision-Making System: A Vocabulary to Enrich Stream Data

Peter Wetz, Tuan-Dat Trinh, Ba-Lam Do, Amin Anjomshoaa, Elmar Kiesling, and A Min Tjoa

Abstract The future of the earth's environmental systems will, to a major extent, be determined in cities, where already more than 50 % of the human population is concentrated. Pervasively available sensors and the data they generate can help to address pressing environmental challenges in urban areas by making crucial information available to researchers and decision-makers. However, environmental data is at present typically stored in disparate systems and formats, which inhibits reuse and integration. Furthermore, the large amounts of environmental data that stream in continuously require novel processing approaches. So far, research at the intersection of environmental sciences and urban data infrastructures has been scarce. To address these issues, we develop a novel framework based on semantic web technologies. We apply data modeling and semantic stream processing technologies in order to facilitate integration, comparison, and visualization of heterogeneous data from various sources. This paper presents the concept of a platform for environmental data stream analysis, and focuses on the design of a new vocabulary to semantically enrich the processed streams. The implemented architecture shall be capable of informing and supporting decision-making by non-expert users. We propose and discuss a three-step framework, present a vocabulary to model environmental data streams, and outline initial results.

Keywords Environmental data streams • Semantic sensor network ontology • RDF data cube vocabulary • Stream processing

1 Introduction

In 2010, for the first time in history, more than half of the global population was concentrated in urban areas [17]. As a consequence of this development, the future of our environment will to a major extent be shaped by policies implemented in cities. To make informed decisions and implement sensible environmental policies,

P. Wetz (✉) • T.-D. Trinh • B.-L. Do • A. Anjomshoaa • E. Kiesling • A.M. Tjoa
Institute of Software Technology and Interactive Systems, TU Wien, Favoritenstraße 9-11/188, 1040 Vienna, Austria
e-mail: peter.wetz@tuwien.ac.at

J. Marx Gómez et al. (eds.), *Advances and New Trends in Environmental and Energy Informatics*, Progress in IS, DOI 10.1007/978-3-319-23455-7_17

decision-makers need a solid and up-to-date understanding of environmental issues. In this context, technologies and methods from computer science have strong potential to improve the understanding of our environment and contribute towards solving environmental challenges [9].

Most harmful developments in urban areas are directly linked to people's behavior, which impairs air and water quality, causes waste disposal problems and noise pollution, and affects the local and global climate. A key motivation of this work is to contribute towards mitigating such negative environmental effects of urbanization through ICT-based (Information and Communication Technologies) methods. In particular, ICT can support city planners' strategic decision making and assist in real-time environmental decision-making, such as, for instance, analyzing traffic and air pollution data streams to dynamically optimize traffic routing. Such applications are rendered possible by the ubiquitous presence of data stream-generating sensors. From an environmental management perspective, this rich data fabric can be seen as a major advantage that cities have over rural areas. Integrating this real-time sensor data may lead to innovative citizen services and may ultimately help to trigger change in how we interact with the environment [15]. At present, however, available means to exploit continuously generated environmental data streams are still limited and existing environmental information systems that monitor air pollution, water quality, or transport systems based on real-time data operate only within the isolated scopes of their domain. Abundant raw data is only a first step; to extract relevant insights, it is necessary to enrich it with contextual information, integrate it with data from various other streams, and carefully process and analyze it.

Another important motivation for this work is to provide public access to environmental information. The European Union (EU) Directive on public access to environmental information [7] mandates public access to and systematic distribution of environmental information through ICT. However, there are serious technical barriers that inhibit citizens from readily accessing environmental information. These barriers include (i) distribution of data among different agencies and lack of a single point of access, (ii) heterogeneous storage without standardized presentation, (iii) focus on static data and neglect of increasingly important real-time data, and (iv) lack of embedding of data within its "context", that is, providing and utilizing additional information based on the surroundings of the data is currently not possible.

Our vision is a Smart City data platform capable of measuring, sensing, analyzing and presenting the environmental "pulse" of a city, i.e., measured via characteristics such as air quality, noise pollution, water quality, and traffic information. This platform should exhibit characteristics such as timeliness, accuracy, usability, scalability, and modularity. Semantic technologies are particularly useful in this context because they facilitate both data integration and query-driven reasoning based on formalized vocabularies. Ontologies ensure data homogeneity and Resource Description Framework (RDF) stream processing techniques facilitate real-time information integration.

In this paper, we conceptualize and outline the architecture of a web-based platform for the semantic integration of heterogeneous environmental data sources in (near) real-time. This platform provides a unifying layer that integrates diverse environmental city data. In particular, we tackle three main challenges, i.e., to (i) integrate data originating from different sources and formats, (ii) facilitate semantic querying of the integrated stream data following Linked Data principles, and (iii) address information needs in the environmental domain in (near) real-time. The research presented shall be seen as "work in progress" since the platform[1] is still in its early stages of development [22]. Nonetheless, we proof our concept by means of initial use cases based on environmental data. This article extends preliminary work already published (cf. [24]) with the motivation, design and description of an ontology for annotating environmental data streams.

The remainder of this paper is organized as follows. Section 2 discusses related work; Sect. 3 motivates and explains the design of a vocabulary for annotating environmental data streams, which lays the foundation for the platform's architecture that is then described in Sect. 4. Initial results are discussed in Sect. 5; finally, we conclude the paper and give an outlook on future work in Sect. 6.

2 Related Work

To provide powerful ontology-based query languages that support continuous queries, efficient means to process semantic data streams are necessary. The amount of research in this field has been expanding rapidly in recent years and several approaches have been developed: C-SPARQL [3], CQELS [13], SPARQLstream [5], EP-SPARQL [1], and INSTANS [16]. Moreover, a W3C RDF Stream Processing Community Group[2] has been formed with the aim of defining a common model for producing, transmitting, and continuously querying RDF streams. However, there is currently no system that supports the whole process from data acquisition to data utilization and enables flexible and efficient use of generic streams.

Balduini et al. [2] present an approach to identify events in a city leveraging a Streaming Linked Data Framework. In contrast to our work, it focuses specifically on social media as a data source. This simplifies the semantic modeling of the data, but also limits the approach to geo-tagged tweets. In particular, the framework does not fuse multiple social streams, but analyzes a single stream at a time.

Lécué et al. [10] predict the severity of road traffic congestion using real-time heterogeneous data streams. The proposed approach is similar to ours, but focuses strongly on the traffic domain and on predictive reasoning, whereas our goal is to provide a generalized system that supports a larger spectrum of use cases.

[1] See http://linkedwidgets.org (accessed 23 February 2015).

[2] See http://www.w3.org/community/rsp/ (accessed 23 February 2015).

Tallevi-Diotallevi et al. [20] present a real-time urban monitoring framework implemented for the city of Dublin. The authors extend CQELS and C-SPARQL to facilitate merging of CSV and RDF streams. Integration of other formats is not supported and explicit semantic enrichment and subsequent use of semantics are not covered.

Initial work on the design of an ontology for the environmental monitoring domain has also been done, in particular by reusing the Semantic Sensor Network Ontology (SSNO) [6] or the RDF Data Cube Vocabulary (QB). However, none of the existing approaches make use of both SSNO and QB in a truly unified manner.

Lefort et al. [11, 12] aim to combine both ontologies in order to convert a historic climate data set to RDF. They provide a web-interface to browse the data while making use of the encoded semantics. However, the vocabularies are coupled more loosely than is the case in the present work. The authors use SSNO to publish metadata and QB for observation data. Their work is based on the premise that the declarations of observed properties in SSNO as classes are not directly compatible with their declarations in QB as properties. In our work, we did not have to tackle this issue, because we model the data structure in a different way (see Subsection "Defining the Measure Property according to QB").

Stocker et al. [19] design an ontology that aligns and specializes the generic concepts of several upper ontologies, i.e., SSNO, DOLCE + DnS Ultralite (DUL), QB, Situation Theory Ontology (STO), GeoSPARQL, Time Ontology in OWL, and the PROV Ontology (PROV-O). However, they also model *ssn:Observation* as a different notion than *qb:Observation*. The former represents a sensor observation that relates to the sensor that made an observation (result of sensing); the latter represents a data set observation, for instance, a line of a textual observation file (result of computations). Since their work aims at modeling a multi-step process of environmental monitoring this separation is suitable in this particular case.

Tarasova et al. [21] reuse QB to capture heterogeneous environmental data sources. Their approach is similar to ours, except that the authors do not operate on data streams. Instead, they process static data sets of historic data. Moreover they do not use SSNO or SWEET (Semantic Web for Earth and Environmental Terminology) [14] ontologies, so many terms they introduce are redundant and actually already modeled in the respective source ontologies.

3 Ontology for Modeling Environmental Data Streams

The design of an ontology for modeling and annotating environmental data streams is a prerequisite for the architecture presented in Sect. 4 and a central contribution of this article. Ontologies form the core of semantic web technologies. They are used to describe data and explicitly define their semantics. Due to the heterogeneity of environmental data, it is not feasible to design a single ontology that covers all environmental domains. Therefore Linked Data—as the data infrastructure of the semantic web—allows for the arbitrary use of multiple coexistent vocabularies to describe data. Our ontology takes these considerations into account and represents an extensible vocabulary for sensor observations in the environmental domain, which can be complemented with external information.

The main contribution of this work is (i) to provide a concise, but complete vocabulary to model, annotate and semantically enrich environmental data streams while (ii) relying on the reuse of already existing ontologies wherever possible. Since both the Semantic Sensor Network Ontology (SSNO) and the RDF Data Cube Vocabulary (QB) have overlaps in their fields of application, i.e., capturing (sensor) observations, this is—to the best of our knowledge—the first approach to combine and align both ontologies while preserving and respecting the concepts' definitions. Our integrated ontology allows for queries that exploit the ideas, notions and use cases of both source ontologies.

The following sections provide a detailed overview on the design process, requirements, reused ontologies, key considerations, and lessons learned. The ontology discussed is actively maintained and available online.[3] In this article we use the prefix *es* (Environmental Streams) when referring to terms of this ontology.

3.1 Requirements

Environmental data streams are heterogeneous and vary along several dimensions, e.g., data type, update frequency, or covered domain. To process them in a homogeneous environment, it hence is essential to define a unifying data model. We choose ontologies as a foundational data model to describe our domain of interest, i.e., environmental data. Ontologies are highly useful for data integration tasks and to enable unified data access.

Incoming data gets converted to a data model that conforms to a described ontology and therefore enables access to the previously separated sources by means of a controlled vocabulary. The conversion process can also be seen as a semantic enrichment of the data, because raw data gets annotated using well-defined domain concepts. In other words, we add meaning to the data which can later be exploited by applications and humans. Since the enriched concepts are both machine- and human-readable, they can be reused by semantic clients as well as by humans, for instance, through defining queries. Semantic enrichment is a fundamental requirement to realize consecutive steps of the presented architecture, i.e., data streaming and stream processing. Other tasks to realize are contextualized sensor discovery, semantic search and query mechanisms, and knowledge extraction from sensor-generated data.

In our work, we design a new ontology for the environmental domain based on the following goals and requirements:

[3] https://github.com/beta2k/environmental-stream-ontology (accessed 23 February 2015).

- *R1 – Data integration*: It is essential for the ontology to allow integration of different data sources that capture environmental observations. Therefore, it should be capable of storing knowledge about phenomena that appear in our environment. Moreover metadata about sensors (update frequency, location, etc.) and observations (unit of measure, etc.) should be stored. The challenge is to define the vocabulary not too broadly or too narrowly.
- *R2 – Reuse of existing ontologies*: reusing knowledge whenever possible is a key concept of the semantic web. This is necessary to avoid the dissemination of terms and concepts with overlapping meaning via different URIs. Moreover, reuse of ontologies saves effort, because one does not have to start development from scratch. This approach also facilitates interoperability between, for instance, other sensor networks using the same vocabularies.
- *R3 – Lightweight*: The result should be a lightweight ontology which is easy to (re)use. Users should be able to start working with the terms quickly and without reading long documentations. Lightweight ontologies are commonly used for data standardization purposes, as is the case here and when the resulting application does not involve complex reasoning tasks based on so-called heavy semantics, which is explicitly not the aim here.
- *R4 – Reusability and extensibility*: Users of the vocabulary should be able to reuse and extend it easily. This requirement is satisfied by this ontology, because terms that should be reused can be picked arbitrarily and users are not expected to reuse every defined term. Similarly, the vocabulary is designed to be generic enough so it can be extended easily with more concrete definitions, if required.
- *R5 – Separation of knowledge*: The problem of conflating concepts arises when the domain to be modeled is not defined precisely and completely. The vocabulary at hand should avoid conflating terms by separating concepts such as *observed property*, *observed feature*, *unit of measure*, etc. Moreover, we aim to separate domain knowledge from operational knowledge when designing the ontology. Domain knowledge shall act as static and stored background knowledge, which supports knowledge extraction. It does not change frequently. Operational knowledge is represented through concrete observations that flow in steadily in the form of data streams. These two layers shall be clearly separated in the model. By preventing conflated concept definitions and ensuring a clear separation of knowledge domains the ontology facilitates semantic interoperability and powerful query mechanisms. It makes the concepts easier to query and to align them to vocabularies of other domains (this is also related to the extensibility requirement R4 above). Moreover, the combination of static knowledge (e.g., geographic maps, point-of-interest data, etc.) with operational knowledge (i.e., dynamic data streams) improves the quality of new knowledge that can be deduced.
- *R6 – Observation Aggregation*: It should be possible to aggregate sensor observation data with typical aggregation functions like MIN, MAX, SUM, or AVG. This requirement can be addressed through appropriate SPARQL queries, but the underlying data model, i.e., the ontology, should also encode the data in a way that allows for such queries. Moreover, it should be possible to combine and

aggregate observations stemming from different data sources along different dimensions. For instance, a query could ask for *"Provide the average temperature of sensors located near Vienna from the last two weeks"*. The ontology has to be designed to allow for aggregation along temporal or spatial dimensions, as well as supporting typical aggregation functions.

- *R7 – Dynamic selection of data streams*: Whereas this is primarily a user interface issue, the ontology should support the definition of new data streams with their accompanying data structure. We expect that data streams will be added or removed over time and their encoding in the data model should be flexible enough to account for this.
- *R8 – Exploitation of hierarchical structures*: Ontologies can store knowledge in hierarchical structures; an ontology for the environmental domain can make use of this characteristic to support convenient query formulation. A potential query could be *"Provide all sensor observations which measure precipitation properties"*. If the observed properties are encoded in a hierarchy, where different types of precipitation such as snowfall or rainfall are subsumed under the same concept, general queries can be formulated conveniently.
- *R9 – Stream and observation discovery:* This requirement is related to R4 and R8; it states that both streams and observations should be discoverable along key dimensions such as spatial, temporal, and environmental characteristics. Given the example that a sensor observes *air temperature*, it should for instance be possible to query sensors based on either observed properties (*temperature*) or features (*air*). This allows for flexible query processing along the axis of properties and features. Moreover, since we also encode temporal and spatial knowledge, these dimensions should be supported for querying as well.

3.2 Reused Ontologies

Table 17.1 summarizes the used vocabularies, their namespaces, prefixes, application domain and how they were reused (either via *owl:imports* statement (*I*) or via simple redeclaration (*R*)). Figure 17.1 depicts how existing ontologies were combined via subclass statements.

Figure 17.2 provides an overview of the classes and their relations via properties from an SSNO-centric point of view, whereas Fig. 17.3 provides an overview over used classes and relationships from a QB-centric point of view. In the former, one can see that we derive *es:Sensor*, *es:Observation*, *es:SensorOutput*, and *es:ObservationValue* from respective classes of SSNO and QB. *Sweet:HumanActivity*, *sweet:Phenomena*, and *sweet:Substance* are aligned with *ssn:FeatureOfInterest* since they contain conceptual domain knowledge which later will be encoded as Features complying with SSNO best practices. Furthermore, we use *geo:hasGeometry* in combination with *geo:Point* to model sensor locations and *ssn:observationResultTime* in combination with *time:Instant* to capture temporal information of observations.

Table 17.1 Reused ontologies, their namespaces, prefixes, application domains, and how they were reused. The prefixes are used also further in the article to describe classes and properties of respective ontologies

Ontology name	Namespace	Prefix	Domain	Reuse
Semantic Web for Earth and Environmental Terminology (SWEET)	http://sweet.jpl.nasa.gov/2.3/	sweet	Environmental terms	R
Semantic Sensor Network Ontology (SSNO)	http://purl.oclc.org/NET/ssnx/ssn#	ssn	Sensor and observation descriptions	I
RDF Data Cube Vocabulary (QB)	http://purl.org/linked-data/cube#	qb	Multi-dimensional observations	I
Time Ontology in OWL	http://www.w3.org/2006/time#	time	Temporal data	R
GeoSPARQL	http://www.opengis.net/ont/geosparql#	geo	Spatial data	R

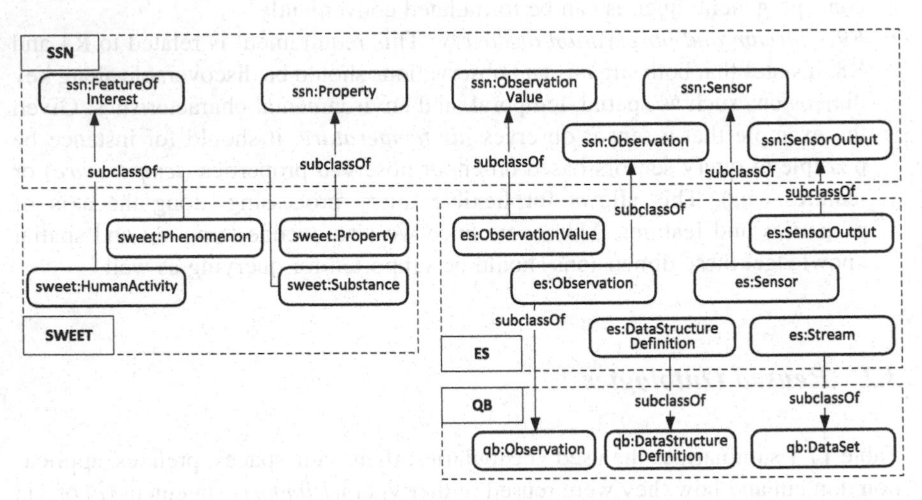

Fig. 17.1 Reuse and combination of external ontologies via *owl:subclassOf* properties. Prefixes are used as defined in Table 17.1, except *es* which is the namespace of the developed ontology

Figure 17.3 shows that *es:Stream* is derived from *qb:DataSet*. Furthermore, we see that units of measurement are encoded as *sweet:Unit* via *sweet:hasUnit* at the data set level. The bottom half of the figure represents instance data, i.e., the actual data structure definition, observation and data stream instances according to QB.

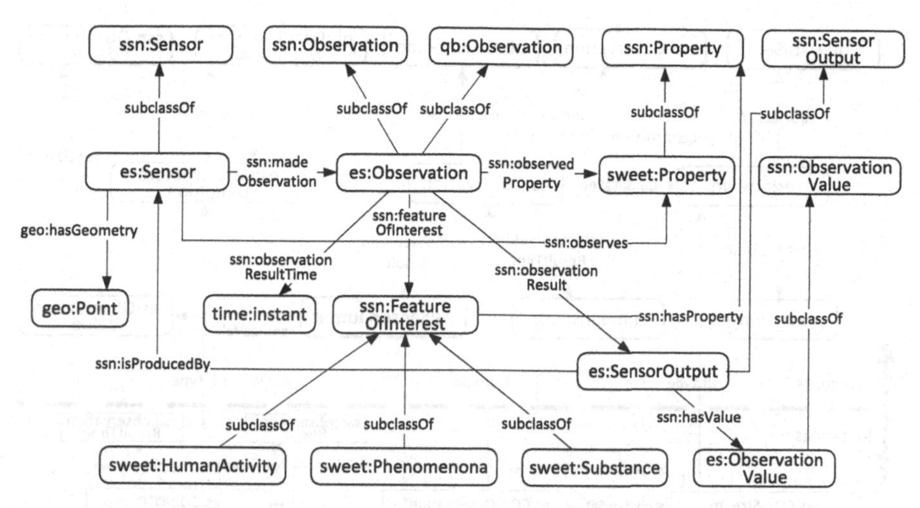

Fig. 17.2 SSNO-centric view over classes and their relations. *Es* (environmental streams) is the prefix for the developed ontology. It defines new classes deduced from upper ontologies and reuses external properties

3.3 *Considerations*

Several considerations and design decisions had to be taken into account while implementing above requirements. They will be described in the next subsections.

3.3.1 Decisions Between Ontology Alternatives

QUDT vs. SWEET Several ontologies are available for encoding certain kinds of data. In our case we want to model properties of observations, e.g., *temperature*, *length*, *speed*. Two ontologies were considered for this purpose: QUDT – Quantities, Units, Dimensions and Data Types Ontologies[4] (prefix: qudt) and SWEET – Semantic Web for Earth and Environmental Terminology[5] (prefix: sweet). We did a non-exhaustive evaluation of environmental properties, which we possibly want to model in our future stream processing application. Table 17.2 shows that QUDT does not support all evaluated properties (*qudt:QuantityKind*) as opposed to SWEET (*sweet:Property*). We also note that QUDT developers state that *height* and *depth* are not encoded as quantity kinds, because they are specific quantities of the quantity kind *length*.

All in all, the terms of the SWEET vocabulary appear to be more targeted towards the environmental domain. This and the fact that units of measure are also encoded

[4] See http://www.qudt.org/ (accessed 23 February 2015).

[5] See https://sweet.jpl.nasa.gov/ (accessed 23 February 2015).

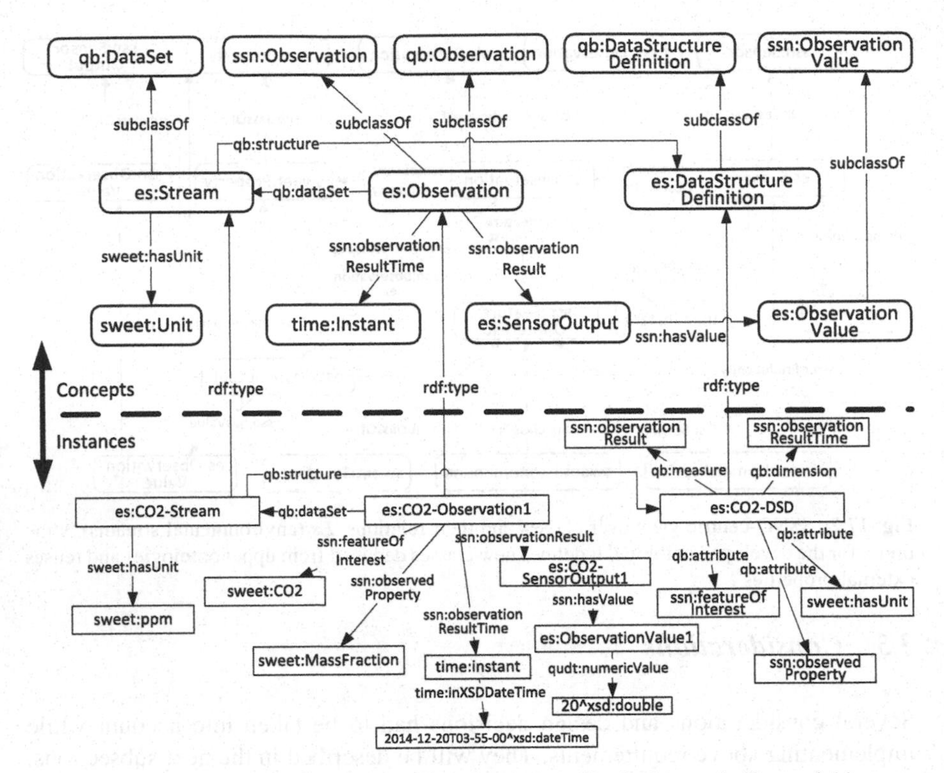

Fig. 17.3 QB-centric view over classes and their relations. On the *bottom half* potential instance data is visualized, i.e., how an actual data stream, observation and data structure definition is modeled based on the created vocabulary. Note the absence of location data (*geo:hasGeometry*) at the observation and the encoding of the unit at the data stream instance (*es:CO2-Stream*). For details see Subsection "Storing Spatial and Temporal Observation Data"

Table 17.2 Comparison of supported observation properties of the QUDT and SWEET ontologies

	QUDT	SWEET
Temperature	Y	Y
Length	Y	Y
Depth	N	Y
Height	N	Y
Pressure	Y	Y
Angle	Y	Y
Direction	N	Y
Speed	Y	Y

in SWEET (and related to *sweet:property*)—which may enable intelligent reasoning and querying mechanisms—led us to the decision to use SWEET.

GeoSPARQL vs. WGS84 Several ontologies for encoding spatial data are available. Due to its simplicity, the WGS84 ontology[6] appears to be very popular. This

[6] See http://www.w3.org/2003/01/geo/ (accessed 23 February 2015).

simplicity, however, comes with the tradeoff of being less expressive. In contrast, the GeoSPARQL ontology[7] is an official Open Geospatial Consortium (OGC) standard and allows for more complex spatial encodings such as lines or polygons. This enables queries such as *"Provide observations from all sensors in the city of Vienna"* given the coordinates of Vienna are described as a polygon. GeoSPARQL also comes with a query language that allows, for instance, queries based on spatial relations.

3.3.2 Reusing Ontologies

There are two options for ontology reuse: either through the *owl:imports* statement, or by redeclaring external classes and properties to be reused.

Using *owl:imports* makes the entire external ontology a semantic part of the new ontology. This may not always be desired behavior, for instance, if large and complex ontologies are imported. Moreover, *owl:imports* is transitive, that is, if ontology *A* imports ontology *B*, and *B* imports *C*, then *A* imports both *B* and *C*. Because we aim to design a lightweight ontology (cf. *R3*), we decided to import small ontologies only. In cases where we want to reuse terms of large ontologies, or if we just pick particular terms, we simply redeclare them.

Avoiding *owl:imports* for large ontologies has the advantage that semantic web tools such as reasoners do not need to load all referenced ontologies when working with our vocabulary. The redeclarations are useful for OWL tools which require proper definitions of used classes and properties. Linked Data applications can use the URIs of the definitions for dereferencing. Overall, *owl:imports* appears to be more suitable for reuse of all axioms of an external ontology to the end of rich inferencing. However, as our goal is to reuse existing and well-known terms to facilitate interoperability, declaring them is enough.[8]

Most importantly, we use redeclarations of the following classes: *sweet:HumanActivity*, *sweet:Phenomena*, and *sweet:Substance*. These classes are furthermore encoded as subclasses of *ssn:FeatureOfInterest* meaning that the respective SWEET classes are used as features of interest for the actual instance data. Similarly, *sweet:Property* is aligned *with ssn:Property*. From the Time Ontology in OWL we declare *time:Instant* and *time:inXSDDateTime* to capture temporal data, i.e., when an observation has been made. From the GeoSPARQL Ontology we reuse *geo:hasGeometry* and *geo:Geometry* to store spatial information, i.e., locations of sensors. *Qudt:numericValue* is used to encode actual sensor measurements. We use *owl:imports* only for QB and for SSNO because we reuse a large fraction of these ontologies.

[7] See http://www.geosparql.org/ (accessed 23 February 2015).

[8] On semanticweb.com different ways of reusing terms when designing an ontology have been discussed and explained (see http://answers.semanticweb.com/questions/18505/ontology-import-vs-owlsameas-in-ontology-design, accessed 23 February 2015).

3.3.3 Storing Spatial and Temporal Observation Data

To store spatial (i.e., locations of observations or sensors) and temporal data (i.e., when an observation has been made), we had to make several decisions.

We initially considered attaching location data, i.e., the location where an observation has been made, directly at the observation level. However, to conform with SSNO, we decided to do this at the sensor level via *geo:hasGeometry* (see Fig. 17.2). This approach is less redundant because as of now we do not capture any data of moving sensors. In other words, the location of sensors is static, hence, there is no reason to attach the same location information for one data stream to each observation again and again. Moreover, the sensor is attached to each observation through *ssn:madeObservation* anyway, meaning that the location of an observation can easily be retrieved.

A similar rationale applies for the encoding of units of measurement. Those are stored at the data set level (*es:DataStream*) via the *sweet:hasUnit* property. Each *es:DataStream* is related to exactly one *qb:DataStructureDefinition*. Therefore, it generates observations of only a single property and the unit of measurement does not change over time. Observations also only observe a single property at a time, hence, there is no need to encode multiple units of measurement for a data stream.[9] The official standardization document of QB also explains a special *qb:componentAttachement* property which can be used to attach certain attributes, for instance the unit of measurement, to the whole data set in order to avoid redundancy.

Generally QB allows to encode multiple measures at a single observation. For instance, an air sensor could observe *temperature* and *wind speed*. The vocabulary provides two mechanisms to implement this, namely the *multi-measure observations* and the *measure dimension* approaches. However, because we want to comply to both the SSNO and QB and because it is not feasible to create observations in SSNO which capture multiple observed properties, we decide to model single property observations.

3.3.4 Defining the Measure Property According to QB

When working with QB, Data Structure Definitions (DSDs) need to be defined. The actual observation data is stored according to the structure defined in the DSD. We have to tradeoff interoperability and flexibility when combining QB and SSNO.

Usually in the DSD the data provider uses *qb:measure* to create a relationship to a class of type *qb:MeasureProperty* which in the observation data will materialize as a relation to the actual measured value. For instance *eg:lifeExpectancy* may be

[9] There has been some discussion about how and if multiple properties can be encoded into a single observation of SSNO which influenced our decision to also only create single property observations (see http://lists.w3.org/Archives/Public/public-xg-ssn/2014Apr/0007.html, accessed at 23 February 2015).

declared as a *qb:MeasureProperty*. An example observation will then, e.g., contain the triple *eg:Observation eg:lifeExcpectancy "80"*. However, to comply to both SSNO and QB, we declare *ssn:observationResult* as a *qb:MeasureProperty* in the DSD. Starting from an observation, the actual measurement can then be queried via the classes *ssn:SensorOutput* and *ssn:ObservationValue*.

4 Architecture

After describing and discussing the fundamental ontology of the presented platform, we continue to present its overall architecture. This platform for data exploitation that the contributions of this paper are built upon is called *Linked Widgets Platform*. The term *Linked Widgets* was introduced by Trinh et al. [22] to describe an extension of standard widgets [4] with a semantic model, following the Linked Data principles. This semantic model describes the input and output graph of widgets and facilitates discovery and composition of widgets into mashups. The current paper extends the architecture of the platform [23] by introducing stream processing mechanisms embodied in *Linked Streaming Widgets*.

Our framework is based upon semantic annotations that describe the data using domain vocabularies that can be used to integrate heterogeneous environmental data. Whereas the design of this platform is domain-agnostic, we focus on real-time environmental data and the particular challenges and requirements that arise in this area.

Figure 17.4 illustrates the architecture, including extensions that make the platform suitable for stream data. The constituent components can be grouped into three stages, i.e., (i) data acquisition responsible for tying in polling- and streaming-based data sources, (ii) data transformation where raw data is converted

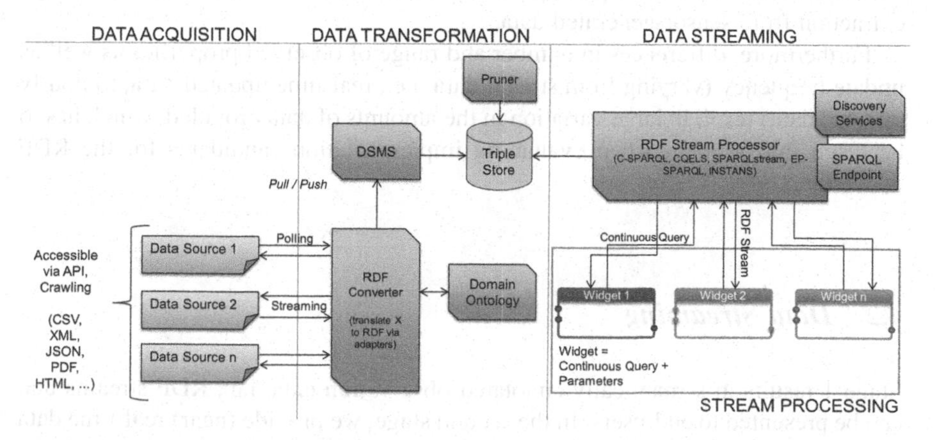

Fig. 17.4 Linked streaming widgets architecture

into time-annotated RDF triples, and (iii) data streaming which provides streams to end-user applications.

Linked Streaming Widgets are used to register continuous queries with user-defined parameters at the RDF stream processor and to handle the resulting data stream.

4.1 Data Acquisition and Data Transformation

Environmental data is available from various repositories, each providing unstructured, semi-structured, or structured data. In many cases, data is presented only on a webpage or via non-standardized interfaces. To allow for timely provision of data via our platform, such data has to be crawled on a regular basis. Data available in (semi)structured formats is more straight-forward to handle, but still needs to be converted into JSON-LD, a recent W3C recommendation [18] that we use as our internal data exchange format.

After conversion, the data is fed into a Data Stream Management System (DSMS) and the triples are stored in a triple store. The RDF converter uses domain ontologies to enrich incoming data sources with semantic knowledge, which later will be utilized to support functionalities such as stream processing or contextualized sensor discovery. The DSMS is dependent upon the RDF stream processor implementation. Currently, we intend to use C-SPARQL as a stream processor which requires a semantic-aware DSMS, hence, the corresponding representation in Fig. 17.4.

Data sources differ in type (rdf, json, xml, csv, pdf, html) and access mechanism (API, file download, manual crawling). Ontologies have been used for many years as a means to overcome the resulting heterogeneity. In the context of our proposed framework they are a valuable tool to define a comprehensive and standardized semantic model which is a prerequisite for semantic search as well as knowledge extraction from sensor-generated data.

Furthermore, differences in number and range of observed properties as well as update frequency (varying from stream data, i.e., real-time updated data, to hourly updated data) result in large variation in the amounts of data provided, which has to be taken into account when evaluating implementation candidates for the RDF converter.

4.2 Data Streaming

Stage 1 results in semantically annotated observation data, i.e., RDF streams that can be presented to end-users. In the second stage, we provide (near) real-time data to the user.

We make use of the publish-subscribe design pattern, which controls what messages are sent by entities that publish data to receiving entities [8]. In the context of the proposed framework, loosely coupled widgets can act as publishers and subscribers. A key advantage of this approach is that through parallel operations, message caching, and routing, this pattern provides the scalability needed to handle flexible stream compositions on our platform. Consequently, it solves the first step in providing environmental data streams to users by allowing clients to subscribe to data streams dynamically.

Furthermore, due to the continuity and large size of data streams, storage is a key issue. To avoid bottlenecks in subsequent procedural steps, we need to define when data becomes outdated and can be deleted.

The architecture supports flexible exploration of the data as depicted in the stream processing section of Fig. 17.4. We achieve this by allowing users to combine widgets in order to answer questions related to environmental data. Via drag and drop, these widgets can be combined into mashups. A mashup can satisfy information needs, e.g., by displaying points of interest that satisfy certain air quality criteria. Widgets leverage the modeled semantics and can be combined in many different ways.

We apply stream reasoning techniques provided through SPARQL extensions, i.e., windowing functions and federation of static data with dynamic streams. Each corresponding data stream is represented by a widget. A web-based graphical interface allows users to assemble these widgets and set parameters for their processing functions. In doing so, users can efficiently explore data streams.

These processing widgets have encoded queries based on stream-specific criteria, e.g., time windows or aggregates (*sum*, *count*, *average*, etc.), and therefore return RDF triples that answer this query, ultimately allowing hands-on combination of data streams. Presentation widgets provide mechanisms to visualize the intended output via, for instance, maps, bar charts, line charts, pie charts, or histograms. This step covers three aspects of leveraging data streams: (i) analyzing via continuous stream queries, (ii) publishing via returning RDF graphs, and (iii) visualizing via corresponding presentation interfaces.

4.3 Semantic Modeling of Stream Data

The semantic model acts as a component which is used to annotate data streams based on appropriate domain ontologies. For the environmental domain, we identify special vocabularies and investigate integration into our framework as follows.

Since ontology reuse is one important principle of the Semantic Web vision, we use existing ontologies in the field of sensors and measurements. Numerous ontologies were proposed with the goal to model sensor observations. Some approaches stand out: First, the Semantic Sensor Network ontology [6], which is the result of the Semantic Sensor Networks Incubator Group at W3C, aims at a top-down approach to model whole sensor networks. Second, the RDF Data Cube

Vocabulary[10] has been widely adopted since its promotion to a W3C Recommendation. This vocabulary is designed for modeling observations and measurements.

In our work, we created a new vocabulary that combines these two modeling approaches. There is some overlap in the available concepts of both ontologies (e.g., *observations* and *properties*) which can be used to link them together. For more details regarding the design of this ontology please refer to Sect. 3.

5 Mashup Based on Linked Streaming Widgets

Figure 17.5 displays an example of a mashup that uses air quality data streams, i.e., carbon monoxide and ozone, as an input. The widgets on the left hand side act as a data source. They are used to register a continuous query at the stream processor (see Fig. 17.4) based on user-defined parameters. The size of the window (range) and the update frequency (step) can be specified. Moreover, the user can decide whether the returned values of the query should be aggregated (min, max, average). The *Stream Merger* is needed to fuse two data streams into a single result stream that can be handled by different visualization widgets, i.e., in this case the *Line Chart* and *Google Maps Widget*. The fusion process can also be used to apply additional processing steps, e.g., transformation, aggregation, or enrichment of the incoming streams.

By applying this approach to environmental data streams, the platform can focus more on the needs and interests of users, e.g., streams can be discovered and used based on contextual information extracted from the stream's semantics. As a result, the user may discover data in his/her proximity, based on his/her interests, time

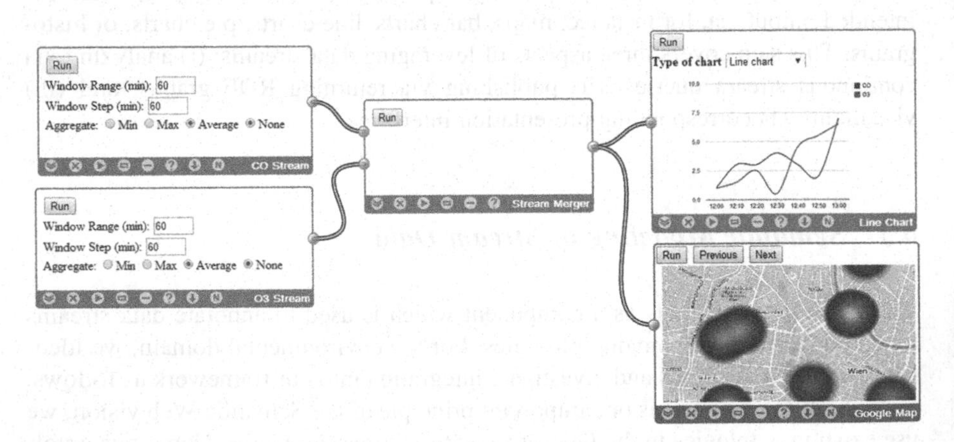

Fig. 17.5 Example of a mashup based on streaming data

[10] See http://www.w3.org/TR/vocab-data-cube/ (accessed 23 February 2015).

constraints, etc. and combinations of these (e.g., air quality sensor observations of the last 30 min within 100 m of the user). Discovery based on current values, aggregates (*sum*, *median*, *mean*, *mode*, *min*, *max*, etc.) or trends (increasing, decreasing, or stagnating) is also possible. For instance, one may be interested in analyzing and comparing pollution data (e.g., air, water, noise) near his/her apartment based on a daily or hourly basis to discover dynamics in the data.

6 Conclusions and Future Work

In this paper, we propose a widget-based framework for the exploration of environmental data streams in an urban context. We divide the approach into three stages and identify important issues that need to be addressed. These include defining a new vocabulary for environmental stream data deduced from already existing and well-adopted ontologies, and applying semantic stream processing methods to facilitate reasoning. We outline the architecture of the platform and discuss a prototypical mashup (i.e., interconnected widgets) example.

We contribute to the state of the art by designing an ontology for the modeling of environmental data streams. We evaluate several potential ontologies for reuse. The resulting vocabulary introduces new concepts by aligning and harmonizing terms of both the Semantic Sensor Network Ontology and RDF Data Cube vocabulary. This enables interoperability and allows to reuse recommended query patterns and best practices which are defined for both ontologies.

Recently, Open Government and eGovernment initiatives, Open Data Portals, and the development towards open provision of public sector information have been growing. As a consequence, many data sources containing environmental information are available to the public.[11,12] Our platform aims to facilitate access to and reuse of these sources. At present, most of these sources provide only infrequent snapshots of static data, we expect that this situation will improve in the future which will facilitate innovative applications in the environmental domain. In the long term, the proposed system could serve as an open data platform for citizens of a "smart city".

The Linked Widgets Platform shall bring together both mashup developers and mashup users. For each of them, it should be as easy as possible to create, (re)use, modify, and execute mashups. By overcoming technical barriers of adoption, citizens will be enabled to interact with the available data sources, e.g., open data, linked data, tabular data while accessing data in different formats. New knowledge can be created by enabling creative (re)combination of available data. The vision is to provide a platform for dynamically building applications that

[11] Task Force Environmental Information: http://www.ref.gv.at/Umweltinformation.1024.0.html (accessed 23 February 2015).

[12] Open Data Vienna: https://open.wien.gv.at/site/open-data/ (accessed 23 February 2015).

leverage semantically enriched environmental data in a timely manner. Ultimately, this could lead to a better understanding of the environment in a local context of a city.

Future work will include implementation of a richer user interface that covers a larger number of use cases. Correspondingly, additional data sources and data input for the platform will be made available and integrated. Furthermore, we will need to find means to combine different types of data. We will also develop mechanisms to decide how long outdated triples will be stored and when they will be pruned. Balancing this tradeoff between being able to compare current values with historic data and the detrimental effects on performance represents an interesting challenge. Discovery Services for finding relevant sensors and data streams will be crucial as well. In addition, as the RDF Stream Processing Group at the W3C is currently making progress towards defining a standard model for RDF stream data, we will follow this process closely.

Future work on the vocabulary will focus on the definition of slices to group observations by fixing the values of certain dimensions. Moreover, the proposed ontology will be evaluated according to criteria of the respective standardization documents. For QB, so-called well-formed cubes have been defined. They conform to a set of integrity constraints that can be evaluated. Similar evaluations need to be found for SSNO. Finally the evaluation will determine the quality and integrity of the proposed alignment.

References

1. D. Anicic et al., "EP-SPARQL: A Unified Language for Event Processing and Stream Reasoning," in *Proc. of the 20th International Conference on World Wide Web*, New York, NY, USA, 2011, pp. 635–644.
2. M. Balduini et al., "Social listening of city scale events using the streaming linked data framework," in *The Semantic Web – ISWC 2013 – 12th International Semantic Web Conference Proceedings Part II*, Springer, 2013, pp. 1–16.
3. D. F. Barbieri et al., "Querying RDF Streams with C-SPARQL," *SIGMOD Record*, vol. 39, no. 1, 2010, pp. 20–26.
4. M. Cáceres, "Packaged Web Apps (Widgets) – Packaging and XML Configuration," W3C Recomm., 2012.
5. J.-P. Calbimonte et al., "Enabling Ontology-based Access to Streaming Data Sources," in *The Semantic Web – ISWC 2010 – 9th International Semantic Web Conference Part I*, 2010, pp. 96–111.
6. M. Compton et al., "The SSN ontology of the W3C semantic sensor network incubator group," *Web Semantics: Science, Services and Agents on the World Wide Web*, vol. 17, 2012, pp. 25–32.
7. Directive number 4 of 2003, *OJ of the EU*, L 41, pp. 26–32, 2003.
8. P. T. Eugster et al., "The many faces of publish/subscribe," *ACM Comput. Surv. CSUR*, vol. 35, no. 2, 2003, pp. 114–131.
9. G. Huang and N. Chang, "The perspectives of environmental informatics and systems analysis," *J. Environ. Inform.*, vol. 1, no. 1, 2003, pp. 1–7.

10. F. Lécué et al., "Predicting Severity of Road Traffic Congestion Using Semantic Web Technologies," in *Proc. of the 11th Extended Semantic Web Conf.*, 2014, pp. 611–627.
11. L. Lefort et al., "A linked sensor data cube for a 100 year homogenised daily temperature dataset," in *5th International Workshop on Semantic Sensor Networks (SSN-2012)*, CEUR-Proceedings, vol. 904, 2012, pp. 1–16.
12. L. Lefort et al., "The ACORN-SAT Linked Climate Dataset," 2013.
13. D. Le-Phuoc et al., "A Native and Adaptive Approach for Unified Processing of Linked Streams and Linked Data," in *The Semantic Web – ISWC 2011 – 10th International Semantic Web Conference Part I*, 2011, pp. 370–388.
14. R. G. Raskin and M. J. Pan, "Knowledge representation in the semantic web for Earth and environmental terminology (SWEET)," *Comput. Geosci.*, vol. 31, no. 9, 2005, pp. 1119–1125.
15. B. Resch et al., "Towards the live city–paving the way to real-time urbanism," *Int. J. Adv. Intell. Syst., vol. 5, no. 3 and 4*, 2012, pp. 470–482.
16. M. Rinne et al., "INSTANS: High-Performance Event Processing with Standard RDF and SPARQL," in *Proc. of the ISWC 2012 Posters & Demonstrations Track*, 2012, vol. 914.
17. H. J. Schellnhuber et al., "World in transition: a social contract for sustainability," German Advisory Council on Global Change, 2011.
18. M. Sporny et al., "JSON-LD 1.0 – A JSON based Serialization for Linked Data," W3C Recomm., 2014.
19. M. Stocker et al., "Towards an Ontology for Situation Assessment in Environmental Monitoring," in *Proc. of the 7th International Congress on Environmental Modelling and Software*, 2014, pp. 1281–1288.
20. S. Tallevi-Diotallevi et al., "Real-Time Urban Monitoring in Dublin Using Semantic and Stream Technologies," in *The Semantic Web – ISWC 2013 – 12th International Semantic Web Conference Proceedings Part II*, 2013, pp. 178–194.
21. T. Tarasova et al., "Semantically-Enabled Environmental Data Discovery and Integration: Demonstration Using the Iceland Volcano Use Case," in *Knowledge Engineering and the Semantic Web*, 2013, pp. 289–297.
22. T.-D. Trinh et al., "Linked Widgets-An Approach to Exploit Open Government Data," in *Proc. of the 15th Int. Conf. on Information Integration and Web-based Applications & Services*, 2013, pp. 438–442.
23. T.-D. Trinh et al., "Open Linked Widgets Mashup Platform," in *Proceedings of the AI Mashup Challenge 2014 (ESWC Satellite Event)*, 2014, p. 9.
24. P. Wetz et al., "Towards an Environmental Information System for Semantic Stream Data," in *28th International Conf. on Informatics for Environmental Protection: ICT for Energy Efficiency, EnviroInfo 2014*, 2014, pp. 637–644.

Chapter 18
How a Computational Method Can Help to Improve the Quality of River Flood Prediction by Simulation

Adriana Gaudiani, Emilio Luque, Pablo García, Mariano Re, Marcelo Naiouf, and Armando De Giusti

Abstract High performance computing has become a fundamental technology essential for computer simulation. Modelling and computational simulation provide powerful tools which enable flood event forecasting. In order to reduce flood damage, we have developed a methodology focused on enhancing a flood simulator minimizing the number of errors between simulated and observed results by using a two-phase optimization methodology via simulation. In this research, we implemented this approach to find the best solution or adjusted set of simulator input parameters. As a result of this, we achieved an improvement of up to 14 % which, for example, represents a significant difference of 0.5–1 m of water level along whole Paraná River basin. In order to find the adjusted set of input parameters, we reduced the search space using a Monte Carlo + clustering K-Means method. Therefore, an exhaustive search over the reduced search space led us to get a "good solution". In summary, we propose add an improvement process on the classical computer model output to improve model quality.

Keywords Flood simulation • Simulator tuning • Optimization methodology • Parametric simulation

A. Gaudiani (✉)
Science Institute, Universidad Nacional de General Sarmiento, Buenos Aires, Argentina

Informatics Research Institute LIDI, Universidad Nacional de La Plata, Buenos Aires, Argentina
e-mail: agaudi@ungs.edu.ar

E. Luque
Computer Architecture and Operating Systems Department, Universidad Autónoma de Barcelona, Barcelona, España

P. García • M. Re
Hydraulic Computational Program, Hydraulic Laboratory, National Institute of Water, Buenos Aires, Argentina

M. Naiouf • A. De Giusti
Informatics Research Institute LIDI, Universidad Nacional de La Plata, Buenos Aires, Argentina

J. Marx Gómez et al. (eds.), *Advances and New Trends in Environmental and Energy Informatics*, Progress in IS, DOI 10.1007/978-3-319-23455-7_18

1 Motivation

Flooding is one of the most common natural hazards faced by human society. Future climate change and its impact on flood frequency and damage makes this problem a serious environmental problem. Flood damage refers to all varieties of harm caused by flooding.

A flood event in natural channels consists of a low or high amplitude flood wave passing through a river reach. Hill slopes and tributary catchments adjacent to the floodplain contribute with received precipitation to flood flows. River modelling plays an essential role in river management decision. In engineering in general and in river management in particular, hydrodynamic models are used to predict flood water levels that occur due to flood wave propagation. The flood wave transmission should be predicted as accurately as possible. River models describe the interactions between bed topography and water motion in a simplified way, but these processes are highly complex [1].

Predictions of simulation flood extents are possible by advances in numerical modelling techniques and the growth of computer power [2]. Nevertheless, a series of limitations cause a lack of accuracy in forecasting, such as the case of uncertainty in the values of the input parameters to the flood model. Hydrodynamic modelling of a fluvial channel involves defining certain parameters as input variables that, for various reasons, may incorporate uncertainties in the results [3]. The parameters' uncertainty has an important impact on the simulation output, which is far from approaching the actual observed data [4].

To overcome this problem, we implemented in our first work [5], a parametric simulation in order to find the best set of parameters. These adjusted parameters will be used as the input set for the underlying flood simulator, emulating an "ideal" flood simulator as much as possible. The main objective of this work is to add an optimization process to the classical prediction approach to tune input parameters, in order to minimize the difference between a real and a simulated result. The optimization method results in a large number of scenarios that allow us to search for the optimal, or suboptimal, set of input parameters. This process requires a huge amount of computations because we need to run as many simulation scenarios required by a parametric simulation as possible, which was carried out with resources in parallel programming provided by high performance computing.

Our work takes advantage of the results performed by the research group of High Performance Computing for Efficient Applications & Simulation at the Autonomous University of Barcelona [6], with the close collaboration of the hydraulic engineering team at the National Institute of Water of Argentina. To conduct the research, we selected the computational model EZEIZA V (Ezeiza). This program is currently used as one of the tools of the Hydrologic Alert and Information System of the National Institute of Water (INA) at Buenos Aires, Argentina [7]. The water resources specialists at INA make use of these computational tools to alert, as early as possible, on the occurrence of extreme water level events in the Paraná River basin, in South America [8]. The main limitations of this computational model are related to the reduced scale of problem resolution (1D) and the inaccuracies in the river geometry representation [9]. The model's challenges are related to the uncertainty reduction in determining the flood peak arrival. This work goes in this direction by providing a better model calibration.

2 The Flood Wave Simulator

2.1 Modelling Overview

A model is a conceptualization of the real system of interest. In other words, a model is a simplification of reality. A good model should achieve a balance between close approximations to the real system, ahead an oversimplified model that ignores important relationships. The conceptual model only takes into consideration conceptual issues. Issues such as mathematical representation and system characteristics propagation are key when the mathematical model is developed. The computational model is the conceptual model implemented on a computer [10, 11].

Computer models discretise the topographical river and estimate flood data (depth, height, velocity). Calculation of flood scope is done using computational models based on forms of shallow water equations. Different numerical schemes for these equations are discussed in [12].

The computational simulations are used extensively as models of real systems to evaluate output responses. In particular, computational models are used to attain a better understanding on inundation events and to estimate flood depth and inundation extent. For these reasons, simulation becomes a powerful tool for predicting flood events and minimizing their environmental effects [13].

2.2 Model Uncertainties

Every phase of modelling and simulation is associated with a source of errors, for example modelling errors and other uncertainties. A full understanding of this problem is crucial for a meaningful interpretation of model outcomes.

One source of uncertainty is the sampling of the cross section area during floods, which can make it difficult to measure the distribution of the wave velocity. Another source of uncertainty are errors in the measuring instruments [14]. Some errors are introduced in the modelling or simulation process, as happens with the modelling simplification of a physical system. A complete identification of river model uncertainties is given in [15]. In [16], the authors' aim is to evaluate the effect of the uncertainty contained in flood risk models, which propagates through the calculation in the final damage estimate.

We mentioned above some sources of uncertainty that affect model parameters and input data. Going into detail, model parameters are measured or estimated in certain particular points, but this is not a direct measurement. Therefore, the value of such parameters involves an estimation error associated with the estimation methodology [3] and must be interpolated to the whole domain. For example, levees' height can be measured in some sections but then it is necessary to estimate the heights for the other sections. The geometry of gauging cross-sections is assumed stable in time, but high flood events cause significant changes. These events prevent discharge measurements and introduce significant interpolation and extrapolation errors, as it is explained in [17].

We assess the overall performance of the flood computer model considering the difference between the predicted model outcome and the observed hydrologic variables. In this paper, we face parameter uncertainties by designing an optimization method to reduce model output error.

2.3 *The Computational Model*

Our work starts using a computer model of a real system such as flood events. The selected software, Ezeiza, is a computational implementation of a one-dimensional hydrodynamic model for a flow net, based on the Saint Venant equations [7].

The Ezeiza software family started its development in the 1970s and the INA staff performs its ongoing updating. This computational model was mainly chosen because it results from the efforts of INA's research team, which aims to achieve flood events prediction and to develop a regional early warning system at La Plata River Basin[1] in Argentina. Ezeiza offers simplicity when exporting results to output files, which can be processed by statistical and/or mathematical software, and for its convenience when running parametric simulations by changing the parameter values in the input files [8]. These features are very useful to take the tuning methodology forward.

An exhaustive study of the Paraná River model performance, which was carried out by INA, underlines the need to improve Ezeiza simulated results [18]. The utility of several efficiency criteria to evaluate hydrological performance model is addressed in [19].

[1] http://www.ina.gov.ar/alerta/index.php

3 Paraná River Model

La Plata basin is one of the most important river systems in the world. The Paraná River is one of the main rivers that form the basin. This is the second longest river of South America and it has a length of 4000 km alongside its major tributary, the Paraguay River (2550 km). The stretch of the Paraná River simulated by Ezeiza extends between the Yacyretá dam (Corrientes) to Villa Constitución (Santa Fe), both in Argentina. The Paraguay River runs from Puerto Pilcomayo (Formosa) to its confluence with the Paraná. Both river basins were divided into a number of sections, to measure river flow or height in each of them.

Large areas of land along the Middle and Lower Paraná margins are frequently subject to extended floods, which cause considerable damage. During the highest floods, monthly discharges at Middle Paraná exceed twice, or even three times, the mean discharge. A complete description of the highest floods at the Paraná basin and the possible climate forcing of such events are shown in [20].

Ezeiza is used to forecast daily water level variations at the Paraná River basin. The data required to define the modelled river system, as shown in Fig. 18.1, are the following:

- Initial conditions: water levels and flow at every point of the river's domain.
- Boundary conditions: time series corresponding to river levels and flow at upstream and downstream points.
- Geometry data: Information on the rivers topography.
- Input Parameters: Manning values and levees' height corresponding to every river section.

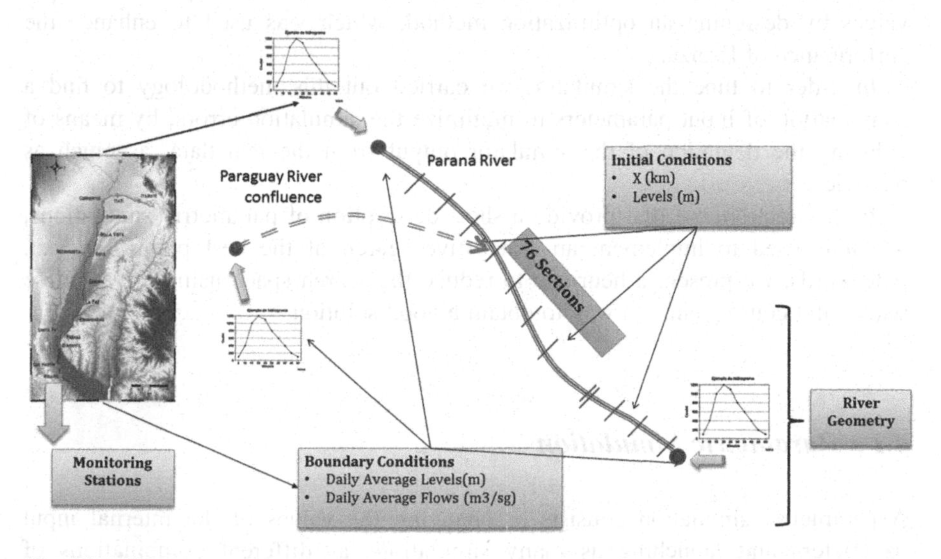

Fig. 18.1 Flow net topology

- Observed data: water heights of Paraná River measured at each monitoring station.

INA provided this information, also including the observed (actual) data from 1994 to 2011. During this period, daily heights were measured at 15 monitoring stations placed along the Paraná River basin.

Despite following 15 control stations, the levels are modelled throughout the length of the river. To define the flooding areas, the information of the modelled height should be crossed with ground levels. In general, every city has predefined levels of warning and evacuation [18].

When a simulation is launched, the simulator Ezeiza returns a time series of height values, which are calculated at each one of the 15 monitoring stations. This simulated data and the observed data, for the simulated period, are compared with each other to determine a *similarity index* (SI) which will be used to measure the simulation accuracy. We will go over this issue further on.

The more sensitive input parameters of the flood routing models are the roughness coefficient (Manning values) and the levees' height. Manning values for flood plains can be quite different from values for channels; therefore, Manning values for flood plains are determined independently from Manning values for channel [21]. Finding an adjusted set of parameters is a key issue for our work and this step is required for developing an optimization methodology.

4 The Optimization Scheme

We overcome the difficulty of providing the model with accurate parameter input values by designing an optimization method, which was used to enhance the performance of Ezeiza.

In order to tune the simulator, we carried out this methodology to find a combination of input parameters to minimize the simulation errors, by means of reducing the deviation of the simulator output from the real data, as much as possible.

In this section, we first provide a short description of parametric simulations, which is used to implement an exhaustive search of the best parameters set. Afterwards, we present a heuristic to reduce the search space using a two-phase search and consequently, a way to obtain a good solution.

4.1 Parametric Simulation

A parametric simulation consists of changing the values of the internal input parameters and launching as many simulations as different combinations of parameter values are possible. This allows to change deliberately the parameter values. A scenario is defined by a particular setting of the set of parameters.

The number of possible scenarios is determined by the cardinality, C_i, for each of the N parameters considered. For every parameter i we define an associated interval and an increment value, which are used to move throughout the interval. The associated domain is determined by an upper and lower bound and a step size to partition such interval, as we show in Eq. (18.1):

$$\left\langle \left[limit_{inf}^{i}, limit_{sup}^{i}\right], Step_i \right\rangle, \forall i \in [0..N] \quad (18.1)$$

In Eq. (18.2), *#Scenarios* represents the calculation of the total number of workload scenarios that result from performing all the possible combinations of parameter values. We show in Eq. (18.3) the cardinality expression for the parameter i. As we perform an exhaustive parametric simulation in this phase, we define each new workload scenario by changing a single parameter, leaving the other fixed.

$$\#Scenarios = \prod_{i=1}^{N} C_i \quad (18.2)$$

$$C_i = \left(\left(limit_{sup} - limit_{inf}\right) + Step_i\right) / Step_i \quad (18.3)$$

Paraná River basin, which represents the model domain, was divided into 76 sections in order to measure river flow and height in each of them. Each section is characterized by a Manning value for the floodplain, another Manning value for the riverbed and a levee height. Here we itemize the parameters' domain:

- Manning values for the floodplain are within the [0.1, 0.2] range, with an ideal step of 0.01
- Manning values for the riverbed are within the [0.015, 0.035] range, with an ideal step of 0.005.
- Levees' height is within the 5–50 m range with a step of 5 m. (The range and step value are set according to the local geography.)

The parametric simulation allows us to carry out an exhaustive search to find an adjusted set of parameters. As the search space is very large, a huge amount of computation is required. We reduce the search process by using an optimization technique.

4.2 Optimization Overview

Optimization is generally defined as the process of finding the best or optimal solution for a given problem under some conditions. Formal optimization is associated with the specification of a mathematical objective function (called f) and a collection of parameters that should be adjusted to optimize the objective function. Mathematically, an optimization problem can be stated as:

$$\begin{aligned} &\max / \min f(x) \\ &\text{subject to } x \in S \end{aligned} \tag{18.4}$$

Where x is the variable; f is the objective function ($f : S \rightarrow \mathbb{R}$); S is the constraint set, and $\exists\, x_0 \in S$ such that

- $f(x_0) \leq f(x) \quad \forall x \in S$ for minimization,
- $f(x_0) \geq f(x) \quad \forall x \in S$ for maximization.

In this work, the optimization process, expressed by Eq. (18.4), can be defined as follows:

$$\vec{x}^{*} = \left[x_1^{\,*}, x_2^{\,*}, \ldots, x_N^{\,*}\right] \tag{18.5}$$

where $\vec{x}^{*}$ is the N-dimensional parameters vector, which optimizes the Eq. (18.4), $\vec{x}^{*} \in S$ and the domain $S \subseteq \mathbb{R}^N$ represents the constraint set defining the allowed values for the $\vec{x}^{*}$ parameters.

In our problem, the search space consists of as many vectors as different combinations of parameter values possible. So, the S-space dimension states the number of possible scenarios. Furthermore, we have to define a process to find a setting for the parameter vector $\vec{x}$, which provides the best value for the objective function $f\left(\vec{x}^{*}\right)$.

When there is no explicit form of the objective function and the parameter settings or design variables are discrete valued, the optimization problems became discrete optimization via simulation problems. We use the results obtained in these subjects by an ongoing project that is being carried out by E. Cabrera et al., at the Research Group in Individual Oriented Modelling (IoM) in the Autonomous University of Barcelona [22, 23].

Our optimization problem is expressed mathematically in Eq. (18.6):

$$\textit{Minimize prediction error} \quad f(Q_{obs}, Q_{sim}) \tag{18.6}$$

subject to	$0.1 \leq MannPlain \leq 0.2$
	$0.01 \leq MannBed \leq 0.035$
	$0 \leq leveesHeight \leq 60$
	$Section \in \{\, selected\ sections\} \subseteq \{76\ sections\}$

The objective function is determined by the real or observed data, Q_{obs}, and the simulated data or simulator output, Q_{sim}. The restrictions must be satisfied by all possible candidate solutions and these conditions define the feasible area; where the optimal or suboptimal solution might be found by optimization techniques.

As mentioned above, a measure of the quality of the simulation outcome is used to implement the optimization methodology, providing a quantitative comparison among different scenarios, which is to say different solutions. We used the root mean square error (RMSE) as a metric to calculate the SI index, in order to evaluate the simulator response for each simulation scenario launched with Ezeiza, and to find the minimum SI. We described in detail the steps involved in this methodology in [5].

4.3 The Search Procedure

Our approach to handle the optimization problem intends to target a search problem within the feasible region. In an initial approach, the search for the optimum was carried out through an exhaustive search technique, even though it implies a lot of search time, it guarantees that the optimum is found, if it exists. Either the exhaustive search was carried out with a parametric simulation that results in a computationally expensive task, in particular when we need a greater search space. When we include more sections in the parameters vector, the computational cost grows exponentially in function of the number of sections considered. Facing this situation, a new alternative technique was explored.

In our previous work [24], we implemented a solution to the "search problem", by using a parametric simulation approach applied to a reduced search space. Running a full simulation under the established conditions lasted 2 min. When we run the 4096 scenarios, the execution time lasted 8192 min (137 h). We used a master-worker approach to parallelize the method and reduce the computing time; however, this solution is not sufficient when the dimension of the parameter vector grows. Therefore, a better approach optimization technique, rather than an exhaustive search, must be used.

The Monte Carlo (MC) heuristic is one of the most popular simulation scheme, even though the yield solution may not be the global optima, but rather an approximate good solution. MC is a statistical sampling method used to approximate solutions to quantitative problems [25]. Our approach, in a first phase, is using a computational process based on an iterative method of MC scheme, which is combined with a K-Means clustering method, in order to identify the regions where the optimum is [26]. The second phase consists in a reduced exhaustive search [23] implemented over a reduced search space. The main idea of this technique is represented in Fig. 18.2.

The selected scenarios that have a better mean SI value than the previous ones are accumulated, in order to reuse past information. The MC program stops when two consecutive iterations are unable to improve the SI value, i.e., it becomes stationary or asymptotic, and the MC stops when the prediction error cannot be improved by the method. During each MC iteration step, a set of 100 simulation scenarios are randomly selected from the whole search space, each of them is used to launch an execution of the simulator Ezeiza. To evaluate the improvement

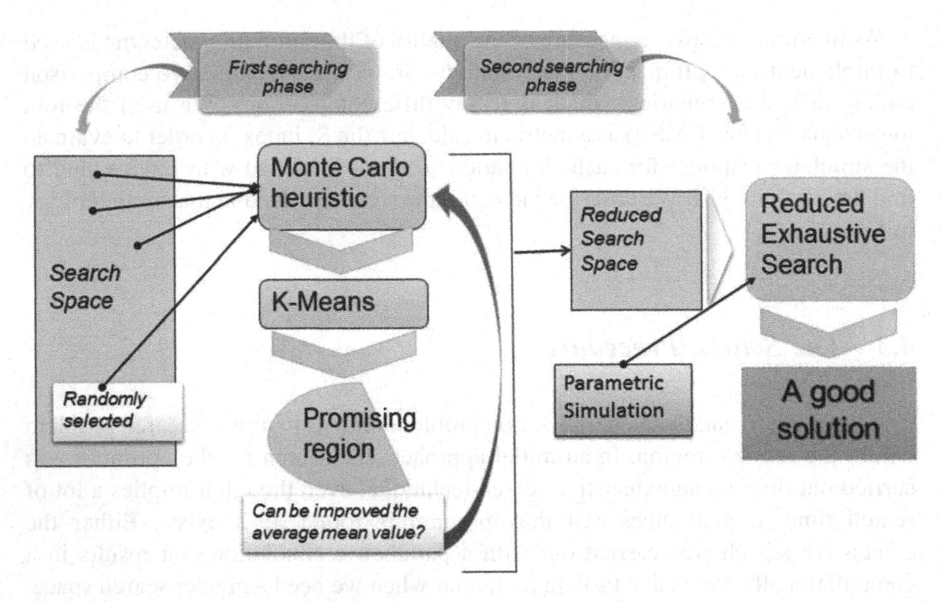

Fig. 18.2 The two-phase searching procedure

achieved by the method, we measure the SI index for each scenario, getting an error measure between simulated results and observed data at each monitory station and taking into consideration the complete time series.

The final SI value is the mean of the RMSE calculated in the 15 output stations. The best SI value is compared with the SI value reached by running Ezeiza with the INA scenario. This scenario configuration is given by the parameters values used currently by INA to forecast the Parana River water levels with Ezeiza. The INA simulated results are the reference point to get the improvement rate achieved from MC method. We calculate this rate in Eq. (18.6).

$$improvement_{station} = \frac{abs(SI_{INA} - SI_{Scenario^*})}{SI_{INA}} \tag{18.7}$$

Where $Scenario^*$ represents the best scenario, $SI_{Scenario^*}$ is the minimum prediction error and SI_{INA} is the INA prediction error.

4.4 Problem Delineation

An exhaustive search always guarantees finding a solution, if the solution exists. In order to determine that solution, it may be necessary to test each possibility and verify whether it satisfies the statement of the problem or not. The computational cost is high because it is proportional to the search space dimension. The whole search space for this model's parameters is determined by: (a) the parameters

Table 18.1 Increasing growth of scenarios depending on the modified sections

Cardinality		Number of sections considered					
M-F	M-R	2	4	6	8	10	76
2	4	64	4096	262,144	1.7E+07	1.1E+09	4.3E+68
3	5	225	50,625	1.1E+07	2.6E+09	5.8E+11	2.4E+89
4	5	400	160,000	6.4E+07	2.6E+10	1.0E+13	7.6E+98
5	5	625	390,625	2.4E+08	1.5E+11	9.5E+13	1.8E+106

Table 18.2 Domain cardinality for Manning values

Manning	Interval	Cardinality	C value
Floodplain	<[0.1, 0.2], 0.1>	((0.2−0.1)+0.1)/0.1	2
Riverbed	<[0.010, 0.04], 0.01>	((0.04−0.01)+0.01)/0.01	4

corresponding to 76 sections along the river basin, (b) considering that each section is divided into 3–5 subsections and (c) the parameters domain, as we described in a previous section.

We combined only the Manning values, leaving aside levee heights. On this basis, if we had implemented an exhaustive search to find the adjusted parameters we would have launched 112^{76} simulations; this means that Ezeiza should have been executed 10^{154} times. Table 18.1 shows the growing number of scenarios according to the parameters cardinality. The parameters considered are floodplain Manning (M-F) and riverbed Manning (M-R) coefficients.

With the aim of reducing the search space we had implemented a parametric simulation algorithm, reducing the domain dimension by combining the possible Manning values in only four sections: 70–72–74 and 76. The domain experimentation cardinality, which is shown in Table 18.2, was calculated using Eq. (18.3). We run the simulator 4096 times: $(4 \times 2)^4$. Even with this reduced setting of the parameter vector dimension, the search space is large. The observed data and the simulated data are a time series of daily river heights at each monitoring station. The simulated period was 365 days (the year 1999).

The SI index provides a metric to select the best scenarios. Each scenario configuration is represented by the objective function, and this function depends on the vectors, whose components are the simulated and the observed data respectively, for each output station and for each simulation day. The restrictions are the possible ranges of values that parameters can take, and they are shown in Eq. (18.6).

5 Experimental Results

The simulator is used as a black box, even though the simulator is more realistic. We used the reduced search space of 4096 scenarios, which is the same domain configuration as the one used in our previous work. The same period was used to

Table 18.3 The best scenarios selected by the first phase of the optimization scheme

	Section 70		Section 72		Section 74		Section 76		Improv.
Scenario	M-R	M-F	M-R	M-F	M-R	M-F	M-R	M-F	Rate (%)
Sce-1	0.02	0.1	0.02	0.1	0.03	0.2	0.03	0.1	14
Sce-2	0.02	0.2	0.02	0.1	0.03	0.2	0.03	0.1	13
Sce-3	0.03	0.1	0.02	0.1	0.03	0.1	0.03	0.1	12
Sce-4	0.02	0.2	0.02	0.2	0.02	0.1	0.03	0.1	11
Sce-5	0.02	0.1	0.02	0.2	0.03	0.1	0.03	0.1	14

Table 18.4 Restrictions to the parameters values for the second phase

	S. 70	S. 72	S. 74	S. 76
M-R	0.02, 0.035, 0.03	0.02	0.02, 0.025, 0.03	0.03
M-F	0.1, 0.2	0.1, 0.2	0.1, 0.2	0.1

carry on the simulation and four sections were selected, which are located at the lower Paraná. We reused the same conditions to evaluate the utility and reliability of this optimization method and to compare both results.

Table 18.3 shows the scenarios resulting from the minimum average of the SI index. These are the scenarios that allow us to reach better simulated results than the INA scenario results, where *M-F* is the Manning value for floodplain, *M-R* for riverbed. We mean that the improvement rate, as we show in Eq. (18.6), is the best achieved. We measure the SI index average at the 15 stations, so we were able to attain rates of up to 14 % average over all the stations at the same time.

An exhaustive search was implemented in the second phase. We show in Table 18.4 the parameters values used to carry out the parametric simulation. This phase was successful. The search returned 4 scenarios with an improvement rate between 20 % and 30 % in 3 stations. We selected the best:

Section 70: (0.03, 0.2), Section 72: (0.02, 0.1), Section 74: (0.035, 0.1), Section 76: (0.030, 0.1)

with an improvement rate of Station Rosario: 21 %, Station San Martín: 25 % and Station Diamante 28 %. We point out that these ratios were achieved with the same scenario.

To illustrate the results, we present the case of Rosario, which is located in the lower part of Paraná River. We show the relationship between observed levels data and those provided by Ezeiza, launched using the best scenario, resulting from our search method, and the INA scenario. Figure 18.3 compares this data throughout the 365 days of simulation.

Taking into account the absolute levels values (in metre), we want to point out in Fig. 18.3, the average improvement obtained between the days 200 and 300 of the simulation time. In this period, we obtained an average improvement of 0.65 m, in relation to the INA values. The difference in values, which range from 0.45 to 0.92 m, may affect forecasting.

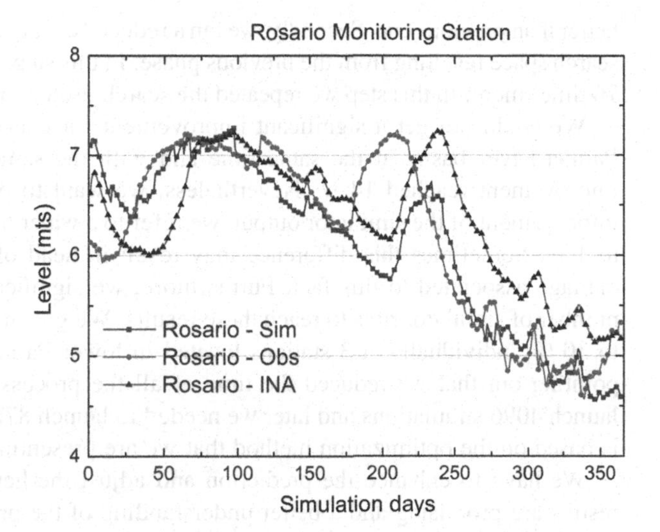

Fig. 18.3 Rosario water level. Time series throughout simulation period

This experimentation was carried out with parallel computing. The optimization method was launched in a Master-Worker application on a Linux cluster of multicore processors. Master-Worker scheme is suitable to parallelize the parametric simulation.[2] The master node distributes the initial workload to each worker node and it carries out a dynamic load balancing of the parametric runs. On the other hand, the execution time in the master is negligible. We took advantage of the K-Means algorithm features to implement a shared-memory parallel program and so to exploit the programming levels of the hardware architecture.

6 Results and Conclusions

In our previous work [24] we ran all the possible scenarios resulting from the parameters values confined to those domains selected to take the parametric simulation forward and to test the reliability of the method. As a result of these experiences, we found 5 scenarios whose prediction error are less than the RMSE reached with INA's current scenario.

Now, in this work, we enhance simulated results and use an extended search space by means of a two phase scheme, implemented taking advantage of parallel computing. Firstly, we get the best (adjusted) set of parameters using an MC asymptotic scheme. MC + K-means arrived to the end in four steps; that means that the simulator was launched 800 times. If the new SI is not lower than the last one stored then there was no need to run MC again. We stopped the process when the new SI value was no

[2] All parallel simulations were carried out using a 56 nodes multicluster Intel(R) Dual Core Xeon (TM) 5030 of 2.66 GHz processors with an infiniband switch.

better than previous one. Secondly, we ran a reduced exhaustive search for the reduced search space resulting from the previous phase. In this step, we only ran the simulator 76 times more. In this step we repeated the search used in our previous work.

We could not get a significant improvement for every station over the whole Paraná River basin, at the same time and with the same scenario. The average improvement reached 14 %. Nevertheless, we want to point out that by a 10 % improvement of the simulator output, we refer to a water height difference from 0.5 to 1 m. Sometimes this difference may result in head of livestock died or other damage associated to this fact. Furthermore, we significantly decreased the total number of simulator runs to reach these results. We got an improvement from 20 % to 30 %, individually at 3 stations located in lower Paraná River, and it is worth pointing out that we reduced the time of all the process. Initially, we needed to launch 4096 simulations and later we needed to launch 876 simulations. This result is based on the optimization method that we are presenting in this paper.

We have to enhance the prediction and adjust the heuristic technique but the results are promising and a better understanding of the problem was achieved. As future work, the MC method + K-Means clustering technique must be tested for all the sections and a huge amount of parameter values should be computed. Just this situation will require strong high performance computing and our work will benefit from and be enabled by this technology. This will be a key computing resource to tune the simulator Ezeiza by achieving more accurate forecasting.

We are working to improve these results and we are making progress in this direction. The best results will be achieved when we consider the monitoring stations located at any one of the three sections of the Parana River.

Acknowledgements This research has been supported by the MICINN Spain under contract TIN2007-64974, the MINECO (MICINN) Spain under contract TIN2011-24384 and it was partially supported by the research program of Informatics Research Institute III-LIDI, Faculty of Computer Science, Universidad Nacional de La Plata. We are very grateful for the data provided by INA and we appreciate the guidance received from researchers at INA Hydraulic Laboratory.

References

1. K. Hansson, M. Danielson y L. Ekenberg, «A framework for evaluation of flood management strategies,» Journal of Environmental Management, vol. 86, pp. 465–480, 2008.
2. L. Hluchy, V. Tran, J. Astalos, M. Dobrucky, G. Nguyen y D. Froehlich, «Parallel Flood Modeling Systems,» de Computational Science ICCS 2002, Berlin, Heidelberg, 2002, pp. 543–551.
3. A. a. S. S. Bárdossy, «Robust estimation of hydrological model parameters,» Hydrology and Earth System Sciences Discussions, vol. 5, n° 3, pp. 1641–1675, 2008.
4. S. Balica, "Parametric and physically based modelling techniques for flood risk and vulnerability assessment: A comparison," Environmental Modelling & Software, vol. 41, pp. 84–92, 2013.
5. A. Gaudiani, E. Luque, P. García, M. Re, M. Naiouf and A. De Giusti, "Computing, a Powerful Tool for Improving the Parameters Simulation Quality in Flood Prediction," Procedia Computer Science, vol. 29, no. 0, pp. 299–309, 2014.

6. M. Taboada, E. Cabrera, M. L. Iglesias, F. Epelde and E. Luque, "An agent-based decision support system for hospitals emergency," Procedia Computer Science, vol. 4, no. 0, pp. 1870–1879, 2011.
7. A. Menéndez, "Ezeiza V: un programa computacional para redes de canales," Mecánica Computacional, vol. 16, pp. 63–71, Septiembre 1996.
8. A. Menéndez, "Three decades of development and application of numerical simulation tools at INA Hydraulics LAB," Mecánica Computacional, vol. 21, pp. 2247–2266, Octubre 2002.
9. F. Saleh, A. Ducharne, N. Flipo, L. Oudin and E. Ledoux, "Impact of river bed morphology on discharge and water levels simulated by a 1D Saint-Venant hydraulic model at regional scale," Journal of Hydrology, vol. 476, pp. 169–177, 2013.
10. R. G. Sargent, "Verification and validation of simulation models," in Winter Simulation Conference, Phoenix USA, 2000.
11. W. L. Oberkampf, J. C. Helton, C. A. Joslyn, S. F. Wojtkiewicz y S. Ferson, «Challenge problems: uncertainty in system response given uncertain parameters,» Reliability Engineering & System Safety, vol. 85, n° 1, pp. 11–19, 2004.
12. M. Saiduzzaman y S. Ray, «Comparison of Numerical Schemes for Shallow Water,» Global Journal of Science Frontier Research, vol. 13, n° 4-F, pp. 1–19, 2013.
13. M. B. Butts, P. J. T., M. Kristensen y H. Madsen, «An evaluation of the impact of model structure on hydrological modelling uncertainty for streamflow simulation.,» Journal of Hydrology, vol. 298, n° 1–4, pp. 242–266, 2004.
14. F. Pappenberger, P. Matgen, K. Beven and L. Pfister, "Influence of uncertain boundary conditions and model structure on flood inundation predictions," Advances in Water Resources, vol. 29, pp. 1430–1449, 2006.
15. J. Warmink, J. Janssen, M. Booij y M. Krol, «Identification and classification of uncertainties in the application of environmental models,» Environmental Modelling & Software, vol. 25, n° 12, pp. 1518–1527, 2010.
16. H. d. Moel, J. Aerts y E. Koomen, «Development of flood exposure in the Netherlands during the 20th and 21st century.,» Global Environmental Change, vol. 21, n° 2, pp. 620–627, 2011.
17. A. Domeneghetti, A. Castellarin y A. Brath, «Assessing rating-curve uncertainty and its effects on hydraulic model calibration,» Hydrology and Earth System Sciences, vol. 16, pp. 1191–1202, 2012.
18. G. Latessa, "Modelo hidrodinámico del río Paraná para pronóstico hidrológico: Evaluación del performance y una propuesta de redefinición geométrica.," INA – UBA, Buenos Aires, 2011.
19. P. Kraus, D. Boyle and F. Bäse, "Comparison of different efficiency criteria for hydrological model assessment," Advances in Geosciences, vol. 5, pp. 89–97, 2005.
20. I. A. Camilloni and V. R. Barros, "Extreme discharge events in the Paraná River and their climate forcing.," Journal of Hydrology, vol. 278, pp. 94–106, 2003.
21. F. Pappenberger, K. Beven, M. Horritt and M. Blazkova, "Uncertainty in the calibration of effective roughness parameters in HEC-RAS using inundation and downstream level observations," Journal of Hydrology, vol. 302, no. 4, pp. 46–69, 2005.
22. E. Cabrera, M. Taboada, M. L. Iglesias and E. Luque, "Simulation optimization for healthcare emergency departments," Procedia Computer Science, vol. 9, pp. 1464–1473, 2012.
23. E. Cabrera, M. Taboada, F. Epelde, M. L. Iglesias and E. Luque, "Optimization of emergency departments by agent-based modeling and simulation," in Information Reuse and Integration (IRI), 2012 I.E. 13th International Conference, Las Vegas, 2012.
24. A. Gaudiani, E. Luque, P. García, M. Re, M. Naiouf and A. De Giusti, "Computational method for prediction enhancement of a river flood simulation," in 28th International Conference on Informatics for Environmental Protection, Oldenburg, Germany, 2014.
25. D. Kroese, T. Taimre y Z. Botev, Handbook of Monte Carlo Methods, New York: John Wiley and Sons, 2011.
26. D. J. C. MacKay, Information Theory, Inference & Learning Algorithms, New York: Cambridge University Press, 2002.

Part VI
Social Media for Sustainability

Chapter 19
A Social Media Environmental Awareness Campaign to Promote Sustainable Practices in Educational Environments

Brenda Scholtz, Clayton Burger, and Masive Zita

Abstract This paper examines the impact of a social media campaign on the environmental awareness of staff members in a Higher Education Institution (HEI). The campaign was designed based on a theoretical model and implemented in a case study of a South African HEI. A centralised website together with selected social media formed the technological foundation of the campaign. Throughout the campaign information about environmental management was distributed by means of these technologies to selected staff members in the case study. Issues related to environmental management as well as suggested strategies to deal with them was also communicated to the staff members. In order to determine the growth of knowledge related to environmental issues and to get feedback on the campaign benefits and problems, surveys and interviews were conducted at regular periods throughout the campaign. The findings revealed that the campaign had many positive benefits, particularly for promoting environmental awareness. The limitations of the study are the small sample size which could be addressed by future research.

Keywords Environmental awareness • Social media campaign • Social media

1 Introduction

Environmental issues are increasingly becoming of concern to the global community [1]. There is thus societal pressure to create environmental awareness which is essential for informing people about the effects of global warming and other environmental hazards. The increase of environmental awareness has risen

---

B. Scholtz (✉) • M. Zita
Department of Computing Sciences, Nelson Mandela Metropolitan University, Port Elizabeth, South Africa
e-mail: Brenda.Scholtz@nmmu.ac.za; Masive.Zita@nmmu.ac.za

C. Burger
Cascade Use, Fakultät II, Universität Oldenburg, Oldenburg, Germany
e-mail: Clayton.Burger@uni-oldenburg.de

J. Marx Gómez et al. (eds.), *Advances and New Trends in Environmental and Energy Informatics*, Progress in IS, DOI 10.1007/978-3-319-23455-7_19

relatively in proportion to the demand for sustainable practices [3]. Environmental responsibility is accepted by the United Nation's Education, Science and Cultural Organisation (UNESCO) as each person's choice and basis towards sustainability. Therefore the decision to mitigate waste, over-utilise fossil fuels and other environmental decisions and factors must be initiated at a personal level in order to affect change [1].

Higher Education Institutions (HEIs) play a critical role in the development of their graduates, and this development should include education regarding environmental awareness and responsibility [16, 30]. Educators and stakeholders should be encouraged to lead by example and through their teaching practice to impart environmentally responsible norms and values. These issues have led to the concept of Earth Stewardship where individuals, as well as organisations, are responsible for being environmentally responsible and promoting sustainable practices [2]. Internationally, pressure from external and internal bodies has resulted in an increasing number of HEIs that are embracing the concept of sustainability in higher education by undertaking green campus initiatives [24].

The promotion of sustainable practices should include creating environmental awareness and a sense of environmental consciousness that will equip students with the norms, values and skills required to make sensible eco-friendly acts [2]. Environmental awareness can be created by developing a personal exploration and discovery of people's surroundings to encourage individuals to participate in achieving a sustainable future [15]. Successful social media campaigns have been shown in some reports [6, 18, 22, 32] to be an effective approach to improving environmental awareness.

Environmental sustainability in higher education is a topic that has been gaining interest in recent years [26]. The large number of research publications, international conferences and declarations held in the last four decades are proof of the growing phenomenon [27]. Although environmental sustainability in HEIs has gained interest, the concept has been particularly prominent in Europe, the United States (U.S.), Asia, Australia, Canada and South America [24, 25] but has had limited penetration into African institutions [4].

This paper addresses a gap in research related to environmental awareness in HEIs, particularly in African HEIs.[1] The purpose of the paper is to propose and implement a model for a social media campaign to improve environmental awareness in higher education. The model is adopted in a case study at the Nelson Mandela Metropolitan University (NMMU) in South Africa. The structure of the paper is as follows. A literature review of sustainability, environmental awareness, environmental education and social media is presented in Sect. 2 and a theoretical model is proposed. The research methods employed in this study to construct, guide and study the campaign are outlined in Sect. 3. This is followed by a discussion in

[1] This paper extends the research published in a related paper "The Use of Social Media as an Enabler to Create Environmental Awareness of Staff in Higher Education" which was published at the EnviroInfo 2014 conference.

Sect. 4 of the results obtained. Finally, recommendations for institutions and subsequent campaigns are outlined in Sect. 5.

2 Literature Review

2.1 Environmental Knowledge and Awareness

The issue of addressing environmental awareness is important due to the increase in the world population and the high lifestyle standards of people because this produces a bigger environmental impact and burden [2]. Several factors contribute to producing sustainable practices (Fig. 19.1). Environmental awareness has been reported as an important factor in increasing sustainable behavior or practices [1, 3, 26] and focuses on knowledge of the environmental concerns happening globally as well as strategies for positively impacting these concerns [1, 16]. The concept extends to public awareness which helps individuals understand and drive other individuals to participate effectively in activities towards achieving sustainable practices [3].

According to Hungerford and Volk [29] environmental awareness provides social groups and individuals with awareness and sensitivity towards the environment and the issues affecting it. Furthermore, there is a positive correlation between environmental awareness and environmental knowledge which implies that an increase in environmental knowledge leads to an increase in environmental awareness [29, 30].

In addition the access to insightful and well-presented information is a key factor for raising and maintaining environmental awareness [3, 10]. Information is often packaged in communication campaigns that reach a larger audience. Environmental awareness is seen as a component to the education process and helps with creating change and improvements in sustainable practices. Yahya and Hashim [8] observe that societal norms and values form an important context for strategies of raising awareness of the natural environment. The existing attitudes that individuals have and their current behavioural patterns similarly contribute to how they adopt sustainable practices when exposed to public awareness initiatives, such as campaigns. Talero [15] also proposes the adoption of campaigns for environmental awareness and states that the most effective way to deliver a campaign depends on the activities and social interaction of the targeted group.

2.2 Environmental Education

Higher Education Institutions (HEIs) are not immune to global environment problems [25]. Awareness of the environmental impact and environmental

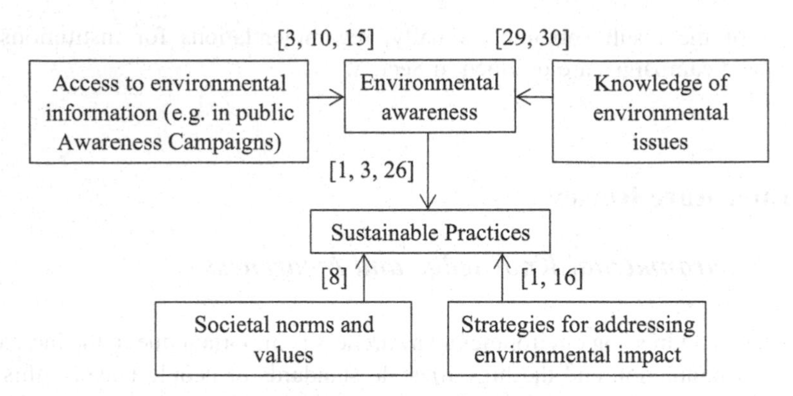

Fig. 19.1 Model for sustainable practices

responsibilities of HEIs is increasing [24, 25, 30]. For this reason many universities internationally are engaging in green initiatives [16, 30] and this is also true in South Africa [4]. In this way HEIs globally are attempting to help reduce their carbon footprints and reduce environmental pressures. South African HEIs are also facing increased pressure to reduce their impact on the environment and to produce sustainability reports since they are obligated to do so by the South African Department of Higher Education and Training (DHET) as specified in the Government Gazette [31]. The gazette defines the report as "an integrated report that conveys adequate information about the operations of a public higher education institution, its sustainability and financial reporting." The report should include a performance review that encompasses economic, environmental and social aspects.

Reporting on sustainability requires individuals to be more aware and informed and the greater the number of aware and informed individuals, the more likely societies will take some form of action to affect environmental change [10]. In HEIs it has been proven that students have a basic awareness of environmental issues but are ignorant on strategies to address these issues. Zsóka et al. [30] state that current university students will have a major impact on the future conditions of the environment therefore it is important to incorporate education of environmental issues into HEIs.

Traditionally, HEIS bear the responsibility of providing a tertiary education to students while providing leadership skills and support for resources to create sustainable practices in their environment [12]. Informing students about environmental issues at universities contributes towards creating environmental sustainability practices [16].

HEIs appoint academics who serve to educate the students of institutions by sharing their knowledge in courses, assessing students and imparting cultural lessons in responsibility and leadership [10]. This approach seeks to create students with the responsibility and knowledge that they need to enable them to become the leaders and experts of the next generation. This highlights the importance of the role that staff members of these institutions play, which forms a crucial context for the delivery of environmental awareness and responsibility in their students.

However, there are several barriers to enabling staff of HEIs to impart environmental awareness to their students, specifically a lack of training, limited organisational support, limited resources, and cultural resistance. These issues cannot be isolated, so a unified approach to training and supporting these staff must be applied, such as through institutional policy and knowledge awareness campaigns.

2.3 Effectiveness of Social Media for Information Sharing

The way individuals interact with each other and within organisations have been changed by social media [28]. The term *social media* is defined as "a group of Internet-based applications that build on the ideological and technological foundations of Web 2.0, and that allow the creation and exchange of user-generated content" [13]. Social media platforms have been shown to be effective tools to communicate and support interaction [14]. There are a plethora of Internet services that can be classified as social media, with Facebook and Twitter being two of the most common platforms in the U.S. [14]. Social media is largely driven by Facebook due to the larger amount of individuals it reaches in comparison to YouTube, Twitter and Instagram in the U.S. [5]. Facebook provides the facilities for users to share content to their network of contacts of various types of content, such as text, links, images and videos. Conversely, Twitter provides limited facilities for sending short messages and embedding images.

Many organisations have adopted social media as a tool for communication and collaboration inside the company because of its numerous usages, and in particular the potential engagement the media provides [7, 21]. Social media access creates an advantage for organisations to use electronic word of mouth as a means of communication and to create brand value [28]. Social media allows individuals to be able to interact, gain access to information faster and provide feedback [20]. The interaction enhances trustworthiness, brand attitude and individuals' commitment [28].

Social media reaches a larger audience than traditional communication mediums and is an interaction medium that is highly accessible and scalable in publishing platforms such as websites [18, 33]. This support for websites allows easy accessibility to the public and enables educators to be able to communicate the content with credible sources. The use of social media in organisations has been shown to enhance brand strategy, build relationships and allow businesses to gain a competitive advantage [20]. The use of social media for marketing, communication and brand management has been reported as having a positive effect on product knowledge and customer relationships [17]. Social media has proven to be a more effective technique for influencing attitudes and behaviours of individuals in marketing communications as compared to traditional communications. Hudson, Roth, Madden and Hudson [5] agree that interaction with brand strategy has a direct effect on the emotional attachment of individuals. This supports the argument that engagement on social media is triggered by marketing communications which play a dominant role in the behavioural outcomes of individuals. Erdoğmuş and Çiçek [33] agree that social media builds brand loyalty amongst clients and facilitates better communication with them. Social media has been reported to change the way information and awareness can be generated, and information retrieved and used in organisations [5].

Marketers are using individuals' comments on social media platforms as a basis for strategic marketing [34]. Manipulating user generated content builds awareness and loyalty [35]. The study of Idumange [18] reported that social media can be used to enable educators to be able to communicate internally and effectively in the workplace and to share important information. Studies [19, 20] have also reported that social media is an awareness channel that can enhance intra-organisational knowledge sharing. Campaigns conducted through social media platforms are recognised as having a positive impact on revenue [20]. Social media campaigns have also been shown to be successful in promoting, changing awareness and behavior.

Tess [9] reports that social media are widely available and mainly accessed at university where they are used by students for communication, collaboration and learning. Social media in higher education learning can improve the effectiveness of communication. Most students that have social media accounts rarely use social media for educational purposes; instead they mostly use it for communicating with friends and career networking. Hussain [23] supports this and confirms that students mainly use social media for enjoyment, accessing their academic information and current affairs.

Environmental awareness campaigns rely heavily on information and seek to provide individuals with the proper knowledge, skills and attitudes to address environmental problems [15]. Mooney et al. [22] confirm that environmental awareness campaigns play a big role in improving an individual's environmental awareness and changing their attitudes towards environmental issues. Numerous environmental campaigns have been conducted using social media, for example social media have been used to foster environmental behaviour, to get petitions signed, to provide news, to provide motivation and to improve awareness [32]. Social media are effective in carrying out environmental campaigns because they have the ability to distribute information quickly and are cost effective. Idumange [18] confirms that social media are able to support environmental initiatives because they are able to reach a wide spectrum of audience. They can be easily accessed; they are easy to use; they enable users to get instant responses, and they allow instant modification of responses by means of comments.

The use of social media in environmental campaigns helps raise awareness by means of informing individuals in an interesting way. Social media campaigns have been shown to be successful when carefully executed [14]. Five best-practice guidelines for conducting social media campaigns were recommended by Tobey and Manore [14]. These guidelines include a combination of participative feedback and proper research done prior to initiating the campaign (Fig. 19.2). The first

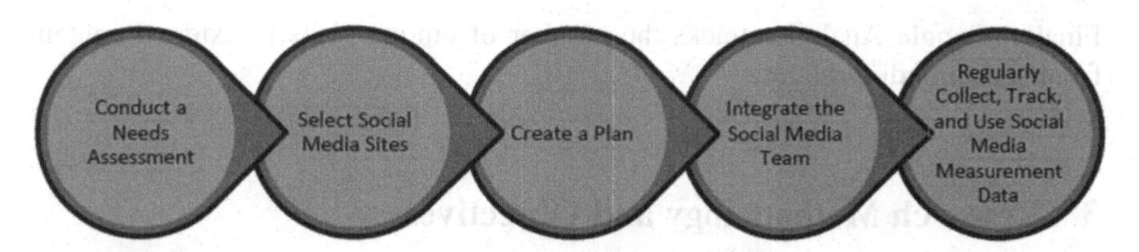

Fig. 19.2 Guidelines to a successful campaign strategy (Adapted from [14])

guideline for adopting an awareness campaign is to conduct a needs assessment. This needs assessment could consist of a literature review of the related research.

The second guideline highlights the importance of selecting the most appropriate social media sites to be used in the campaign. There are many different types of social media tools that can be used to raise awareness of environmental issues [18]. Popular social media include Blogs, Facebook, Twitter, Google Plus and LinkedIn. According to Tobey and Manore [14] Twitter encourages marketing strategies such as word-of-mouth and discussions through short messages whilst Facebook encourages social networking and accessing of collected social network information. Facebook provides the facilities for users to share content to their network of contacts. Text, links, images and videos are examples of the types of content shared. Facebook can also be used to discuss numerous environmental issues and allow users to share their opinions and assist each other on how to conserve environmental resources [18]. Twitter can be used to carry out discussions that can contribute to solving environmental problems that are being faced. YouTube provides users with visual representations and audio and can be used to address environmental issues such as increases in carbon footprints.

The third guideline to a successful campaign strategy is to create a plan [14]. The use of engaging content and a conversational approach to fresh educational, meaningful content is the key to a successful campaign on social media. To be able to reach a campaign's objective and the target audience, campaigns require a set structure. The structure explains what the campaign will focus on and how it will reach the audience. The structure serves as a driver to influence the participants to interact with the website and spread awareness.

The fourth guideline to a successful campaign involves communication and integration of the social media team. All members of the team should meet regularly to share objectives and vision and to communicate the strategy of the campaign to all those involved. The last guideline is to collect, track and use social media measurement data on a regular basis. This measurement data can be obtained by measuring tools such as Google Analytics, Sprout Social and social media tracking buttons. These tools help monitor and assess the usage of social media and can be used to provide key insights into campaigns that are driven by social media. Examples of these tools are Facebook *likes* and *shares* and Twitter *tweets*, *retweets* and *followers*. Social Sprout allows the analysis of demographics of geographically distributed content, while social media tracking buttons provide an aggregated summary of social media interactions for a specific piece of content.

Finally, Google Analytics tracks the number of unique visits to external content from social media sources.

3 Research Methodology and Objectives

The increase in global concerns, has led to education playing a fundamental role in empowering responsible sustainable development. An environmental awareness campaign is needed to support university staff to be knowledgeable and informed about environmental issues. The environmentally aware staff body can then share their awareness with students to promote sustainability, both in practice and in education. The main objective of this study is to design and implement the model for a social media campaign for improving environmental awareness (Fig. 19.1). A case study research strategy was used and the case was a South African university, the Nelson Mandela Metropolitan University (NMMU). The theoretical model was used to design the environmental awareness campaign which was then adopted at the university for a period of six weeks.

The structure of the Environmental Social Media Campaign (EcoSafe) Model consisted of three cycles (Fig. 19.3). The guidelines recommended by Tobey and Manore [14] were used to plan the campaign (Fig. 19.2). The first guideline recommends conducting a needs assessment and was incorporated into cycle one. The needs assessment included an extensive review of the literature to identify best practice. In cycle one the planning of the campaign included identifying the target market and the population, designing the content and activities of the campaign and an initial exposure of participants to environmental content. As the focus of the campaign was to raise awareness in a higher education context, a purposive sample was employed to attract participants that are currently employed by the institution. The sample of NMMU staff which took part in the campaign consisted of 30 active participants who were selected on a voluntary basis. These participants were encouraged to share information with their respective social media contacts as well. The participant sample consisted of academic, professional and support staff from the spectrum of departments at the institution, drawing from Computing Sciences, Engineering, Student Governance, Media Studies, Journalism, Mathematics, Development Studies, Finance and Examinations departments.

The second guideline followed was to determine which social media were preferred and to select the social media which would be used in the campaign. The most popular social media used in the United States are Facebook, Twitter and Pinterest [14]. According to the South African Social Media Landscape 2014 study [11], Facebook has the highest number of active users in South Africa (9.4 million). However in educational environments this could be slightly different and, for this reason, a survey was undertaken prior to the commencement of the campaign to determine the preferences of the participants.

The questionnaire contained questions regarding which social media platforms were used by each participant and how they accessed the platforms. The

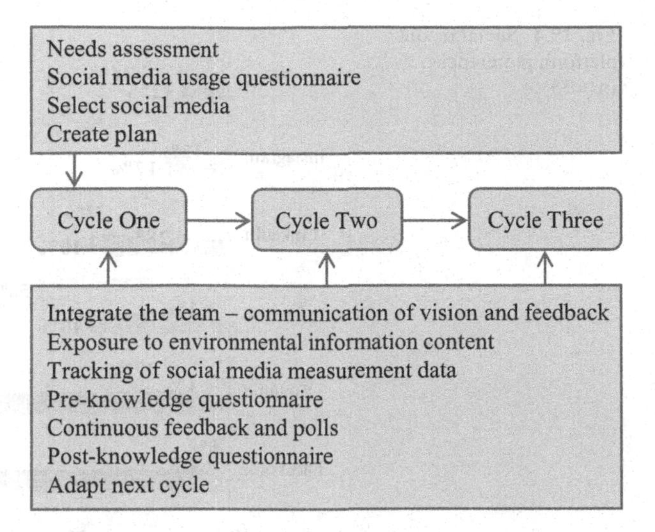

Fig. 19.3 Environmental social media campaign cycles

participants were given a set of options for some of the most widely used social media platforms to indicate their preference and 25 participants completed the survey correctly. The results (Fig. 19.4) showed that the majority (96 %) of the participants who completed the social media usage questionnaire ($n = 25$) indicated positively (Agree or Strongly Agree) that they used Facebook and the remaining 4 % were neutral. Therefore none of the participants were negative about their usage of Facebook. YouTube had the second highest positive score with 79 % of participants reporting that they Agreed or Positively Agreed that they used YouTube. Twitter and LinkedIn both had the same percentage of participants Agreeing or Strongly Agreeing that they used these platforms; however Twitter was selected to use in the campaign rather than LinkedIn. The reason for this decision was that LinkedIn and Facebook are similar types of social media (providing facilities for users to share various types of content to their network of contacts), whereas Twitter would provide a different type of social media to the campaign, one which provides limited facilities for sending short messages and embedding images.

Instagram and other blogs were rated lowest in the survey and were therefore not included into the campaign. In addition a website was created which enabled links to these social media and also allowed participants to share content related to environmental issues on the web site to increase the reach of the information.

The participants were asked how much time they spend on their preferred social media platform and what hampers their access to the platform. The majority of the participants indicated that they spend between one and two hours daily on their platform to network with peers and engage with media. The constraints to usage of social media platforms listed were time (43 %), Internet access (35 %) and knowledge (22 %).

The second and third cycles consisted of engagement with the campaign, interaction with the website and social media, observing and assessing the

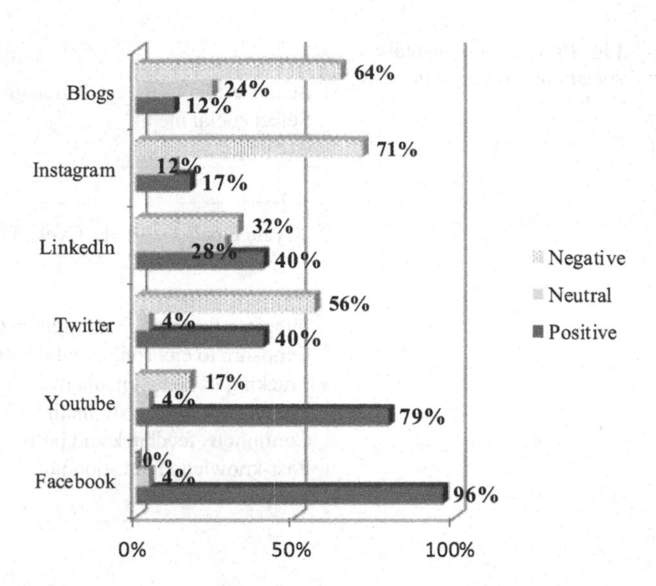

Fig. 19.4 Social media platform preferences (n = 25)

participants' environmental knowledge, as well as assessing the participants sharing of environmental information. The campaign was driven through shareable content that was posted on the central campaign website three times a week. Participants were required to visit the website regularly to ensure exposure to the provided content. All content was peer-reviewed prior to posting to ensure relevance to the campaign and accuracy of content.

Examples of some of the posted content themes were:

- Sustainability initiatives and projects, such as renewable energy initiatives;
- Recycling facts and tips; and
- Popular sustainability initiatives, such as Earth Hour.

Examples of some of the posted questions were:

- *What are some of the ways to save electricity at university?*
- *Which sources of energy and electricity are renewable (if harvested)?*
- *Which diseases can indoor pollution cause?*

During all three cycles, a *continuous feedback survey* was made available whereby participants could provide feedback about the campaign. Various *polls* were provided as part of the content to determine participation levels. Finally, *a post-campaign questionnaire* was provided to determine if the participants had shared the content provided through their social media networks, which networks were preferred and if their environmental awareness had been improved.

All data gathered was anonymous to protect the identities of participants. Similarly, the actual proliferation of the content was not measured as this could pose an ethical issue to the participants as their social media profiles would need to be tracked. A limited number of *semi-structured interviews* were also conducted with participants to verify the results of the questionnaires and polls and to discuss the campaign structure and content at the end of the campaign.

4 Findings

4.1 Growth in Environmental Knowledge

The aim of the campaign was to evaluate participants' environmental knowledge and spread the environmental information to create environmental awareness. To determine the level of environmental knowledge of participants before and after the campaign, a *pre-campaign questionnaire* and *post-campaign questionnaire* was administered to each participant. The questions focused on general environmental knowledge as well as knowledge of institutional specific environmental initiatives. A total of 10 questions were presented in each of the questionnaires. The comparative results indicate an increase in knowledge in each cycle of the campaign (Fig. 19.5). The cycles cannot be compared as the participant sample varied slightly in each cycle.

In cycle one the participants' knowledge increased by 2 %. This slightly low increase could be due to the fact that cycle one was the first exposure to the campaign and this type of content and participants were still familiarising themselves with the website. The knowledge growth for cycle two was much higher at 45 % and this large increase from cycle one to cycle two was probably due to the fact that participants had now become more familiar with the content and the campaign. Participants became more comfortable with the website, and shared the environmental information with additional participants. In cycle three there was a knowledge growth of 33 %, which is slightly lower than the growth in cycle two. This could be due to the fact that as identified by the collected social media measurement data and reported by the participants there was more activity in cycle three than the previous cycles.

Throughout the duration of the campaign, participants were requested to complete *continuous feedback surveys* and *polls*. The aim of the poll surveys was to encourage participants to give their opinions on environmental matters and to provide strategies on how to implement sustainable practices as recommended by [1]. Three continuous feedback surveys were presented over the course of the campaign. The role of the continuous feedback surveys was to gain insight into the experiences and preferences of participants and allowed the researcher to change the type of content published to align with the media that the participants preferred the most. The participants indicated that they prefer fresher content comprised of current topics and visual media, such as images and video. For example one participant stated that he preferred "*To see the different content, such as the videos*" and another participant stated that he liked "*To be kept informed*". The participants indicated that they were surprised at many of the

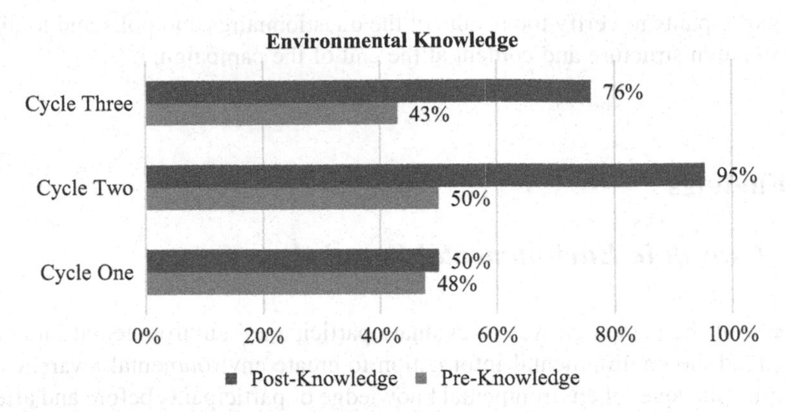

Fig. 19.5 Assessment of environmental awareness before and after the campaign (n = 30)

statistics shown on the website and that they had gained an appreciation of topical environmental factors, such as rhino poaching in South Africa and subsequent biodiversity loss. The increased knowledge thus contributed to increased awareness which confirms other similar findings [29, 30].

From the results, it can be deduced that the campaign facilitated an increase in environmental knowledge during the campaign. In the second and third cycles participants became more active with the social media and in sharing the environmental content, likely due to the adaption of the content provided to more shareable content such as video. This played a role in the increase of their environmental knowledge. It was observed that participants may have found it difficult to adapt to the campaign model during the first cycle.

4.2 *Social Media Campaign Usage*

The sharing analytics gathered from the EcoSafe campaign website revealed a mean of 82 hits per day throughout the course of the campaign. Many of these hits could be due to external search engine hits or web crawlers, but several were from IP addresses that belong to the institution. The majority of participants (90 %) agreed that the EcoSafe Campaign provided access to different information types (Fig. 19.6). More than half (60 %) of the participants who interacted in the EcoSafe campaign stated that it promoted flexibility of discussions.

Semi-structured interviews were held with three participants in order to verify the questionnaire results. The interviews revealed that participants regularly shared content that appeared in the campaign with the social network of peers when they felt that it would be appreciated by their peers. The most popular shared content amongst the interviewees was video content as it can be easily accessed.

The prevailing opinion voiced by the participants was that both the environmental and NMMU content was relevant and interesting, but that there was not an

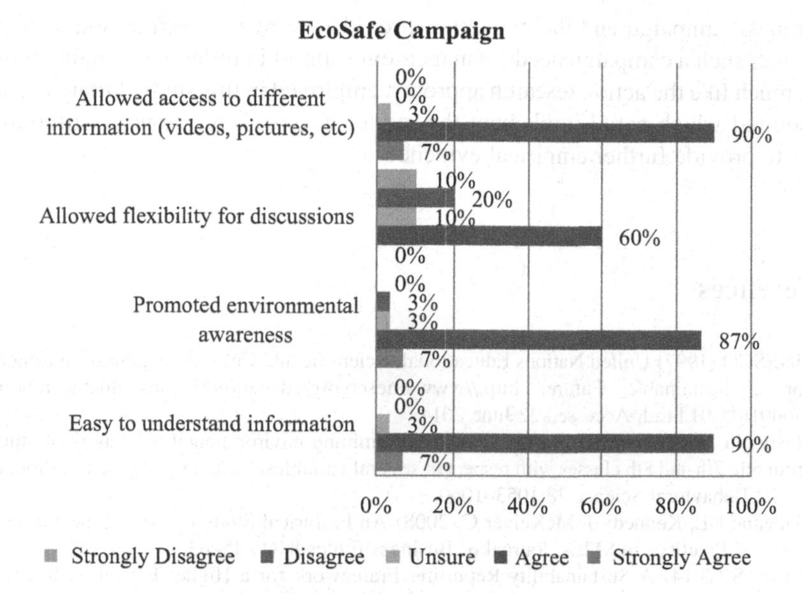

Fig. 19.6 EcoSafe campaign benefits (n = 30)

institutional culture for sustainability. They emphasised that they would share their newly gained environmental knowledge and values with their students in order to strengthen the graduate profile of the students while encouraging Earth stewardship. One participant stated that it is "worthwhile to take action in striving to reduce the environmental issues". Participants indicated that to create sustainable development as an organisation more sustainability awareness campaigns are needed.

5 Recommendations and Conclusion

The results of this study suggest that the staff participants of the higher educational institution used in the study felt that they benefited from the campaign and that they had engaged with the content provided. There was an increase evident in the environmental knowledge of the participants. Key findings indicate that institutions, such as the university investigated, should invest in awareness campaigns regarding their environmental initiatives, and make them accessible to staff members. This was previously lacking in the university case study investigated. These awareness campaigns need to provide engaging content and well-presented information in order for staff to become environmentally aware. The limitation of this study is in the small sample size. However, it is a qualitative study which forms part of a larger, ongoing study. In spite of this limitation, the theoretical model proposed and initial findings provide a valuable contribution to the research field of environmental awareness and the use of social media for promoting these campaigns. Social media were shown to be a valid means of conducting an environmental awareness campaign and the guidelines provided were successfully implemented. However such a campaign needs management support in order to maintain momentum, much like the action research approach employed in this study. Future research is required which could implement the model in other educational institutions in order to provide further empirical evidence.

References

1. UNESCO (1997) United Nations Educational, Scientific and Cultural Organisation Educating for a Sustainable Future. http://www.unesco.org/education/tlsf/mods/theme_a/popups/mod01t05s01.html. Accessed 22 June 2014.
2. Haşıloğlu MA, Keleş PU, Aydın S (2011) Examining environmental awareness of students from 6th, 7th and 8th classes with respect to several variables: 'sample of Agri city'. Procedia – Social Behavioral Science 28:1053–1060.
3. Gadenne DL, Kennedy J, McKeiver C (2008) An Empirical Study of Environmental Awareness and Practices in SMEs. Journal of Business Ethics 84(1):45–63.
4. Bosire S (2014) A Sustainability Reporting Framework for a Higher Education Institution. Dissertation, Nelson Mandela Metropolitan University.
5. Hudson S, Roth MS, Madden TJ, Hudson R (2015) The effects of social media on emotions, brand relationship quality, and word of mouth: An empirical study of music festival attendees. Tour. Manag 47:68–76.
6. Ali MSS (2011) The Use of Facebook to Increase Climate Change Awareness among Employees. In: International Conference on Social Science and Humanity, Singapore, 2011.
7. Ngai EWT, Tao SSC, Moon KKL (2015) Social media research: Theories, constructs, and conceptual frameworks. Int. J. Inf. Manage 35(1):33–44.
8. Yahya WK, Hashim NH (2011) The Role of Public Awareness and Government Regulations in Stimulating Sustainable Consumption of Malaysian Consumers. In: Proceedings of the International Conference on Business, Engineering and Industrial Applications (ICBEIA 2011), Kuala Lumpur, Malaysia, 5–7 June 2011.
9. Tess PA (2013) The role of social media in higher education classes (real and virtual) – a literature review. Computers in Human Behavior 29(5):60–68.
10. Thomas I (2004) Sustainability in tertiary curricula: what is stopping it happening? International Journal of Sustainability in Higher Education 5(1):33–47.
11. World Wide Worx (2014). South African Social Media Landscape 2014. http://www.worldwideworx.com/wp…/Exec-Summary-Social-Media-2014.pdf. Accessed 15 January 2015.
12. Nicolaides A (2006) The implementation of environmental management towards sustainable universities and education for sustainable development as an ethical imperative. International Journal of Sustainability in Higher Education 7(4):414–424.
13. Kaplan AM, Haenlein M (2010) Users of the world, unite! The challenges and opportunities of Social Media. Business Horizons 53(1):59–68.
14. Tobey LN, Manore MM (2014) Social media and nutrition education: the food hero experience. Journal of Nutrition Education and Behavior 46(2):128–33.
15. Talero G (2004) Environmental Education and Public Awareness. http://worldfish.org/PPA/PDFs/Semi-AnnualII English/2nds.a.eng_F2.pdf. Accessed 25 July 2013.
16. Ralph M, Stubbs W (2014) Integrating environmental sustainability into universities. Journal of Higher Education 67(1):71–90.
17. Jucaitytė I, Maščinskienė J (2014) Peculiarities of Social Media Integration into Marketing Communication. Procedia – Soc. Behav. Sci. 156:490–495.

18. Idumange J (2012) The Social Media as a Platform for Creating Environmental Awareness in the Niger Delta Region. Paper presented at the 3rd Environment Outreach Magazine Public Lecture and Environmental Awards, Effurun, Delta State, 28 September 2012.
19. Vuori V, Okkonen J (2012) Knowledge sharing motivational factors of using an intra-organizational social media platform. Journal of Knowledge Management 16(4):592–603.
20. Parveen F, Jaafar NI, Ainin S (2014) Social media usage and organizational performance: Reflections of Malaysian social media managers. Telematics and Informatics 32(1):67–78.
21. Stieglitz S, Schallenmuller S, Meske C (2013) Adoption of Social Media for Internal Usage in a Global Enterprise. Paper presented at the 27th International Conference on Advanced Information Networking and Applications Workshops, Barcelona, Spain, 25–28 March 2013.
22. Mooney P, Winstanley A, Corcoran P (2009) Evaluating Twitter for Use in Environmental Awareness Campaigns. http://www.cs.nuim.ie/~pmooney/websitePapers/MooneyWinstanley Corcoran070709-FINAL.pdf. Accessed 1 February 2014.
23. Hussain I (2012) A Study to Evaluate the Social Media Trends among University Students. Procedia – Social and Behavioral Sciences 64:639–645.
24. Disterheft A, da Silva Caeiro SS, Ramos MR, de Miranda Azeiteiro UM (2012) Environmental Management Systems (EMS) implementation processes and practices in European higher education institutions – Top-down versus participatory approaches. Journal of Cleaner Production 31:80–90.
25. Velazquez L, Munguia N, Platt A, Taddei J (2006) Sustainable university: what can be the matter? Journal of Cleaner Production 14:810–819.
26. Alshuwaikhat HM and Abubakar I (2008) An integrated approach to achieving campus sustainability: assessment of the current campus environmental management practices. Journal of Cleaner Production 16(16):1777–1785.
27. Lozano R (2006) A tool for a Graphical Assessment of Sustainability in Universities (GASU). Journal of Cleaner Production 14(1):963–972.
28. Dijkmans C, Kerhof P, Beukeboom CJ (2015) A stage to engage: Social media use and corporate reputation. Tour. Manag. 47:58–67.
29. Hungerford H, Volk T (1990) Changing learner behavior through environmental education. Journal of Environmental Education 21(3):8–22.
30. Zsóka Á, Szerényi ZM, Széchy A, Kocsis T (2013) Greening due to environmental education? Environmental knowledge, attitudes, consumer behavior and everyday pro-environmental activities of Hungarian high school and university students. Journal of Cleaner Production. 48:126–138.
31. Department of Higher Education and Training (2014) Regulations for Reporting by Public Higher Education Institutions, Government Gazette 37726.
32. Willson J (2010) A look at how non-profits are using social media to raise environmental awareness. http://www.helium.com/items/1952299-how-environmental-groups-are-using-social-media. Accessed 15 April 2012.
33. Erdoğmuş IE, Çiçek M. (2012) The Impact of Social Media Marketing on Brand Loyalty. Social and Behavioral Sciences, 58:1353–1360.
34. Bright LF, Kleiser SB, Grau SL, (2015) Too much Facebook? An exploratory examination of social media fatigue. Computers in Human Behavior 44:148–155.
35. Dahnil MI, Marzuki KM, Langgat J, and Fabeil NF. (2014) Factors Influencing SMEs Adoption of Social Media Marketing. Social and Behavioral Sciences, 148:119–126.

Chapter 20
Supporting Sustainable Development in Rural Areas by Encouraging Local Cooperation and Neighborhood Effects Using ICT

Andreas Filler, Eva Kern, and Stefan Naumann

Abstract The following contribution presents an approach how information and communication technology (ICT) can be used to support local cooperation and neighborhood effects. The presented approach maps typical real-world processes in the area of local cooperation between stakeholders from schools, universities, and companies to an Internet platform with the motivation to support the communication between the aforementioned audiences and thereby reduces demographic problems in the model area in a sustainable, long-time perspective. The contribution describes the process from the conceptual application of the idea as well as the prototypical realization of the technical prototype *vitaminBIR*. This encompasses, from the conceptual perspective, the definition of the main goals of the platform supported by a survey collecting needs and preferences of all addressed audiences and the conceptual phase of the platform itself. Furthermore, from the implementation perspective, the prototypical implementation is supported by different software libraries, a concept for data self-management, and an iterative usability study to also check against the prior evaluated needs and preferences of the audiences, is addressed. A first usage analysis is presented at last.

Keywords Local cooperation • Sustainable development • Neighborhood effects

---

A. Filler (✉) • S. Naumann
Trier University of Applied Sciences, Environmental Campus Birkenfeld, Institute for Software Systems (ISS), P.O. Box 1380, 55761 Birkenfeld, Germany
e-mail: a.filler@umwelt-campus.de; s.naumann@umwelt-campus.de

E. Kern
Trier University of Applied Sciences, Environmental Campus Birkenfeld, Institute for Software Systems (ISS), P.O. Box 1380, 55761 Birkenfeld, Germany

Leuphana University of Lüneburg, Scharnhorststr. 1, 21335 Lüneburg, Germany
e-mail: mail@nachhaltige-medien.de

J. Marx Gómez et al. (eds.), *Advances and New Trends in Environmental and Energy Informatics*, Progress in IS, DOI 10.1007/978-3-319-23455-7_20

1 Demographic Problems in Rural Areas in Europe

The emigration of young people to the bigger cities and the brain drain related with this fact [1, 2], as well as the general aging of the society [3], are big demographic problems especially in rural areas in Germany, and in other European countries, nowadays [4]. Companies observe a shift from a "buyer's market" to a "seller's market" regarding employees for open position [5]. Local authorities need to handle the population loss on several levels, e.g., spatial development [6], cultural development [4], or tourism and leisure activities [7].

For these reasons, it is apparent that strategies for the sustainable development, especially for rural areas, need to be found. An important aspect hereby is to show up young people the potentials of their hometown and region [8] and related with this, the possibilities for qualifications and jobs. As job search and recruiting is usually intrinsically motivated by people and companies, we present a community-based approach to encourage all persons involved to cooperate in that matter with an intended advantage for both parties.

This publication extends our former work-in-progress paper [9] by going more into detail regarding the completion of the developed website and app and provides first results of our ongoing usage analysis of the same. In the following section we will therefore describe the basic idea of our approach (Sect. 2.1), followed by the conceptual development (Sect. 2.2) and prototypical implementation (Sects. 2.3 and 2.4). We will proceed with the presentation of the results from our usage analysis (Sect. 2.5) and close with a brief outlook (Sect. 3).

2 Supporting Local Cooperation Using ICT

The approach presented in this contribution mainly focuses on the idea to support local cooperation and neighborhood effects by information and communication technology (ICT), in particular by providing an interactive Internet platform. In detail, our goal is to map typical real-world processes of local cooperation and neighborhood effects, i.e., interactions between people from different involved audiences, e.g., schools, companies etc., to an ICT system. Therefore, we focus on the following aspects:

1. Easy understandable transfer of the real-world processes to the navigation structure of the system to achieve a good ease of use
2. Screen and information design that is attractive for all involved audiences
3. High level of information quality that exceeds the expectations of the user
4. Mainly self-managing system to reduce runtime service costs

All information provided in the system is, by concept, restricted to information that follows the guideline to work against the aforementioned demographic problems, e.g., job offers only from companies inside the specific local area, educational events provided by the local university, etc.

2.1 Conceptual Development

As part of the pilot project *LandZukunft* we apply the described approach on the cooperation processes of audiences in the District of Birkenfeld in Germany. The pilot project *LandZukunft*, which is funded by the *German Federal Ministry of Food and Agriculture*, focuses on the sustainable support of rural areas during their endeavor to develop regional value chains and protect local jobs. Our contribution focuses on the opportunities ICT can provide to support cooperation in rural areas, while a corresponding project followed the same approach by using face-to-face communication with the involved focus groups. Hereby, we are able to provide a holistic approach to address the prevalent problems on a technical channel available at any time on the one hand and a personal contact person on the other hand.

While applying the presented approach, we focus on the processes of cooperation of schools, universities, and companies on the organizational side, as well as the neighborhood effects between pupils, students, and human resources departments of local companies on the social side. By supporting these audiences and the restriction of the information on the platform to a specific local area, we specifically follow the concept to reduce emigration of young people as well as the mentioned brain drain. When we are successful in this, we expect to have positive impacts on the sustainable development of the appropriate local area. With the application described in the following, we furthermore plan to create a solution that can be transferred to other rural regions with similar problems, after a successful prove of our approach in the District of Birkenfeld. Since *LandZukunft* is a pilot project, this transfer of knowledge and technology is also one of the main concepts of the funding. We therefore focused on essential problem, by addressing the most relevant audiences and problems in the first step. It's planned to extend the concept by integrating further audiences and services in the future after a successful validation of the presented approach, but in the first step we focused on the most relevant audiences and most relevant services.

It is proofed that peer networking has (1) a high impact on the job search or recruiting and its factor of success for students as well as for companies and that (2) these neighborhood effects can be supported by ICT [10–12]. Therefore we decided to design an Internet platform, which supports these peer networking aspects with the goal to help pupils and students to find good apprenticeships and jobs in companies near their residence as well as companies to find appropriate job candidates. To complete the general idea, the platform also encourages offering and searching for side jobs between pupils and students among themselves, e.g., private tutoring. We hereby follow a holistic approach by not only providing job offers, but also possibilities to check public transport connections to the appropriate location. On a higher organizational level, the platform motivates company leaders and human resources authorities as well as headmasters of schools to find and agree to long-term partnerships. Long-term partnerships in this context can be partnerships between schools and companies with periodic practical events, workshops as preparation for apprenticeships, or appointments for networking between pupils and companies. Also practical events or job fairs organized by a university in collaboration with companies etc. belong to this category. Summarized, we offer a platform to encourage different audiences to long-term cooperation between each other. We expect that these changes lead to positive demographic effects and with this to a reduction of brain drain to bigger cities as well as to positive effects regarding the other connected aspects already mentioned, e.g., spatial development, cultural development etc.

2.1.1 Analysis of Existing Regional Platforms

In order to define requirements for a new regional platform we analyzed existing platforms of other regions that follow a similar concept. Based on an explorative web analysis we selected three German platforms: *PFIFF*,[1] *Ems-Achse*[2] and *Jobwunderland*.[3] We analyzed these platforms in terms of usability, content-related and technical implementation. Following, we list the key findings of the analysis:

- Usability: The usability is of high relevance for regional online platforms. In general, the usability of the analyzed platforms is given. Indeed, in two of three cases the information flood is too high in order to be able to gather relevant information quickly. Thus, this should be avoided in case of *vitaminBIR*. Here, usability tests can be helpful (see Sect. 2.2).
- Content: Both the job types as well as the target groups are broad in scope. These aspects depend on the addressed region. The usage of the platform functionality is free in most of the cases. In order to reach the specific target group and her needs in the region of Birkenfeld, we conducted an assessment survey that will be described in the next section.
- Technical implementation: Overall, the implementation of the analyzed platforms is based on open source software. Apart from general job searches all functionalities require a registration on the platform. Regarding social media, all of the platforms provide a fan page on Facebook. Neither interfaces to other job portals nor an application for smartphones exist. Before starting to implement *vitaminBIR* we will analyze existing open source solutions and the linking to Facebook to announce our platform as well as job data supplier to extend our job offers.
- Unique features: In contrast to the analyzed regional platforms *vitaminBIR* will directly involve schools, pupil, universities, students, and companies. To date pupil are rarely addressed, universities mostly act the single part of the platform operator.

[1] http://pfiff-sachsen-anhalt.de/
[2] http://emsachse.de/
[3] http://jobwunderland.com

Detailed results of the requirement analysis for *vitaminBIR* based on similar existing regional platforms can be found in [13].

2.1.2 Assessment Survey

In order to identify the processes that should be supported by the platform as well as the information required by the audiences for the same, we conducted a paper and web based assessment survey before we started into the conception phase. The survey (survey data: n = 304, 138 pupils (μ = 16.64 years), 150 students (μ = 26.23 years), 16 company representatives) was, in the beginning, prepared as one generic questionnaire for all audiences and afterwards elaborated as three different versions for the specific audiences. That way, we were able to create questionnaires optimized for each audience on the one hand regarding language, e.g., youth or business language, and question selection, e.g., by adding additional questions for specific audiences. On the other hand it enabled us, to relate the answers on specific question between the audiences, e.g., regarding expectations of responsibilities or preferences of specific audiences on the same topic. The survey helped us to evaluate expected features, responsibilities and information. These three aspects will be presented a bit more in detail in the following:

1. The survey results enabled us to gain an objective perspective on the steps that are seen as most relevant to be supported by such a platform, to be able to give them a higher importance in the platform concept.

   Exemplary item: The expectations of all audiences from which side a communication regarding an open job position should start.
   Result excerpts (all results can be seen in Fig. 20.1):

   - Students: 54.4 %: applicant contacts company, 36.2 %: no preference
   - Pupils: 6.2 % applicant contacts company, 68.2 %: no preference
   - Companies: 68.8 %: applicant contacts company, 25.0 %: no preference

   Result interpretation: The majority of the participants over all audiences expect the applicant to contact the company, not vice versa.
   With these results we can especially support the preferred steps on the platform.

2. We determined the information that is required and expected by the users of the platform to support the aforementioned steps.

   Exemplary item: Which information do companies expect from applicants when applying for an open position.
   Results excerpts: Most relevant information from company perspective as evaluated by the survey: Name, residence, age, motivational letter, employment type, branch, earliest starting date, language knowledge, and photo.

3. Additionally, we identified some technical preferences by the survey.

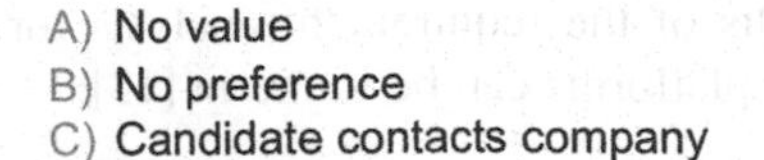

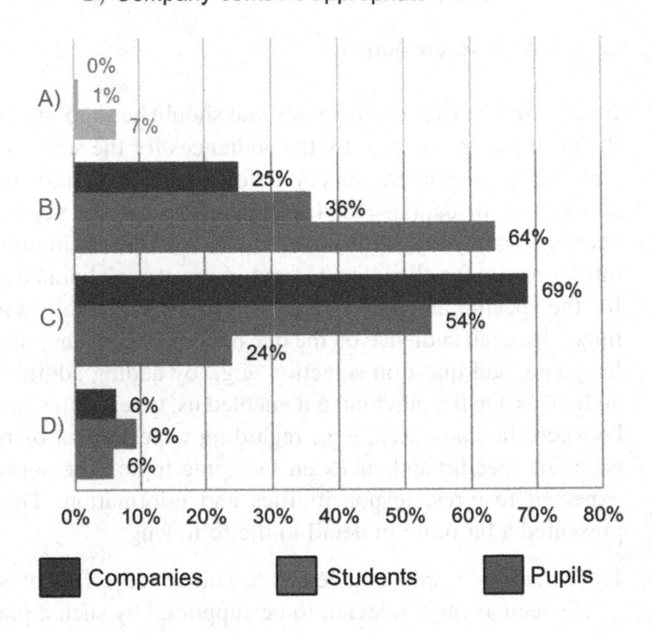

Fig. 20.1 Results of exemplary item of assessment survey

Exemplary item: Does the user want to use existing or separately created user accounts to access the personalized or private sections of the platform.

Result interpretation: All audiences prefer the creation of own accounts instead of reusing existing Facebook, Twitter, Google, Microsoft or XING accounts.

2.1.3 Screen Design Concept

Based on these survey results, we elaborated the first concept for the Internet platform, which can be seen as first screen design concept below (see Figs. 20.2 and 20.3). The concept follows a combination of a target group and intention based navigation approach. That way, the information a user is searching for on the platform, can be found in not more than two mouse clicks. If possible, the user can also find the information without giving any personal information in advance, i.e., it is not necessary to request user information to provide the information.

Summarized, users from all audiences can come to a page containing the relevant information with only two steps and will, if it is appropriate for the particular information, be able to filter the given information to find the best-fitting results afterwards. An additional top menu supports direct navigation over all pages for returning, experienced users who want to skip the two-step process.

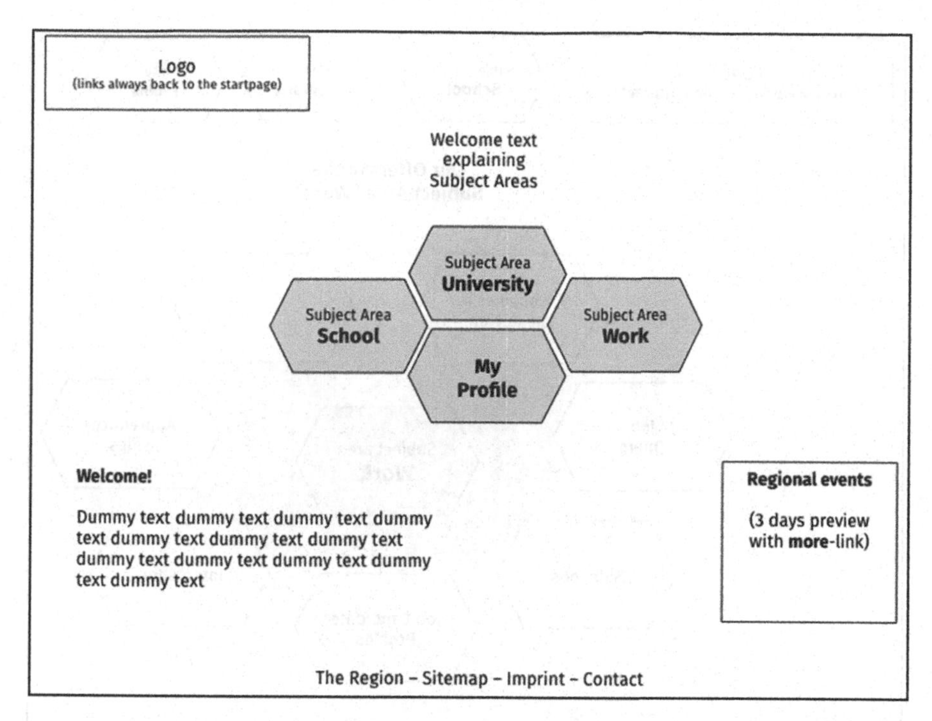

Fig. 20.2 Concept draft of the homepage of the internet platform (Translated from German)

The screen design concept drafts showing the exemplary results page of a two-step process of a student searching for a side job offer as well as the top menu (see Fig. 20.4) can be seen in the following. By focusing on the aforementioned two-step process while developing the concept, we hope to reduce the bounce rate[4] to a minimum.

2.1.4 Use Case Development

A parallel process beside the development of the first screen design was the description of the use cases. The use cases elaborated in this step are a collection of all functionalities provided by the Internet platform. As usual, they contain information about all involved stakeholders and the functions provided by the platform. They also depict how the stakeholders can use the functions, i.e., the general interplay of all functions as well as dependencies between the functionalities. They can be used to continuously check the ongoing implementation against the concept. Altogether the first version of the Internet platform realizes 28 use cases.

[4] The bounce rate is the number of users leaving a platform directly after the first page view.

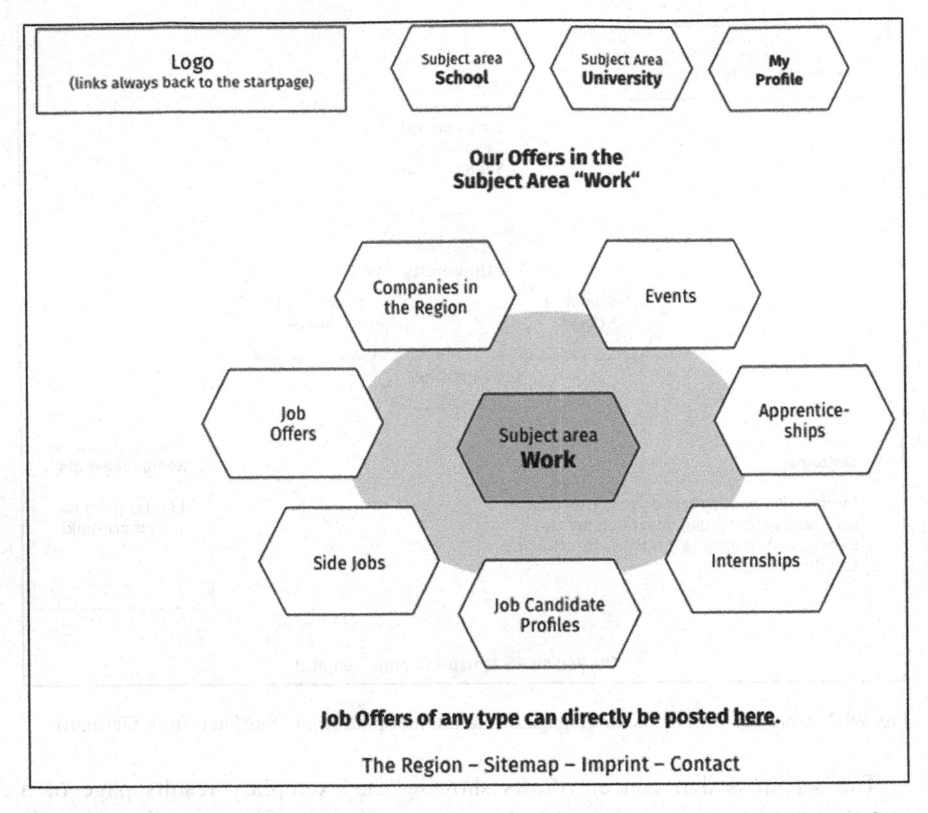

Fig. 20.3 Concept draft of the subject area "work" of the internet platform (Translated from German)

An excerpt from the use case overview showing some of the use cases and their dependencies between each other can be seen below (see Fig. 20.5). Solid lines in the graph describe usual transitions between two cases, while dotted lines describe optional or alternative ways to use the platform. To exemplify the process, the implementation result of *UC09: Password reset* can be seen in the following figure (see Fig. 20.6).

2.2 Prototypical Implementation

Our prototype, *vitaminBIR* (www.vitamin-bir.de), is developed as an interactive Internet platform based on the content management system *TYPO3*[5] for easy content management and own extensions to provide specific interactive

[5] http://typo3.org

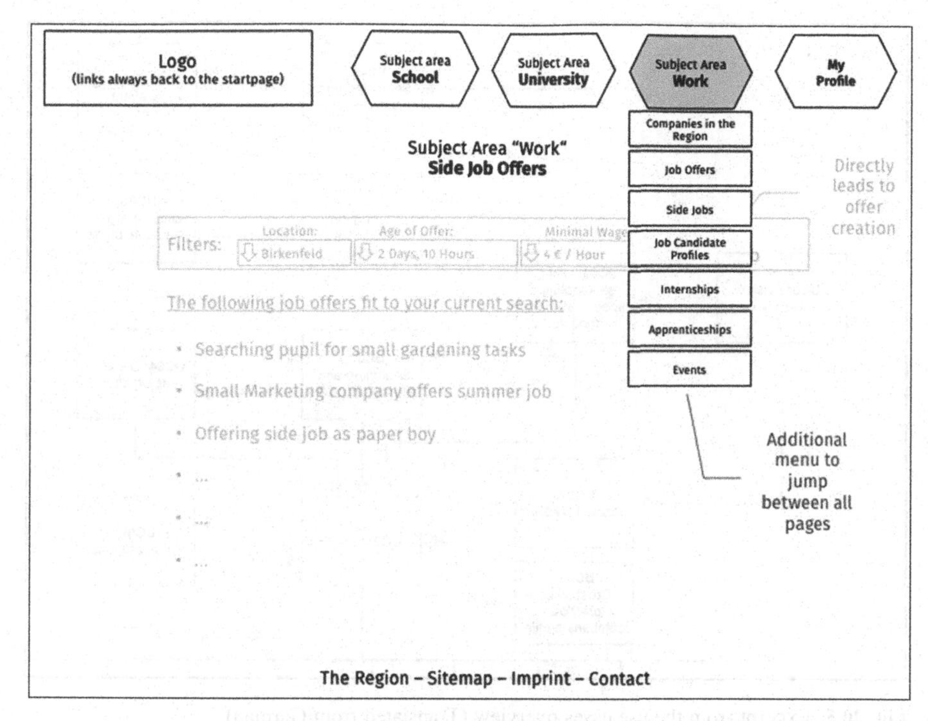

Fig. 20.4 Concept draft of the additional top menu on the platform (Translated from German)

components on the platform, which directly support the use cases addressed. On the one hand, when designing *vitaminBIR* a major aspect of our technical concept was also a self-managing design, to reduce management effort for the developed platform to a minimum. On the other hand, we always focused on simplicity and low entry barriers for all involved audiences to motivate the users to use the cooperation features provided by the platform. This process is also supported by an iterative usability and adaption cycle before the release of the Internet platform. Both aspects will be elaborated a bit more in the following paragraphs. Details on the mobile version of the platform will be given in the last paragraph of this section.

2.2.1 Data Self-Management

When delivering a service like the aforementioned Internet platform, one of the major problems is the data management, in specific the effort to keep the collected information correct and up to date. To provide such a huge amount of different types of information about schools, companies, jobs, apprenticeships, job candidate profiles etc. in a high information quality, it is necessary to automatize most of the data management steps to keep the effort on a minimum. An approved approach to solve this problem when delivering a service is, to involve the user into these processes and to reduce the administrational effort for the administrator hereby as

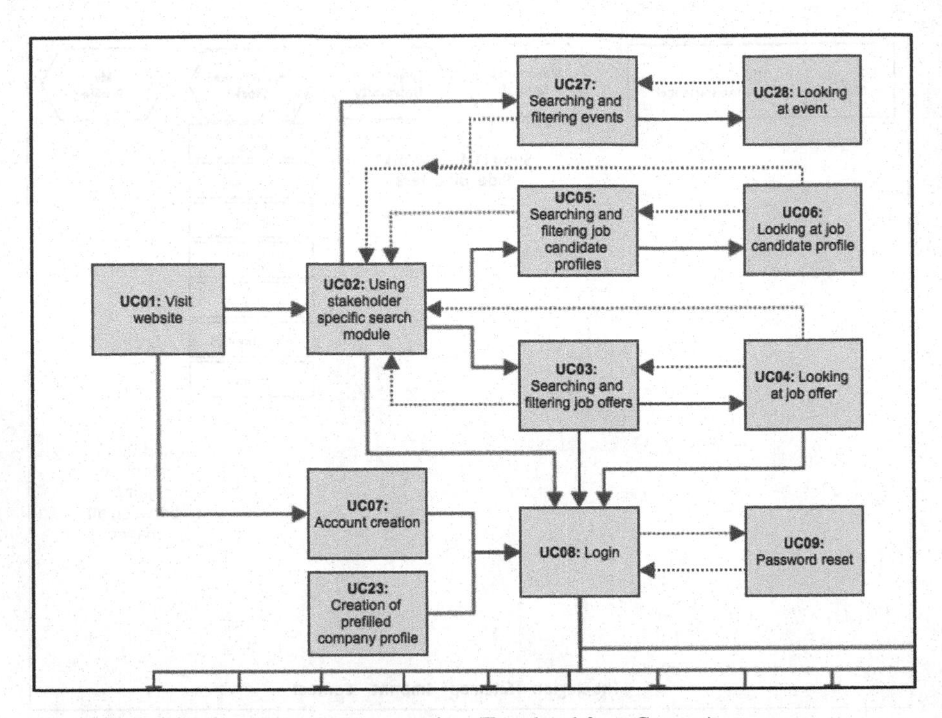

Fig. 20.5 Excerpt from the use cases overview (Translated from German)

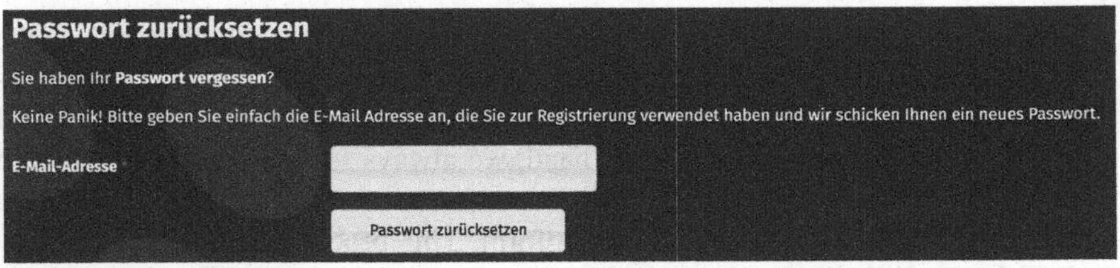

Passwort zurücksetzen

Sie haben Ihr **Passwort vergessen?**

Keine Panik! Bitte geben Sie einfach die E-Mail Adresse an, die Sie zur Registrierung verwendet haben und wir schicken Ihnen ein neues Passwort.

E-Mail-Adresse

Passwort zurücksetzen

Fig. 20.6 Exemplary implementation of use case *UC09: password reset* (Available in German language only)

much as possible [14]. Thus, the effort of data management is almost completely shifted from the platform provider to the user. When designing and implementing *vitaminBIR* we followed this approach in many ways:

1. The management of all personal and institutional account data is completely outsourced to the user, while the system validates the input when the user tries to store the same to only allow consistent data states to be saved. The validation of the data is done by a combination of the internal validation capabilities provided by the *TYPO3 Extbase* technology and program logic implemented in PHP. The required data model and controller classes are realized as standardized *TYPO3*

vitaminBIR Model extension. An example for the internal *TYPO3 Extbase* validation of an email address can be seen in the following source code snippet:

```
/*
 * @var string
 * @validate NotEmpty, EmailAddress
 */
protected $email = '';
```

Beyond using the validators provided by the *TYPO3 Extbase* framework, also own validators have been implemented for specific data attributes like German postal codes or currency amounts.

2. All data-based lists on the platform, e.g., job position listings, school listings, company listings, events etc., are also managed by the responsible person directly on the platform. Using the *TYPO3 Fluid* template technology these listings are automatically created from the data objects. The conditional functionalities of *TYPO3 Fluid* hereby help to be able to deal with optional fields and empty field values in a professional way. The templates are also part of the aforementioned *TYPO3 vitaminBIR Model* extension. An example for a *TYPO3 Fluid* condition to show the application deadline of a job offer on the platform can be seen in the following source code snippet:

```
<f:if condition="{jobOffer.applicationDeadline}">
  <f:then>
    <f:format.date format="d.m.Y">
      {jobOffer.applicationDeadline}
    </f:format.date>
  </f:then>
  <f:else>
    not provided
  </f:else>
</f:if>
```

Some listings on the platform were created in a very general manner to be reused on several pages of the platform under many different circumstances, for example job listings or contact information.

3. To enable the platform to provide a high amount of job offer results (especially in the starting phase) or to enable companies to import their own job offers automatically without any further management effort, the job database of the *German Federal Employment Agency*[6] is periodically filtered and imported into the platform. Based on several filtering steps regarding completeness of

[6] http://www.arbeitsagentur.de/web/content/EN/index.htm

information, regional jobs, assignment of jobs to companies on the platform it is ensured that only relevant job offers are imported. The import not only encompasses job offers, but also apprenticeships and internships and also synchronizes changes, which are performed on the data source later. From the technical perspective, the data preparation is performed by a combination of a proprietary import script for the data exchange client by the agency, the *BaseX*[7] XML database and a *Java*[8] application. Afterwards the prepared and pre-filtered data is imported into *TYPO3* using a scheduled task.
4. The administrator can define self-management rules in the backend of the platform to motivate the users to keep their information up to date and to automatize the process of checking and updating information using email, i.e., a user can be asked by email, when a data object has not been changed for a while, if the information is still correct. The user can confirm that the information is correct or adjust it on the platform. The system realizes if the user reacted on a mail or not and proceeds accordingly based on further rules. An exemplary rule-based self-management flow can be seen below (see Fig. 20.7). The administrator has only to act, if the user does not react at all. To provide such functionality another own *TYPO3 vitaminBIR Maintenance* backend extension has been developed, which enables an administrator to setup maintenance messages with placeholders that will be sent under specific rule-based circumstances.

By combining these approaches we hope to minimize the effort of data management on the platform by still having a high level of information quality. Especially for the last mentioned approach of self-management (4) a wide-ranging documentation has been developed, which is the basis for the setup of the rules in the backend of the platform. The current set of rule encompasses 17 adjustable rules and 31 possible maintenance messages with user-specific placeholders. Different messages can also be prepared depending on the type of stakeholder, i.e., school representatives, company representatives, or ordinary users.

The described integration of the job database of the *German Federal Employment Agency* has to be seen as a proof of concept. Further services, e.g. integration of public transport or event calendars, are planned to become or are already integrated as well. The idea is to reduce the data management effort not only for the platform provider, but also for the user, by providing services for automatic data aggregation and integration from several sources that are relevant for the audiences of the platform.

[7] http://basex.org

[8] http://www.oracle.com/technetwork/java/index.html

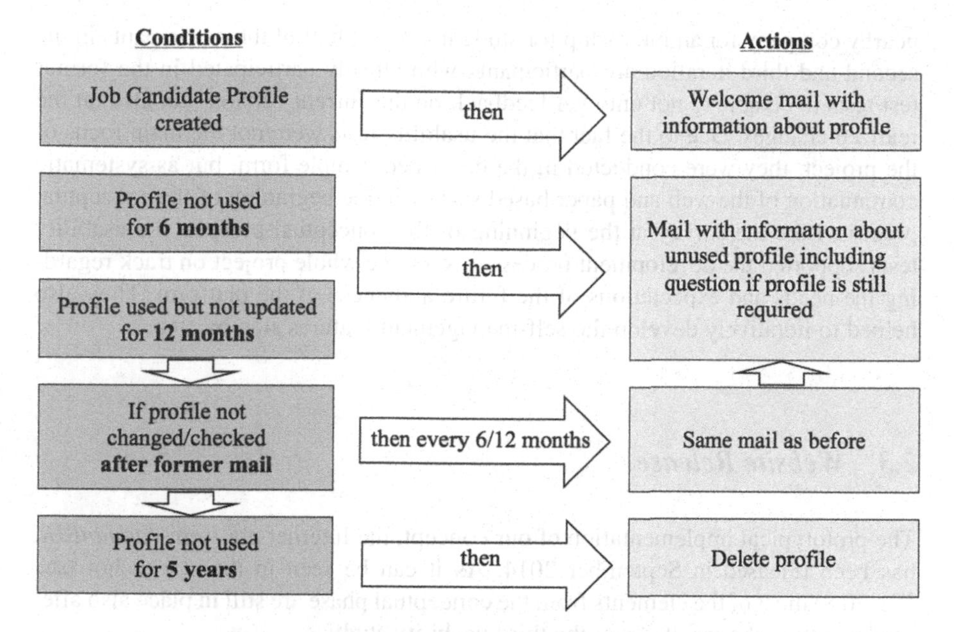

Fig. 20.7 Exemplary rule-based self-management flow for job candidate profiles

2.2.2 Usability and Adaption Cycle

Another problem when developing a service for such a broad variety of users and processes is the potential loss of the formerly aimed simplicity as well as the potential fail of providing not the processes that were expected by users. Therefore we conducted an iterative usability study with adaption cycles to keep on track regarding the former problems. The study was split into three usability tests with about five participants from each user group, i.e., three appointments for tests with five pupils, five students, and five employees from human resources departments of companies in the District of Birkenfeld.

The first test took place using an early Alpha Version, the second test some weeks later using a revised Beta Version and the last test has been performed on the Release Candidate Version. The resulting Release Version contains all improvements and changes, because the time between these tests was used to adjust the platform based on the feedback from the former tests. The tests were conducted as qualitative usability tests with *Concurrent Probing* [15], which means that the participants follow some tasks they have to do on the prototype platform while the supervisor analyses the behavior and interrupts in case of unexpected behavior of the user. Starting with the third iteration half of the participants were not interrupted in case of problems to check if they will find another or any solution for the given tasks by their own since the real users do prospectively not have a helping hand on their side.

The tasks given to the participants were typical tasks for the appropriate audience, i.e., creating a job candidate profile for a holiday job for pupils or finding a nearby company for an internship for students. About half of the participants in the second and third iteration are participants who already participated in the former test/tests to be able to not only get feedback on the current version, but also on the realized changes. Due to the fact that the usability tests were not the main focus of the project, they were conducted in the described, simple form, but as systematic continuation of the web and paper based survey in the beginning of the conceptual process. Like the survey at the beginning of the conceptual process, the usability tests supported the development process to keep the whole project on track regarding the needs and expectations of the future audiences of the platform. They also helped to iteratively develop the self-management features step by step.

2.3 Website Release

The prototypical implementation of our concept, the Internet platform *vitaminBIR*, has been released in September 2014. As it can be seen in the screenshot (see Fig. 20.8) most of the elements from the conceptual phase are still in place also after implementing the results from the third usability study.

Compared to the first conceptual approach (see Figs. 20.2 and 20.3) six mid-size changes can be seen in the Release Version. These changes will be addressed in the following.

1. The profile button has been changed to an "offers" (German: "Angebote") button. Some users expected the username to be clickable after logging in to an own account, while others expected to find a full feature list in the main menus. Therefore we made the username clickable and leading to the profile page, while the free button has been used to provide a submenu for all offers available on the platform, i.e., job offers, apprenticeships, appointments etc.
2. A short introduction video promotes and motivates the functionalities provided by the platform.
3. The concept missed an area where currently interesting functionalities of the platform can be promoted to the user. Therefore the image link section "in focus" (German: "Aktuelles") has been added. Based on the current week of the year special offers are promoted in this section, e.g., internships for pupils one week before the summer holidays, theses offers for students at semester start.
4. The steps of selecting a subject area of interest, e.g., "school", and a primary platform functionality in the proximate step, e.g., "list of schools in the region", has been replaced by a new one-page module with an optional stakeholder selection, e.g., "I'm interested in offers for pupils" (German: "Mich interessieren Angebote für Schüler"). By doing so, all major functions of the platform can be reached with one click in the tag cloud area of the element, while preselecting a stakeholder filters and highlights topics specifically for the selected stakeholder.

Fig. 20.8 Screenshot of the first release version of the *vitaminBIR* platform (Available in German language only)

The color scheme of a stakeholder is hereby reused for highlighting the offers in the tag cloud, as it can be seen in the following figure (see Fig. 20.9).

5. The stakeholder selection has been extended for "founders" (German: "Gründer") due to an additional project funding and focus extension. Also founders can have high impact on the positive development aspects of a rural region. Therefore this extension has already been implemented for the Release Version in a later phase of the project.

2.4 Mobile Application Release

By concept, a website is available for a broad audience and can be viewed on several operating systems and with different browsers. During the conceptual development of *vitaminBIR* we realized that building the platform highly compatible for many operating systems and browsers is only one important aspect to reach our audiences: Especially for young people, in our case pupils and students, an important information source are mobile applications on their smartphones. To provide professional mobile applications for several common platforms on the one hand, while having a minimal additional development effort on the other side, we

Fig. 20.9 Tag cloud with optional stakeholder selection providing access to the platform features (Available in German language only)

decided to provide *vitaminBIR* also as web-based application using the *Apache Cordova*[9] framework for the *Android* and *iOS* operating systems.

The *vitaminBIR* mobile application (see Fig. 20.10) only provides a reduced subset of the use cases of the Internet platform. This subset reduces the app to read-only searching functionalities to motivate the user to use the website to access the more complex functionalities without the hassles of editing contents on small device screens.

The *vitaminBIR* mobile application has been released on the *Google Play Store*[10] by the end of December 2014. The *iOS*[11] followed in January 2015.

2.5 Usage Analysis

In addition to the assessment survey and the usability tests, we collected usage data to address our research question if ICT can support regional cooperation using *Piwik*[12] and *Google Analytics*.[13] Additionally, we wanted to find out the criteria for such an Internet platform. The data approve different aspects of our assumption that will be presented in the following.

---

[9] http://cordova.apache.org/

[10] https://play.google.com/store/apps/details?id=de.ucb.vitaminbir

[11] https://itunes.apple.com/us/app/vitaminbir/id940521967

[12] http://piwik.org/

[13] http://www.google.com/analytics/

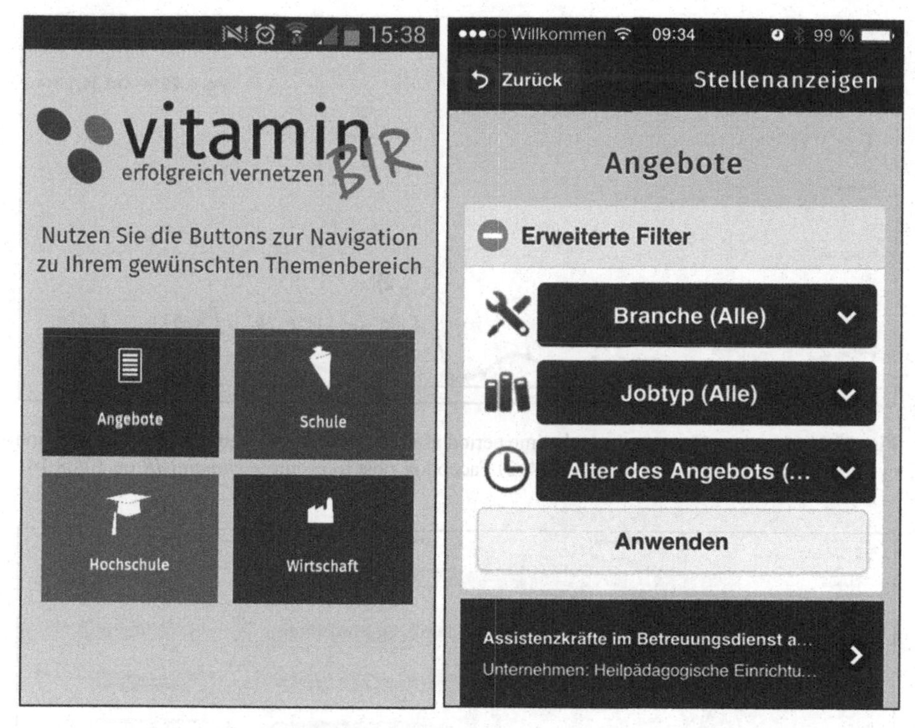

Fig. 20.10 Screenshots of the start screen (Android) and the job offer filters (iOS) of the *vitaminBIR* mobile application (Available in German language only)

1. **Personalized and individual.** The results of the survey in the beginning of the project show that it is not enough to offer a solitary information platform to keep it attractive for users. The users want to be informed about personal news using media they are used to. The usage data of *vitaminBIR* show that the page impressions rise after promotion events. Hence, next to the technical development and automatized maintenance, the personal information about *vitaminBIR* or rather regional offers in general is especially important in the beginning. Apart from information events different advertisement channels can be used, e.g., we used social media platforms that are already known by our target group. An extract of the influences of personal and individual promotion of *vitaminBIR* to the hit rates of the platform is depicted in Fig. 20.11.
2. **Regional**. The focus of a regional Internet platform like *vitaminBIR* is very important in order to use an Internet-based instrument to support a region in a sustainable, long-term way. We set the focus on local full time and side jobs, i.e., the kind of jobs the main target group, pupil and students, is looking for. The users flow (see Fig. 20.12) shows that the first interactions of most of the user are "full time job offers" (German: "Stellenanzeigen"), "internships" (German: "Praktika") and "side jobs" (German: "Nebenjobs"). Additionally the users are interested in "regional companies" (German: "Unternehmen in der Region").

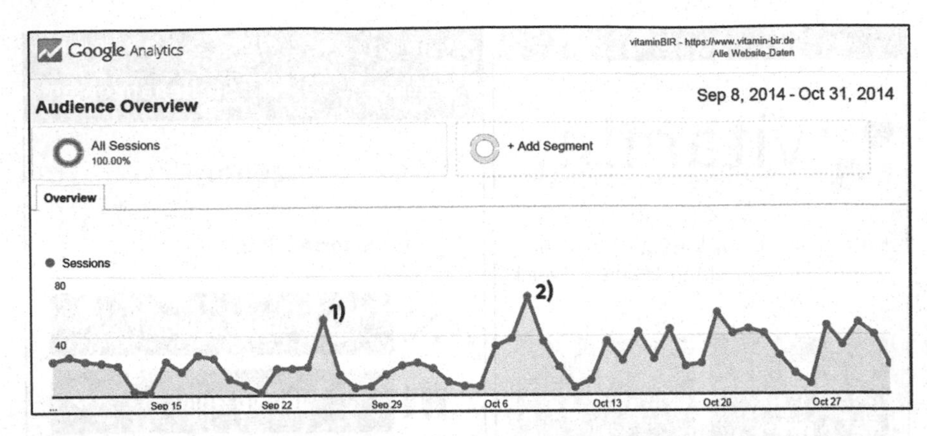

Fig. 20.11 Usage analysis in the early time period after the release; marked events: *1)* information event for school headmaster on 09/25/14, *2)* Facebook post to promote *vitaminBIR* on 10/08/14

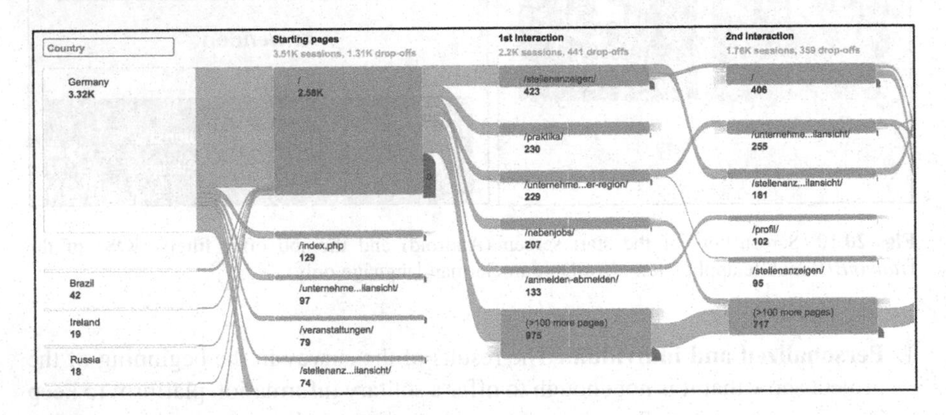

Fig. 20.12 Users flow on *vitaminBIR* (09/08/2014–12/31/2014)

3. **Interactive and direct.** On the homepage of *vitaminBIR* especially the section "in focus" (German: "Aktuelles") and event information catch the users interest (see Fig. 20.13). That shows that the user focuses on direct information. The module with the stakeholder selection is not analysed in this case due to technical incompatibilities. In comparison with the static website *Made in BIR*[14] that offers similar information about companies of the District of Birkenfeld the number of users of the interactive Internet platform *vitaminBIR* is higher.

[14] http://www.made-in-bir.de/

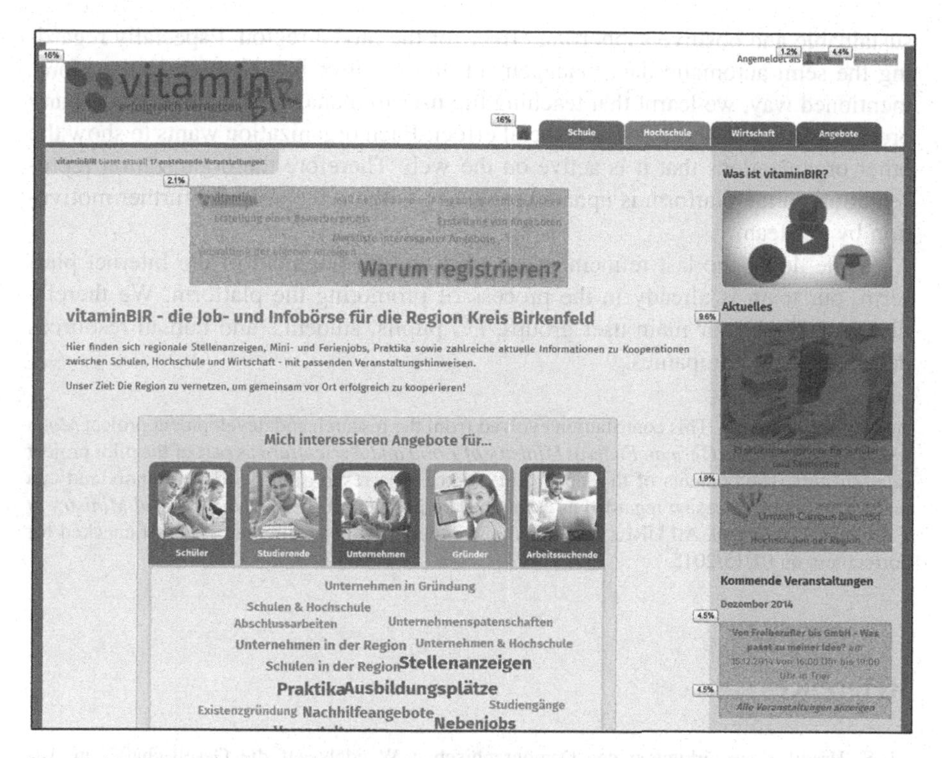

Fig. 20.13 InPage analysis of www.vitamin-bir.de (09/08/2014–12/12/2014)

3 Summary and Outlook

As presented above, the elaborated steps to support sustainable development in rural areas by encouraging local cooperation and neighborhood effects can be transferred into an ICT system. The advantages of ICT systems for this matter are evident: ICT systems, in our case the Internet platform, are available for a large number of people, without time restrictions and hold only a very low entry barrier. Compared to short-time collaboration projects, such a system can also consist for several years with a relatively low managing effort, to sustainably support development of a specific region, not only during the runtime of the current funding or developing project.

Summarized, with *vitaminBIR* we try to provide an ICT system that helps the District of Birkenfeld, as well as other regions in the future by transferring the concept and the software, to reduce the previously mentioned demographic problems, by efficiently supporting real-life processes which reduce the influencing factors of emigration of young people from rural areas.

During the whole conception and realization process of *vitaminBIR* it was always important to continuously check the developed platform against the needs and expectations of the audiences, to ensure that the platform can really have a sustainable and positive supporting effect for the model region. Especially regarding the semi-automatic data management and the user involvement in the aforementioned way, we learnt that teaching the user to manage the school or company profile herself leads to a positive social effect: Each organization wants to show the other organizations that it is active on the web. Therefore the organization representation on the platform is updated regularly without the need for further motivation by our team.

While doing the last refinement steps in the development of the Internet platform, our team is already in the process of promoting the platform. We thereby directly address our main user groups, i.e. pupils, students, and human resources departments of companies.

Acknowledgements This contribution evolved from the research and development project *Made in BIR*, funded by the *German Federal Ministry of Food and Agriculture* as part of the pilot project *LandZukunft*. The contents of this document are the sole responsibility of the authors and can under no circumstances be regarded as reflecting the position of the *German Federal Ministry of Food and Agriculture*. All URLs mentioned in footnotes of this document were last checked for correctness on 01/15/2015.

References

1. S. Hradil, "Auswirkungen des Demographischen Wandels auf die Gesellschaft," in *Alte Gesellschaft, Neue Gesellschaft?*, S. Hradil and J. Weingarten, Eds. Mainz: Zukunftsinitiative Rheinland-Pfalz (ZIRP), 2010, pp. 15–24.
2. B. Müller, "Demographischer Wandel und die Folgen für die Städte – Einführung und Übersicht," *Dtsch. Zeitschrift für Kommunalwissenschaften*, vol. 2004/I, pp. 5–13, 2004.
3. L. Böckmann, T. Kirschey, J. Stoffel, and M. Völker, "Statistische Analysen No. 25 2012, Rheinland-Pfalz 2060." Statistisches Landesamt Rheinland-Pfalz, Bad Ems, 2012.
4. J. Weingarten and S. Zahn, "Kultur, Wirtschaft und Demographie – Chancen der rheinland-pfälzischen Kultur- und Kreativwirtschaft im gesellschaftlichen Wandel," in *Neue Potentiale für Wirtschaft und Beschäftigung*, J. Rump and J. Weingarten, Eds. Zukunftsinitiative Rheinland-Pfalz (ZIRP), 2010, pp. 77–126.
5. H. Schwager, "Demographischer Wandel – Risiko und Chance für Unternehmen," in *Neue Potentiale für Wirtschaft und Beschäftigung*, J. Rump and J. Weingarten, Eds. Mainz: Zukunftsinitiative Rheinland-Pfalz (ZIRP), 2010, pp. 15–18.
6. B. Müller and S. Siedentop, "Wachstum und Schrumpfung in Deutschland – Trends, Perspektiven und Herausforderungen für die räumliche Planung und Entwicklung," *Dtsch. Zeitschrift für Kommunalwissenschaften*, vol. 2004/I, pp. 14–32, 2004.
7. A. Schloemer, "Tourismus und Freizeit – Herausforderungen und Chancen des demographischen Wandels," in *Neue Potentiale für Wirtschaft und Beschäftigung*, J. Rump and J. Weingarten, Eds. Zukunftsinitiative Rheinland-Pfalz (ZIRP), 2010, pp. 62–68.
8. A. Baier and V. Bennholdt-Thomsen, "Der 'Stoff', aus dem soziale Nähe ist," in *Aktivierung durch Nähe*, T. Kluge and E. Schramm, Eds. Frankfurt am Main: Institut für sozial-ö kologische Forschung (ISOE) GmbH, 2003, pp. 12–21.
9. A. Filler, E. Kern, and S. Naumann, "Supporting Sustainable Development in Rural Areas by Encouraging Local Cooperation and Neighborhood Effects using ICT," in *EnviroInfo 2014. Proceedings of the 28 th International Conference on Informatics for Environmental Protection, Oldenburg*, 2014, pp. 621–627.

10. M. Pellizzari, "Do Friends and Relatives Really Help in Getting a Good Job?" 2003.
11. D. Marmaros and B. Sacerdote, "Peer and social networks in job search." 2002.
12. Y. M. Ioannides and L. D. Loury, "Job information networks, neighborhood effects and inequality," *J. Econ. Lit.*, pp. 1056–1093, 2004.
13. E. Kern, B. Luther-Klee, S. Naumann, and A. Filler, "Kriterien für Webportale zur Unterstützung nachhaltiger Regionalentwicklung am Fallbeispiel 'vitaminBIR,'" in *INFORMATIK 2014. Big Data – Komplexität meistern. Lecture Notes in Informatics (LNI) – Proceedings, Volume P-232*, 2014, pp. 1923–1934.
14. J. A. Fitzsimmons, "Consumer Participation and Productivity in Service Operations," *Interfaces (Providence).*, vol. 15, pp. 60–67, 1985.
15. J. R. Bergstrom, "Moderating Usability Tests," 2013. [Online]. Available: http://www.usability.gov/get-involved/blog/2013/04/moderating-usability-tests.html

Printed in the United States
By Bookmasters